Fundamentals of Manufacturing

Supplement

Fundamentals of Manufacturing
Supplement

Philip D. Rufe, CMfgE
Editor

Society of Manufacturing Engineers
Dearborn, Michigan

987

Library of Congress Catalog Card Number: 2004117508
International Standard Book Number: 0-87263-747-6

Additional copies may be obtained by contacting:
Society of Manufacturing Engineers
Customer Service
One SME Drive, P.O. Box 930
Dearborn, Michigan 48121
1-800-733-4763
www.sme.org

SME staff who participated in producing this book:
Rosemary Csizmadia, Production Editor
Frances Kania, Administrative Coordinator

Printed in the United States of America

Table of Contents

Acknowledgments

AUTHORS

Chapter 48—Personal Effectiveness: Philip Rufe, CMfgE
Chapter 49—Machining Processes Analysis: Gary Vrsek
Chapter 50—Forming Processes Analysis: Gary Vrsek
Chapter 51—Joining and Fastening Analysis: Gary Vrsek and Philip Rufe, CMfgE
Chapter 52—Deburring and Finishing: Gary Vrsek
Chapter 53—Fixture and Jig Design: Tracy Tillman, Ph.D., CMfgE, CEI
Chapter 54—Advanced Quality Analysis: Thomas Soyster, Ed.D. and Robert Chapman, Ph.D.
Chapter 55—Engineering Economics Analysis: Philip Rufe, CMfgE
Chapter 56—Management Theory and Practice: Philip Rufe, CMfgE
Chapter 57—Industrial Safety, Health, and Environmental Management: Teresa Hall, Ph.D., CMfgE

TECHNICAL REVIEWERS

Chapter 48—Personal Effectiveness: Tracy Tillman, Ph.D., CMfgE, CEI and William D. Karr, CMfgT
Chapter 49—Machining Processes Analysis: F. Zafar Shaikh and William D. Karr, CMfgT
Chapter 50—Forming Processes Analysis: F. Zafar Shaikh and William D. Karr, CMfgT
Chapter 51—Joining and Fastening Analysis: F. Zafar Shaikh and William D. Karr, CMfgT
Chapter 52—Deburring and Finishing: F. Zafar Shaikh and William D. Karr, CMfgT
Chapter 53—Fixture and Jig Design: Gary Vrsek and William D. Karr, CMfgT
Chapter 54—Advanced Quality Analysis: Kathy Kustron, Ken Zimmer, Dan Fields, Ph.D, and William D. Karr, CMfgT
Chapter 55—Engineering Economics Analysis: Poch Bombach, Jihad Albayyari, Ph.D., Laura Powers, CPA, and William D. Karr, CMfgT
Chapter 56—Management Theory and Practice: Tracy Tillman, Ph.D., CMfgE, CEI and William D. Karr, CMfgT
Chapter 57—Industrial Safety, Health, and Environmental Management: Vince Runde, Dan Hehr, Steven Hall, CPSM, and William D. Karr, CMfgT

Preface

This book was designed to be used in conjunction with Chapters 11-47 of *Fundamentals of Manufacturing*, Second Edition to provide a structured review for individuals planning to take the Certified Manufacturing Engineer (CMfgE) examination. The topics covered are the result of a 1999 survey of manufacturing managers, manufacturing technologists and engineers, and manufacturing educators. Its purpose was to identify fundamental competency areas required of manufacturing technologists and engineers in the field.

While the objective of this supplement is to help prepare manufacturing engineers for the certification process, it is also a useful reference for manufacturing professionals and individuals with limited manufacturing experience or training.

Introduction

MANUFACTURING CERTIFICATION

Manufacturing is concerned with energy, materials, tools, equipment, and products. Excluding services and raw materials in their natural state, most of the remaining gross national product is a direct result of manufacturing.

Modern manufacturing activities have become exceedingly complex because of rapidly increasing technology and expanded environmental involvement. This, coupled with increasing social, political, and economic pressures, has increased the demand for highly skilled manufacturing engineers and managers.

The principal advantage of certification is that it proves an individual possesses not only the minimum academic requirements, which are based upon a specific set of standards, but more importantly, the practical experience required of a manufacturing engineer or manager on the job.

Many persons currently employed in industry can successfully measure themselves against the set of standards, but they cannot provide documentation concerning their ability. SME's certification program is designed to provide successful candidates with documentary evidence of their abilities. The designations Certified Manufacturing Engineer (CMfgE), Certified Manufacturing Technologist (CMfgT), Certified Engineering Manager (CEM), and/or Certified Enterprise Integrator (CEI) are bestowed upon successful candidates.

Philosophically, the purpose of manufacturing certification is to gain increased acceptance of manufacturing engineering and management as a profession and to ultimately improve overall manufacturing effectiveness and productivity.

PURPOSE AND OVERVIEW

The purpose of this supplement, in conjunction with Chapters 11-47 of *Fundamentals of Manufacturing*, Second Edition, is to provide the manufacturing engineer or manager with a structured review in preparation for taking the Certified Manufacturing Engineer (CMfgE) examination. It is also a useful reference for manufacturing professionals or individuals with limited manufacturing experience or training.

The major areas of manufacturing reviewed in the text include personal effectiveness, manufacturing processes, fixture and jig design, quality, engineering economics, manufacturing management, and industrial safety, health, and environmental management.

Sample problems and questions are included at the conclusion of each chapter for practice. Answers are included for all questions, but it is recommended that questions be attempted before reading the answers.

EXAMINATION SPECIFICS

The Certified Manufacturing Engineer (CMfgE) examination is a three-hour open-book exam consisting of 150 multiple-choice questions. The exam consists of 110 general manufacturing engineering knowledge questions and 40 questions from preselected focus areas within the "body of knowledge." The focus areas are processes, management, integration and control, joining and assembly, support operations, and quality.

Each major area and its relative emphasis in the exam are listed as follows:

Mathematics/Applied and Engineering Sciences/ Materials	14.6%
Product Design	13.8%
Manufacturing Processes	12.7%
Production Systems	21.1%
Automated Systems and Control	5.2%
Quality	13.1%
Manufacturing Management	13.3%
Personal/Professional Effectiveness	6.2%

ADDITIONAL INFORMATION

The bibliography at the end of this book contains appropriate resources for more detailed information on the topics covered. Additional study resources for the Certified Manufacturing Engineer (CMfgE) exam include, but are not limited to, a pencil and paper practice exam and a Windows® -based self-assessment program. The self-assessment program (based on the same topics as the exam) will help to determine your strengths and weaknesses. Built-in bibliographic references suggest study materials for topics where additional help is needed.

For more information regarding the exam or additional resources, please contact the Society of Manufacturing Engineers at 313-271-1500 or e-mail: training@sme.org. Information also can be obtained on SME's website, www.sme.org/certification.

Any questions or comments regarding this supplement are welcome and appreciated. Please direct questions and/or comments to training@sme.org.

Chapter 48

Personal Effectiveness

One of the largest competency gaps for engineers and managers is interpersonal skills. College graduates typically are proficient in areas such as mathematical reasoning, engineering mechanics, materials, quality, processes, etc. However, they generally lack proficiency in skills such as:

- written and oral communication,
- listening,
- negotiating,
- confrontation and conflict management,
- meeting management, and
- creativity and innovation.

48.1 COMMUNICATION

Communication is the process of sending and receiving messages. It is the exchange of information and ideas. Effective communication occurs when the receiver receives and interprets the message as intended by the sender.

There are two basic forms of communication: nonverbal and verbal. Examples of nonverbal communication include facial expressions, gestures, posture, clothes, grooming, and punctuality. Nonverbal communication is usually more spontaneous than verbal communication. People typically do not formulate their facial expressions, whereas they would give more thought to formulating a sentence or paragraph to express an idea.

Verbal communication, in the form of written or oral communication, conveys more complicated and structured messages than nonverbal communication. Written communication can express highly detailed and complicated messages. It also is a form of communication that can be saved and/or stored. For example, meeting minutes often are stored for later reference if needed. Oral communication is sometimes more convenient than writing an e-mail or report. Nonverbal communication can also be combined with oral communication to increase the message's meaning.

LISTENING SKILLS

Listening is one of the most important communication skills for personal and organizational success. Listening accounts for a major part of the time spent communicating. Poor listening skills can result in miscommunications and costly errors in terms of lost time and money.

To be a good listener a person must first be motivated to listen. People are generally motivated if the communication is job related and/or if there is some interest in the communication. Generally, higher motivation results in better listening.

Being a good listener requires a person to focus on the content as opposed to the delivery style. This is referred to as content listening (Bovee 2000). Regardless of whether the presenter's communication style is entertaining or dry, it is the content of the message that counts. For example, a person with poor communication skills may convey an important message, whereas a person who is entertaining may be easy to listen to but has nothing important to say.

Another important part of being a good listener is not jumping to conclusions, thereby deferring evaluation to the end. Often, a speaker may arouse emotions within the listener. This may cause the listener to stop listening and start formulating a response before the entire message is absorbed. Premature evaluation can seriously inhibit the listener's ability to understand the true meaning of the message.

A good listener gives nonverbal or verbal feedback to the speaker. For example, making eye contact with the speaker can mean the listener acknowledges the message. No eye contact can mean the listener has lost interest or disagrees with the message. Verbal feedback can express recognition, agreement, and disagreement. A listener may verbally indicate receipt of the message with agreement or disagreement.

In general, feedback should be constructive, timely, coherent, and impersonal (Bovee 2000). Feedback needs to be helpful rather than harmful, and practical. It has to relate to something within the speaker's realm of control. Feedback needs to occur soon after the message is sent or it is not very effective. Also, feedback loses its meaning if it is viewed as personal. If a speaker views feedback as personal, he or she may ignore it completely.

48.2 WRITTEN COMMUNICATION

There are many forms of written business communication such as letters, resumes, memos, reports, and proposals. The writing process generally follows a series of steps. First, what is the purpose of the message? Is it informative or persuasive? Second, who is the audience? Is the audience comprised of co-workers or strangers, for example? Is the audience familiar with the topic and what will their probable reaction be? It is generally advisable to anticipate the audience's questions and provide enough information as practical. Third, the appropriate communication medium is selected based on the purpose and audience. The most common forms of written communication are memos, letters, reports, and proposals. Memos are typically internal documents that communicate a brief message. They are typically one-page documents, which are normally directed to many people; however, the audience could be a single person. Figure 48-1 illustrates a generic memo.

Letters are more formal documents that are typically sent to people outside the company or organization. They are generally one to two pages in length and printed on company letterhead. Figure 48-2 illustrates a generic letter.

Reports are a form of written communication conveying facts, data, and other technical information. They can be written for readers internal or external to a company. Reports can communicate the results of a project or experiment. In the case of mutual funds, a report may disclose the yearly earnings and stock holdings to fund shareholders.

Proposals are similar to reports. However, they are generally persuasive in nature as opposed to objectively communicating factual information. Proposals can be a few pages to several hundred pages in length. A proposal typically asks for approval or support of an idea or project. The proposal's audience is typically internal to a company or organization; however, the audience also can be external.

After the communication medium has been selected, the author needs to organize and compose the message. In the organization process, the author identifies the main points of the message to be communicated. In other words, the author identifies what information he or she wants the audience to receive and remember. It is generally advisable for the author to develop an outline starting with a main idea and supporting

Southern Community Center

Inter-office Memo

Date: 6/10/04

To: Center Employees

From: Bob Wikert, Human Resources

Cc: Jim Layhan, President and CEO

RE: Parking

On July 15 the employee parking lot on the east-side of the building will be repaired. Repairs will be completed by July 17. During this time please park in the guest parking lot located on the west-side of the building. If you have any questions please contact me at extension 204.

Thank you.

Figure 48-1. Example memo.

ideas. A coherent outline forms the framework for a coherent message. A well-organized and -crafted message will be easier and shorter for the audience to read, thereby making the audience more likely to accept it as opposed to dismissing it.

It is important to remember that the audience will infer the author's tone and mood from the message. The author does not have the luxury of reading the message and using nonverbal communication. Due to an author's poor choice of words or writing style, readers often infer the wrong impression from a message. This happens frequently with e-mails. E-mails are generally drafted quickly and little attention is given to how the reader interprets the words and writing style. For example, the reader may think the author is angry when he or she is not.

An author should choose the level of words and style appropriate for the audience and message. Avoid clichés, such as "time is money," for example. Words should be chosen carefully and composed in a sentence structure that creates the desired tone.

There are three basic sentence structures: simple, compound, and complex. A simple sentence has one subject and one predicate, such as "Car prices have risen in the past year." A compound sentence consists of at least two main ideas or clauses. For example, "Car prices have risen in the past year and mortgage interest rates have fallen." A complex sentence has one main idea or clause and subordinate ideas or clauses. For example, "As a result of falling interest rates, more homeowners have refinanced their homes."

Coherent and fluid paragraphs increase the communication's effectiveness. Paragraphs are typically a group of sentences that all relate to one main idea. They contain a topic sentence, which is supported by the remaining sentences in the paragraph. The topic sentence, which can be a question, provides a summary of what the paragraph will

Southern Community Center

August 1, 2004

Mr. John Smith
Raven Community College
1100 Genoa Avenue
Ypsilanti, MI 48197

Dear Mr. Smith,

The Community Center's administration greatly appreciates your willingness to participate in our career fair. We would sincerely appreciate your attendance at our annual career fair on October 26, 2004 from 12:00–3:00 p.m. in the main lounge of the Community Center. Parking will be provided in the guest parking lot, which is located on the west side of the building. Please bring your parking ticket to the meeting for validation.

Enclosed is an agenda and other related material. We are looking forward to your critical analysis, sound input, and insightful decisions. As a college professional, your participation is essential for the success of the career fair.

Please RSVP by phone or e-mail.

Sincerely,

William Owens
Southern Community Center
734-478-2141
wowens@scc.org

/wo
Enclosure

Figure 48-2. Sample letter.

discuss. For fluid communication, paragraphs must be arranged coherently to prevent confusion. There also must be smooth transition from one paragraph to the next. The transition can be a word, sentence, or paragraph.

The final step in the writing process is editing. Editing consists of rewriting, grammar and spell checking, and proofreading. The key to good writing is rewriting. Depending on the type of writing, a few revisions may be adequate or 10–20 revisions may be necessary. The communication quality increases with the number of revisions. Good writers are not always naturally gifted.

They rewrite several times to perfect their message. Good writers are also typically good readers. In other words, people who read frequently usually are better writers than people who do not.

Word processing and e-mail software packages can check for spelling mistakes. However, there are situations where software will not catch a mistake such as "to or too." Checking grammar can be more difficult, but many word processing software packages come with some form of grammar checking. Some common grammar errors to be aware of are subject-verb agreement, voice, and tense. A singular subject, for example, requires a singular verb, such as "the car travels fast" not "the car travel fast." The two types of voice, referring to verbs, are active and passive. For example, in the active voice, "teenagers like to drive fast cars." In the passive voice, "fast cars are typically driven by teenagers." The voice needs to be consistent throughout the message. Tense refers to past, present, and future. Examples of present and past tenses are respectively: "I am watching the game" and "I watched the game."

VISUAL AIDS

Visual aids can express some messages more simply and clearly than words. They can highlight specific points and, if appealing to the eye, can be persuasive. Visual aids generally consist of tables, charts, pictures, and maps. They provide another avenue for communicating a message.

Tables are generally applicable for displaying data, particularly numerical data. It is cumbersome to read data in paragraph form. Data formatted in a table is much easier to read as illustrated in Table 48-1.

Table 48-1. Example table (Rockwell hardness results for heat treating)

Reading	Air Quenched (HRC)	Oil Quenched (HRC)	Water Quenched (HRC)
1	8	35	56
2	10	38	60
3	6	35	52
Average	8	36	56

Line charts and bar charts are typically used to illustrate data with respect to time such as stock market results. These charts also can be used to compare two or more variables over time. Figure 48-3 illustrates an example line chart. Bar charts, also called histograms, can also illustrate a set of data without respect to time as seen in Figure 48-4.

When comparing subsets of data relative to the complete set, a pie chart is typically used as illustrated in Figure 48-5.

Flow charts typically illustrate a sequence of steps or processes. Figure 48-6 illustrates a flow chart of required program courses and, based on prerequisites, the sequence in which they should be taken.

Radar charts, as illustrated in Figure 48-7, are used to graphically illustrate the value of each variable or indicator with respect to a maximum value such as 10. The chart allows the reader to quickly see the overall performance of a project or team, for example.

Regardless of the type of table or chart included in the communication, there are several universal rules for using them. First, the table or chart must be labeled. The reader must be given a description of the chart and the variables being displayed, including their respective units, if applicable. Second, the table or chart must be large enough to read without straining. Third, the information in the table or chart must enhance or complement the written information. If it does not, there is no reason to include it. Finally, the table or chart must be coherent and esthetically appealing. In some cases, a logical format, boldface letters, cross-hatching, or other formatting, including color, can enhance the message.

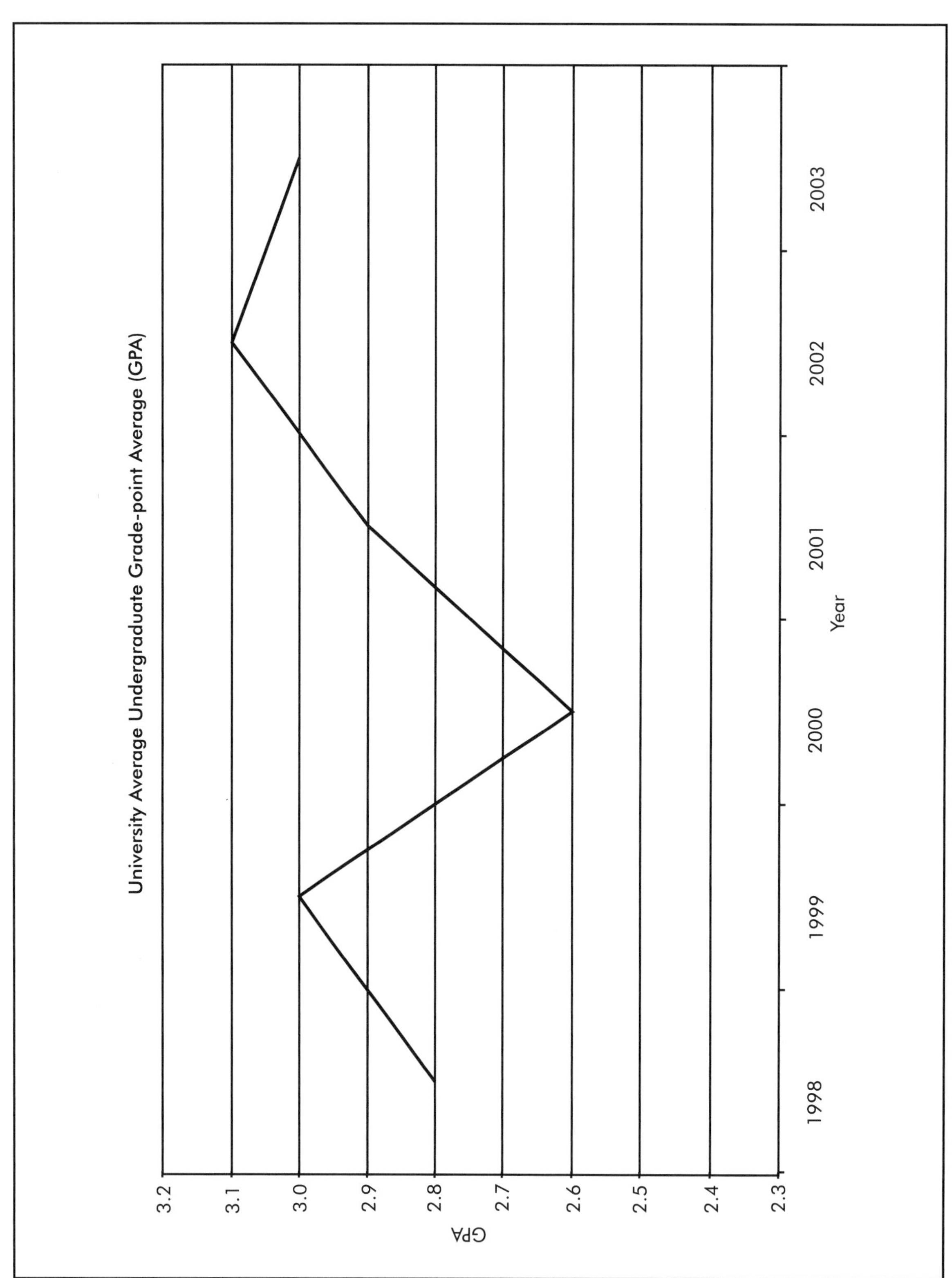

Figure 48-3. Example line chart.

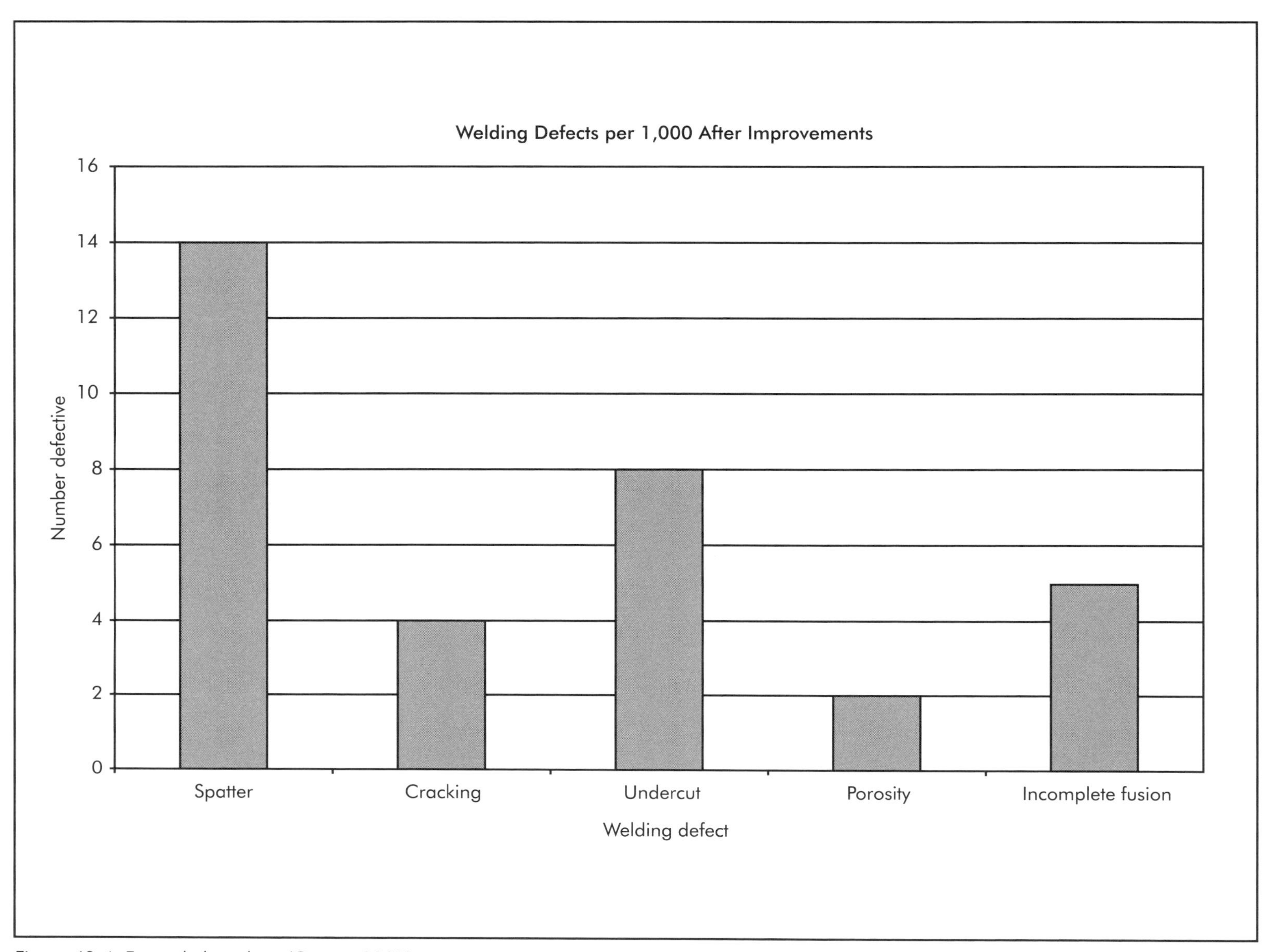

Figure 48-4. Example bar chart (Conner 2001).

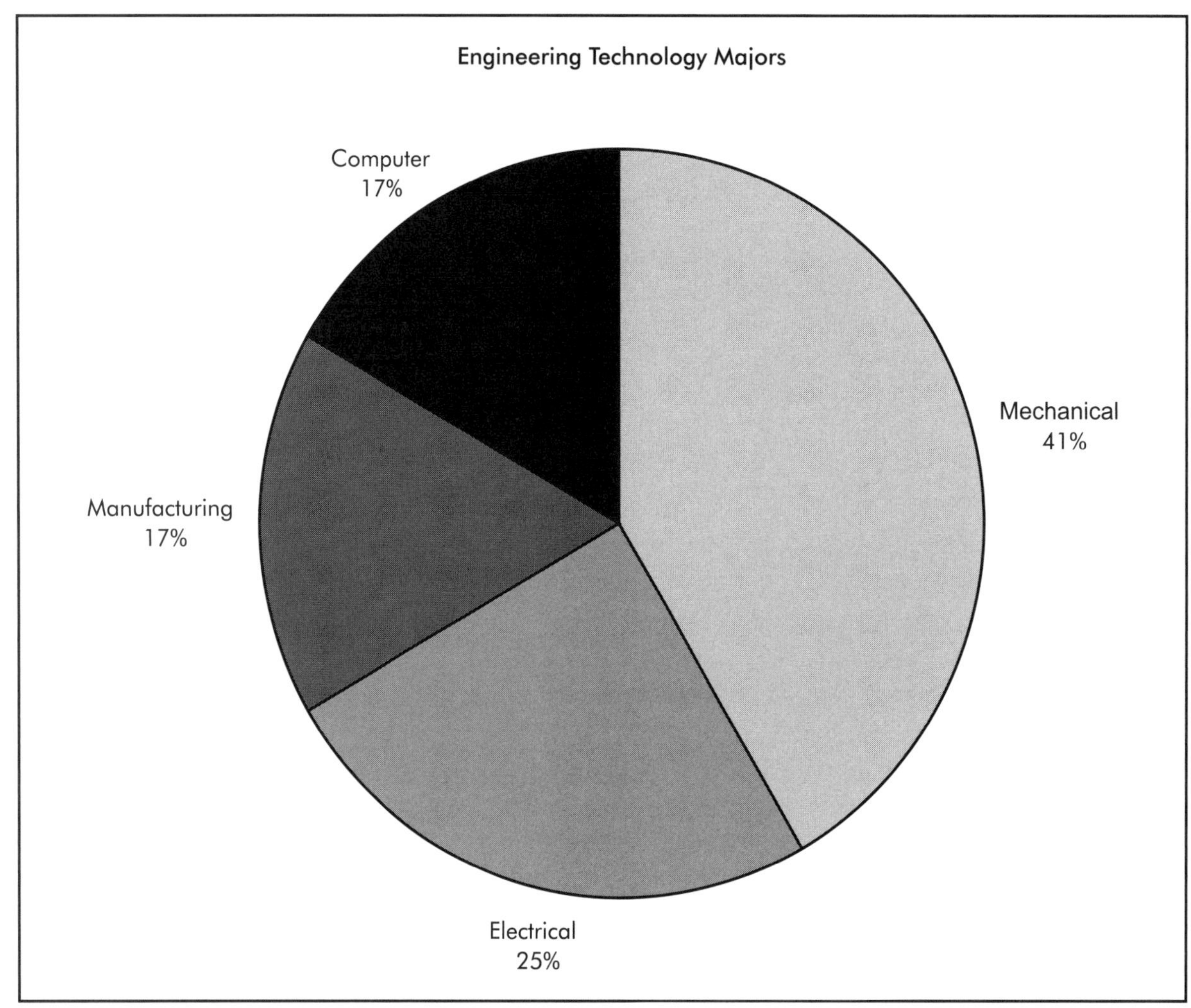

Figure 48-5. Example pie chart.

48.3 SPEECHES AND ORAL PRESENTATIONS

Speeches and oral presentations are integral parts of business communication. Public speaking allows an audience to place a face with a name. It also provides an opportunity for individuals to "stand out" and advance in their career.

Similar to written communication, presentations require careful planning and preparation. Planning a speech or presentation encompasses several steps, including determining the goal and topic, analyzing the audience, developing the main statement, organizing the flow of topics and support material, and determining the length and style.

Presentations generally have two main goals: to inform and/or persuade. In most situations the presentation's goal and topic are already established by the need for the presentation. For example, a project leader may need to give an informative report to

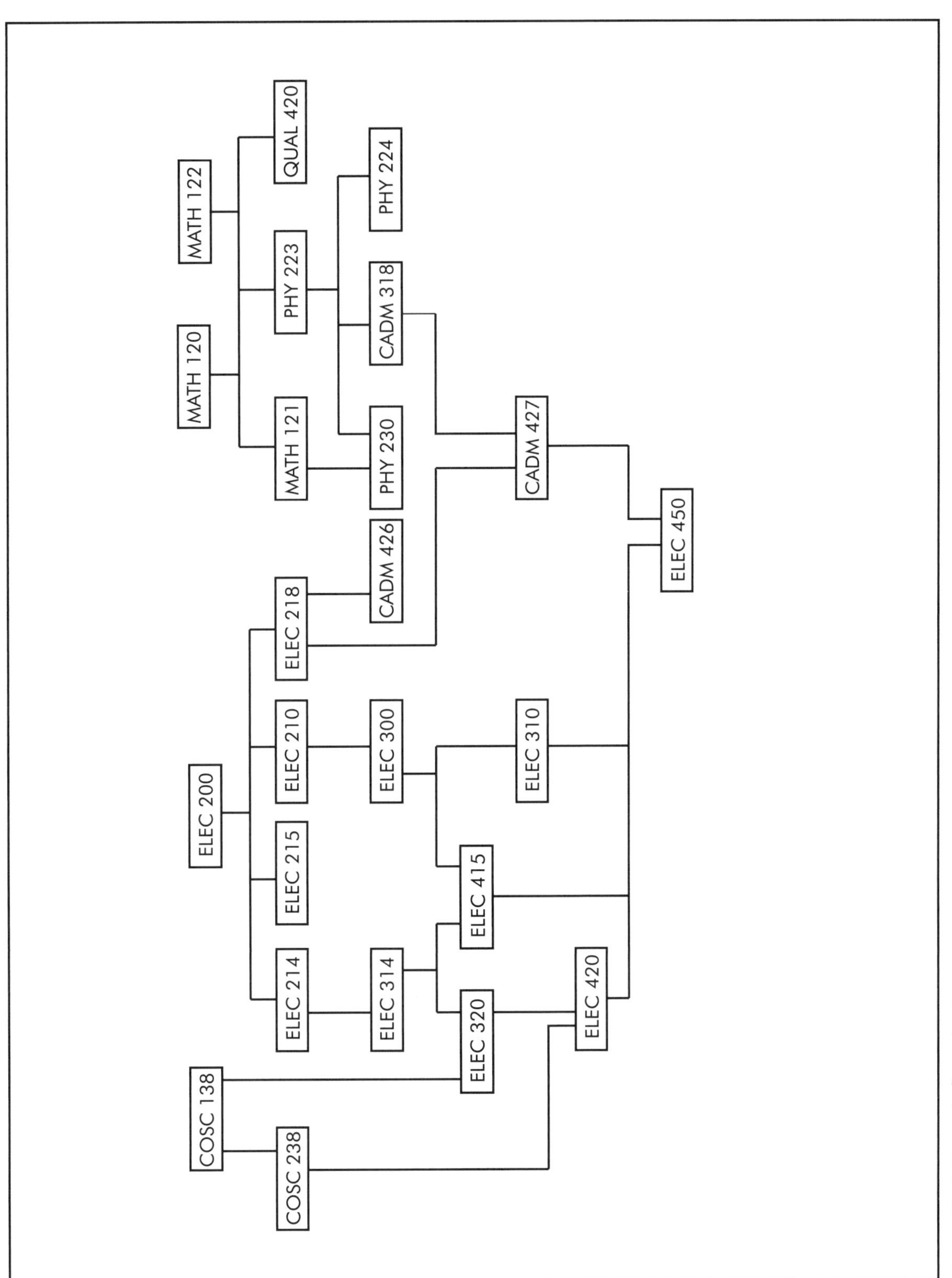

Figure 48-6. Example flow chart.

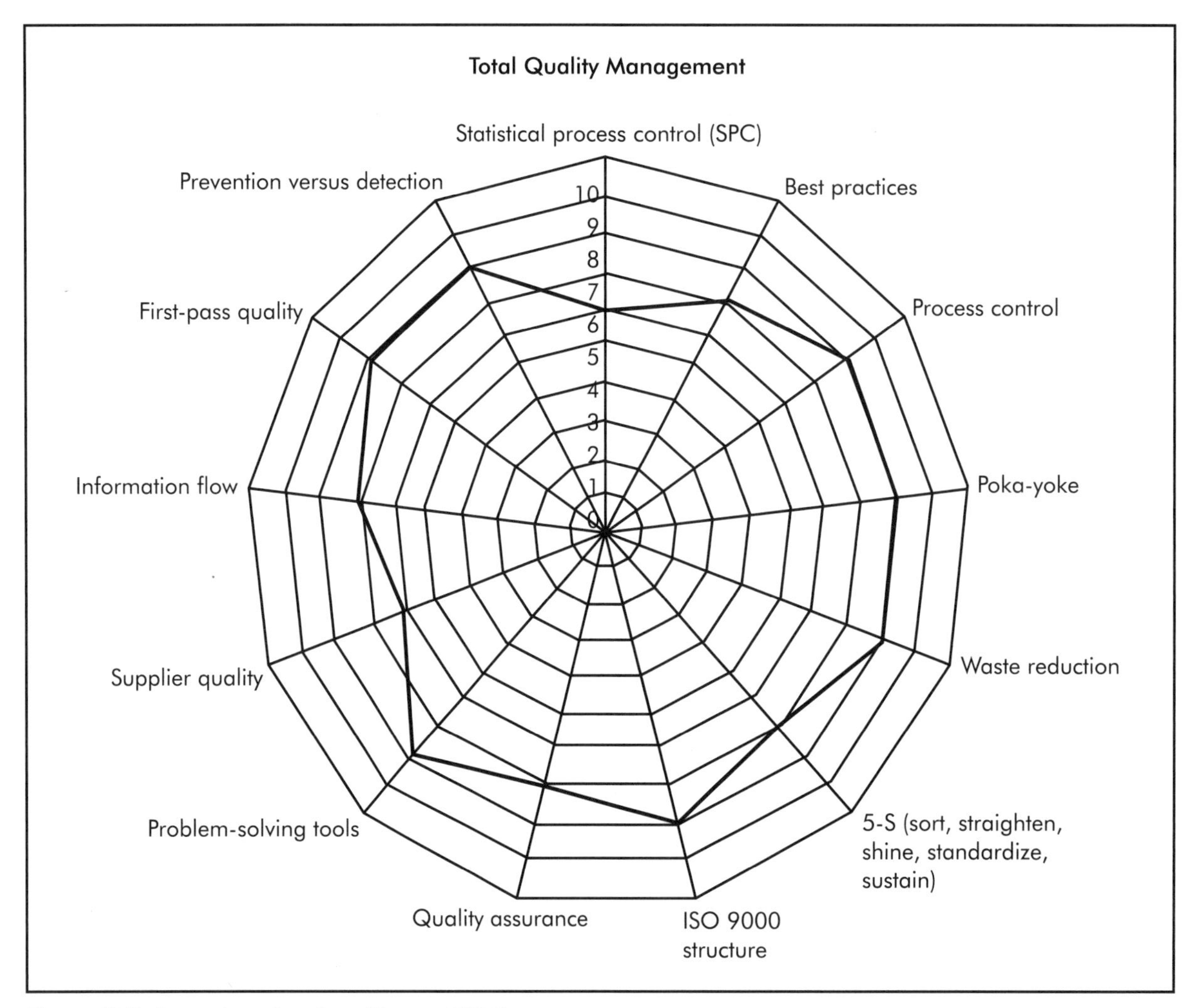

Figure 48-7. Example radar chart (Conner 2001).

his or her superiors. In this case, the presentation is informative and the topic is the project summary.

The next planning step is to evaluate the audience by addressing the following criteria. How big is the audience? Large audiences generally require a more formal presentation without much opportunity for audience interaction. Smaller audiences can be less formal and more interactive. What is the audience's composition, such as gender, age, and ethnicity, for example? What is the audience's motivation for attending? Is attendance required or optional? What is the audience's probable reaction to the presentation? If the presentation is about company downsizing, then the probable audience reaction will be fear and anger. Addressing the probable reaction in the planning stages increases the chances of a successful presentation. What is the level of knowledge of the audience? The level of detail and complexity given in the presentation should correlate to the audience's level of interest and motivation. If the audience is attending the presentation for a project summary, a detailed analysis is not necessary even though the audience may have the appropriate technical background. Finally, what is the relationship between the audience and the speaker? The presentation tone will be different if the audience is comprised of immediate co-workers and friends versus supervisors and/or strangers.

The main statement should inform the audience of the presentation's topic and scope. In the beginning, the speaker needs to identify the presentation's goal. In other words, where is the speaker taking the audience and why?

To guide the audience, the presentation topics should be sequenced logically. The presentation also should be interesting to hold the audience's attention. This can be accomplished by explaining why the presentation topic is important and how the topic personally affects the audience.

Finally, the presentation should close with a brief summary of the main points and a restatement of the purpose. If the audience is expected to formulate a decision or incorporate the presentation material into future actions, the speaker should provide the audience with some guidance as to what to do next.

Prior to the speech or presentation, supporting material or data pertinent to the purpose of the speech or presentation should be compiled. Convincing or supporting information is helpful in communicating the message.

How long should the speech or presentation be? If there is a predetermined time limit, then the speech or presentation must be designed accordingly. However, being allotted 30 minutes, for example, does not mean the presentation or speech must necessarily consume the entire time. An audience is likely to react negatively if a speech or presentation that lasts 30 minutes could have been done adequately in 20 minutes.

Which presentation style should be employed? The style is dependent on the purpose and environmental setting for the presentation. If the purpose, for example, is to present a technical paper at a national conference, then a formal style would be appropriate. If the audience is small and/or comprised of co-workers, an informal presentation style may be appropriate. An informal style also would be appropriate if the presentation involves audience interaction with the presenter. A formal presentation style would be appropriate for larger audiences and/or when the audience is comprised of superiors or strangers. In a formal presentation, the speaker may stand behind a podium and possibly on a stage. The speaker in an informal presentation may simply stand in front of or mingle with the audience.

48.4 NEGOTIATION

Negotiating is the process of working with others to determine a mutually acceptable solution. Negotiations of all types and sizes take place every day and are an integral part of any organization. They can be on a small scale, such as between co-workers, or on a much larger scale, such as between management and labor unions, for example. For all the successful negotiations that occur there are an equal number of those that are unsuccessful. Unsuccessful negotiations generally result in one party conceding more than it needs to or, in the case of labor unions, a strike. Strikes may be considered by some people to be a necessary or acceptable part of negotiating. However, strikes can be detrimental to individuals, families, and the economy. The damage caused by a strike or unsuccessful negotiation, in terms of poor attitudes and hurt feelings, for example, can require a long time to repair.

In negotiating, one party tries to advance its own interests as much as possible and persuades the other side to agree. From that point, compromises and concessions begin. The best case scenario in any negotiating situation is to persuade the other side to agree to your terms because it is in their own best interest (Sebenius 2001).

It is important to understand personal interests and limitations or no-deal options. However, it is equally important to understand the other side's limitations and perspectives and not dismiss them. For example, asking for a substantial raise in the next contract may seem reasonable. However, from management's perspective, the company may be on the verge of bankruptcy. A raise would ensure the company's demise.

Negotiated agreements are seldom based strictly on price. Most people will not accept a deal if it is not equitable and fair. In other words, something is not always better than nothing. For example, an employer may offer employees a choice between a 10-cent raise or nothing at all. Theoretically, 10 cents is better than nothing. However, the employees may view the raise as unfair and insulting, thereby rejecting the offer. Building a healthy relationship between the negotiating parties is equally important and, in some countries, more important than the deal negotiated. The lesson is to not ignore the negotiating process itself. The process must be viewed as personal, respectful, and fair.

In negotiations there are issues, positions, and interests (Sebenius 2001). For example, if employees are asking for raises because childcare costs are rising, the issue is pay raises. The employees' position is that pay raises are mandatory and management's position is strictly no pay raises. Alternately, if the opposing positions converge, then there is an agreeable solution. If not, then there will be a stalemate. What about the interests of both sides? Employees are interested in quality childcare and management is interested in maintaining profits and satisfying shareholders. Is there a deal that can satisfy the interests of both sides? Company-sponsored, in-plant childcare may be less expensive than providing raises and it would satisfy the employees' interest. There are obviously other issues with in-plant childcare. However, the example illustrates how satisfying interests is an alternative to converging positions. Conventionally, negotiating seeks to provide a win-win solution based on common ground between two parties. Common ground is always helpful; however it is not the only source for solutions (Sobenius 2001). In the previous example, the agreeable solution was not based on common interests.

The best alternative to a negotiated agreement (BATNA) can be a powerful negotiating tool (Sebenius 2001). The threat of walking away from a tentative negotiated agreement provides leverage. If management offers an employee a 5% raise but a competitor is willing to hire the same employee at 10% higher salary, then the employee has some leverage to improve the deal. Both sides need to fully understand and appreciate their respective BATNAs.

Finally, for successful negotiations, each side must keep an open mind and not demonize the other side. It is easy to view one's own side as better, smarter, and more honest. It is also easy to view the other side as less moral and dishonest. The perceptions of the other side may become self-fulfilling prophecies. Contending the other side is dishonest, for example, in some cases may actually cause the other side to be dishonest.

The negotiating concepts discussed in this section do not constitute an all-inclusive list. There are other issues to be considered such as communication, timing, personalities, etc.

48.5 CONFRONTATION AND CONFLICT

In the workplace a variety of situations and personal habits can annoy others. In many circumstances annoying habits or actions are not intentional but rather unintentional. Often people are unaware that their actions are being construed as irritating. The following section describes how some of these irritating issues and people can be confronted constructively without conflict.

Confrontation and conflict are not the same thing (Craumer 2002). *Confrontation* implies addressing or bringing issues to another person's attention. *Conflict* implies disagreement, arguing, and hurt feelings. It is possible to confront someone without making them defensive. Not all issues require confrontation. However, the absence of confrontation can result in stress, resentment, and lower productivity.

Before confronting someone, make sure you have enough information and are objective (Craumer 2002). In some cases what you find annoying is unique to your personal biases and most people may not find it annoying. For example, the wording and punctuation of e-mails may be misinterpreted by the reader. One person may view an exclamation point as humorous and another may find it offensive; the interpretation is entirely subjective.

Before confronting someone be sure about the expected outcome and its reasonability. If someone uses a body gesture you find irritating, what do you expect him or her to do about it? It is not fair to confront someone unless there is a reasonable solution. Additionally, if you do not work with this person daily it may be awkward to confront him or her and the situation could become defensive. In this instance you may have to foster a relationship prior to confrontation.

Prior to a confrontation think about the approach, words, and language to use. Be conscious that certain approaches and language may make the other person defensive. Be polite; do not accuse or judge. Do not make the issue personal; focus on the problem (Craumer 2002). Another advisable technique is to discuss joint responsibility and solutions while confronting the other person. In other words, indicate that you may be part of the problem by the way you are interpreting the other person's actions. Also, how can you be part of the solution? What can you do together to solve the problem?

Finally, choose the best place and time for the confrontation. Normally, a confrontation is a private matter. However, if unsure how the other person will react, having a third person in the vicinity or doing it in public may be a good idea. Make sure the setting is conducive to confrontation. Confronting someone in their office or your office may make them defensive. Confronting someone in the lunchroom or away from the workplace may make the other person less defensive.

Finally, remain objective and emotionally detached from the situation (Craumer 2002). When emotions are involved, people can make inappropriate comments or body gestures, resulting in defensiveness and conflict.

CONFLICT

Some conflict within an organization or team is normal and unavoidable. In a team or workplace conflict typically involves:

- manpower resources,
- capital expenditures,
- technical opinions and trade-offs,
- priorities,
- scheduling,
- personality clashes, and/or
- administrative procedures.

Without conflict or without enough conflict, a phenomenon called *group think* can result. Group think occurs when group members do not express their personal opinions but rather willingly submit to what the group as a whole thinks. Group think can lead to bad decisions and inappropriate actions.

Too much conflict, however, can have negative effects. When an excessive amount of conflict exists, teams cannot cooperative effectively and making decisions or taking action is difficult.

There are five basic modes of managing conflict: withdrawal, smoothing, compromising, forcing, and confrontation. Some people tend to withdraw from conflict rather than participate in it, thereby reducing the tension.

Smoothing results from de-emphasizing or avoiding areas of difference and emphasizing areas of agreement. For example, if someone is upset about a decision, the conflict can be reduced by discussing the many areas of agreement that outweigh the one area of conflict.

Compromising involves bargaining for solutions that bring some satisfaction to all parties involved. Compromising works if the parties are willing to give in for something in return. If only one party gives in they feel like the losers. Even if all parties give in they may feel like losers if they are not getting completely what they want.

Forcing is exerting one's viewpoint at the potential expense of another's (win/lose situation). This means one person argues their viewpoint until the other person gives in and abandons their viewpoint. Forcing can result in a poor decision or action. The viewpoint that is forced may not result in the best decision or action to solve the particular problem.

Finally, confrontation involves facing the conflict directly. It is a problem-solving approach whereby affected parties work through their disagreements. The problem-solving approach allows people who firmly believe in their individual viewpoints to work together to create a mutually acceptable solution.

The best mode of conflict management depends on the situation. In general, compromising and confrontation are usually the best modes of conflict management because each party's viewpoint formulates the final decision or idea. Withdrawal, smoothing, and forcing eliminate one party's viewpoint and rely exclusively on the other's viewpoint.

48.6 MEETING MANAGEMENT

Meetings consist of a number of people gathered together to exchange information, generate ideas, and develop collective decisions. Meetings can be a contentious issue among employees. It is not uncommon to hear people say they could get their work done if it were not for all the meetings they attend. Meetings with appropriate objectives and planning can be very useful. The following section discusses how to plan and execute good meetings, and the common pitfalls of poor meetings.

MEETING PREPARATION

Prior to a meeting the person facilitating it must determine the objectives and if the potential outcome will be worth the time and money invested in the meeting. Typical meeting objectives include information sharing, idea generation, and assigning responsibilities (Jay 1976).

Information sharing is an important meeting objective, but there are other modes of sharing such as e-mails and memos. A meeting is not advisable if the information could be communicated effectively through an e-mail or memo. A meeting is necessary if the information needs to be disseminated by a particular person or the information is complicated and requires further explanation. If a particular problem within the group needs to be solved or a new policy generated, a meeting would be productive. Some idea-generating techniques such as brainstorming require live interaction as opposed to e-mails or memos. Group discussion allows collective decision making on problem solutions and policies.

Finally, a meeting is a forum for assigning responsibility for implementing any ideas or policies generated in the meeting. In other words, "who is going to do what?" Individual roles and responsibilities should be clear to everyone in the group.

Meetings are not free and therefore should have a good return on their investment. The cost of a meeting includes wages or salaries and time. Is it worth paying someone to participate in a meeting (investment) rather than for working on his or her own responsibilities? The meeting outcome (return) needs to be of some value since the participants are being paid for their contributions to the meeting and not performing their individual responsibilities. The value of the outcome is sometimes hard to quantify. Certainly cost-cutting ideas generated by the group, for example, can be quantified. However, outcomes such as policy changes that affect employee morale are more difficult to measure.

In addition to the return on investment of the overall meeting, the same cost/benefit analysis should be applied to each agenda item. In other words, do not spend 95% of the allotted meeting time on an agenda item or items that likely have a minimal outcome.

If a meeting is justified, further preparation is required, such as determining the meeting length and participants, and preparation of the agenda and other written materials as needed. In general, a meeting is more successful and valuable if the number of participants is small. Larger meetings can be successful; however, more intensive preparation is necessary. The length of the meeting is also a factor in its success. Approximately two hours is the maximum length for an effective meeting (Jay 1976). The most valuable meeting outcomes are derived from the first half of the meeting. After this time period, the participants get tired and their attention and focus diminishes. In some cases, it is a good idea to assign a time limit to the meeting and time limits to each agenda item to keep the group focused.

The agenda, which has been mentioned several times, is the outline or blueprint for the meeting's facilitation. All potential items should be evaluated for appropriateness prior to placement on the agenda. All issues to be discussed should be on the agenda. The inclusion of "other business" or "other items" as the last agenda item is an invitation to waste time (Jay 1976). If it is important enough to discuss in the meeting, then it should be distinctly listed on the agenda. Additionally, for the facilitator to be in complete control of the meeting there cannot be any surprise items raised by a member of the group. Topics for discussion should be solicited from group members prior to the meeting. If the topic warrants discussion it can be placed on the agenda.

Finally, any proposal or lengthy written material to be discussed should be distributed before the meeting. If members are expected to intelligently discuss the material they need time to review it beforehand. If members are expected to read material during the meeting, it obviously should be kept short.

MEETING EXECUTION

Conducting a meeting effectively and efficiently is the facilitator's responsibility. The facilitator guides the participants toward achievements they could not make individually. A facilitator should not impose his or her ideas on the group. In measuring personal achievement, being busy or spending time in meetings does not mean someone is effective. Activity for the sake of activity is not necessarily productive.

It is the facilitator's job to keep the meeting on schedule and the participants focused. The facilitator should make the meeting objectives clear and make sure the participants understand all the issues. The facilitator should politely and objectively prevent participants from making long speeches and defer issues that will not be concluded in the time allotted to the next meeting.

The facilitator can lead group discussions by adhering to the following technique (Jay 1976). First, at the beginning of discussion of each agenda item, the facilitator makes sure the reason for the item is clear. Second, the history of the agenda item is discussed. For example, how long has this problem existed? Third, what is the current status of the problem? Diagnosis is the fourth step. What is the root cause of the problem? Finally, the last step is the derivation of a solution. How can the problem be solved?

Additionally, it is the facilitator's job to control those in the meeting who take a long time to say very little and to encourage those who are silent to voice their opinions and/or facts. The facilitator should encourage differing viewpoints and ideas while maintaining control of the discussion. It is important that everyone in the meeting feels their input is welcome and that their ideas will not be immediately dismissed. This is one reason, if possible, to call on the more senior people last (Jay 1976). If the senior participants speak out first, then the less senior participants may feel the need to publicly agree with them. The less senior participants may be reluctant to contradict the senior participant's ideas.

POST MEETING

Following the meeting, the minutes taken during the meeting should be distributed to all meeting members and others if appropriate. The minutes should contain the following information:

- time, date, and location of the meeting;
- members present;
- agenda items and summaries of their respective discussions;
- issues that were voted on with the voting results;
- time the meeting was adjourned; and
- if appropriate and available, the time, date, and location of the next meeting.

48.7 CREATIVITY AND INNOVATION

Creativity and innovation are important for competing in the global economy. Creativity and innovation also help individuals become personally successful. If it is critical for success, how does a person become creative and innovative? How does a person generate original thoughts and ideas? To begin, creativity is not limited to people with creative talents. To be creative, a person must first decide to be creative and then emulate the acts of creative people (Zelinski 1998). In other words, do not give up on being creative because you do not have any creative talent. Also, intelligence is not a strong indicator of creative ability (Gary 1999). If people have enough intelligence to understand the field they are working in, then their intelligence will not limit their creativity.

Young people are generally thought to be creative. The psychology of a "creative" type person is similar to that of a three-year old or preschool-age child. They are frequently inquisitive and ask the question "why." With age, people tend to become more accepting, less inquisitive, and more conforming. Though subject to more paradigms, older, more experienced people can rely on their experiences to be creative and innovative.

Being creative and innovative involves risk taking. With the development of anything new there is always the risk of failure and criticism. It is safer not to have an original thought or idea. However, the benefits of successful ideas and innovations can be enormous in terms of professional and economic success. Without sharing creative ideas the physical transformation into innovation is impossible. In fact, the merits of creative thinking can only be proven through implementation.

Finally, being creative and innovative is hard work. Successful ideas and corresponding innovations occur after long periods of experimentation, testing, and failures. In the case of patents, invention is defined as the combination of conception and reduction into practice. Conception is the "eureka, I've got it" moment and reduction into practice is the process of making the concept functional. Reduction into practice could take hours, days, or years.

How can people become creative or more creative? To start with, a person should become more open-minded, less complacent, and more willing to take risks or risk being wrong and/or criticized. Look for the unexpected, which usually is found in failures (Drucker 1998). Innovations arise from the process of implementing creative ideas. Following industry and market changes can be helpful in creatively thinking about good ideas for the future. Also, demographic changes can influence creative thinking. Take advantage of the perception of others. For example, if the public perceives fuel economy to be very important, then focus creative energy on fuel economy related ideas such as hybrid automobiles. Finally, creativity and innovation involving completely new knowledge and technology can involve a fairly long implementation period. Creativity and innovation can, however, originate from existing knowledge and technology. For example, it is common for individuals and companies to improve upon patented products and technology.

REVIEW QUESTIONS

48.1) Describe content listening.

48.2) How are proposals different than reports?

48.3) Name the two types of charts typically used to illustrate data with respect to time.

48.4) What does BATNA refer to?

48.5) How is confrontation different than conflict?

48.6) What mode of conflict management is being used when someone refuses to participate in a conflict?

REFERENCES

Bovee, C.L. and Thill, J. V. 2000. *Business Communication Today*, Sixth Edition. Upper Saddle River, NJ: Prentice Hall.

Conner, Gary. 2001. *Lean Manufacturing for the Small Shop*. Dearborn, MI: Society of Manufacturing Engineers.

Craumer, M. 2002. "Confrontation Without Conflict." Harvard Management Communication Letter, June.

Drucker, P. F. 1998. "The Discipline of Innovation." Harvard Business Review. November-December.

Gary, L. 1999. "Beyond the Chicken Cheer: How to Improve Your Creativity." *Harvard Management Update*, July.

Jay, A. 1976. "How to Run a Meeting." *Harvard Business Review*, March-April.

Sebenius, J. K. 2001. "Six Habits of Merely Effective Negotiators." *Harvard Business Review*, April.

Zelinski, E. J. 1998. *The Joy of Creative Thinking*. Berkeley, CA: Ten Speed Press.

Chapter 49

Machining Processes Analysis

49.1 CUTTING TOOL MATERIALS

In cutting, the goal is to achieve the appropriate accuracy, surface finish, and maximum productivity. Whether it involves production or prototyping, cost, timing and quality are the key drivers. Traditional machine technology has been constrained by the limitations of the cutting tools themselves. Regardless of how fast machines can operate or how sophisticated the software is, if the cutting tool is inadequate, the job will be done incorrectly. Workpiece materials are also becoming more sophisticated, presenting new challenges to the machining industry.

"Advanced" cutting tool materials refer to those materials needed in cases where higher cutting speeds are used, which generate higher tool temperatures. The exact material composition of the advanced cutting tools available is difficult to define given that it is a very competitive industry and the compositions are proprietary. This section describes some of the key materials: carbide, coated carbide, and cermet. Ceramic and diamond cutting tool materials were discussed in sufficient detail in Chapter 20 of *Fundamentals of Manufacturing*, Second Edition.

CARBIDE

The carbide cutting-tool family comprises a powdered metal formulation that has been very popular for tooling. Carbide tooling dates back to the 1930s when it was developed as a means to cut at higher speeds. Carbides grew in popularity due to their minimal cost increase over high-speed steel (HSS) tools, with the added capability of typically a three to five times speed increase. In fact, carbide tools perform well only at high speeds.

The use of carbide tools became widespread after the development of the powdered metal sintering process, which allowed difficult materials, such as tungsten, to be fused together into net-shape tools. Carbide tools are also referred to as "cemented carbide" because in the sintering process tungsten carbide (WC) powder is "cemented" by using a small amount of cobalt (Co) as the binder. Sintering consists of heating the mixture of WC and Co powder to a temperature of about 2,800° F (1,538° C).

The binder, a ductile material, provides strength and increases toughness and shock resistance; however, it also reduces hardness and wear resistance. Nickel (Ni) is sometimes used as a binder to improve chemical resistance and if magnetic properties are a concern. Titanium (Ti) and tantalum (Ta) also can be added to modify the properties of the tool. Grain size is an important factor. An increase in grain size will increase toughness but reduces the hardness and wear resistance of the tool.

Tungsten carbide tools offer toughness, which is necessary for making interrupted cuts. Figure 49-1 shows the superior hardness of carbides over a wide variety of temperatures versus most other tools. These so-called straight WC-Co (tungsten carbide with a cobalt binder) carbide tools are well-suited for machining most cast irons, nonferrous metals, and nonmetallic materials. They provide an increase in cutting speed capability plus the ability to cut harder materials with improved efficiency.

The key benefits of carbide are:

- high hot hardness,
- high elastic modulus for toughness,
- good thermal conductivity, and
- low thermal expansion.

Carbide's wear characteristics are also good. However, they are still not as good as the newer ceramic materials. The softer carbides are tough, but the harder grades have higher wear resistance and are better for finish cuts. However, they are more brittle.

Cutting tools made from WC-Co are not suitable for machining steels. The tool's incompatibility results in rapid wear in the chip contact area. This is evidenced by crater wear at the top of the cutting edge. For cutting steel, titanium or tantalum is added to the carbide because these additives are harder and more stable than tungsten. Titanium is added to resist cratering and tantalum is added to improve hot hardness and thermal shock resistance, thereby providing better nose and edge wear. This is especially helpful in heavy, rough cutting operations.

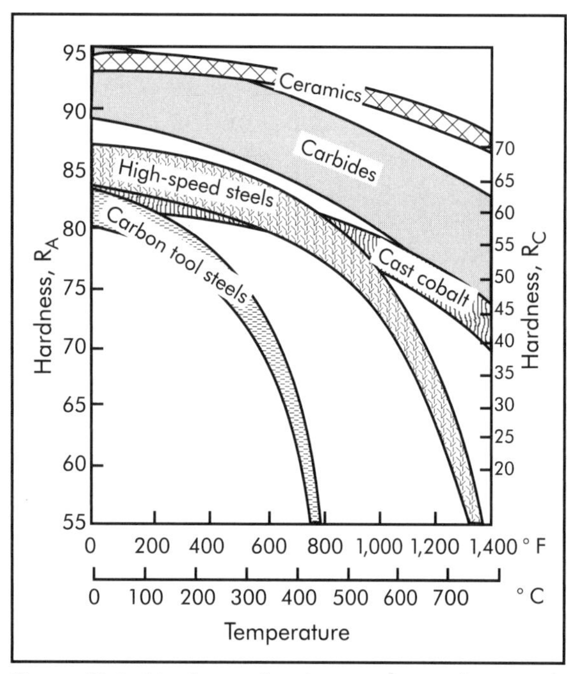

Figure 49-1. Hardness of various cutting-tool materials as a function of temperature (Drozda and Wick 1983).

Micrograin carbides are an extension of the conventional WC-Co grades. These tools maintain wear resistance or hardness because of their extremely fine microstructure, while the increased binder levels they contain maximize strength and shock resistance. The major advantage of micrograin carbides is their higher toughness compared to conventional grades of carbide with equal hardness but lower binder content. They are used for machining applications requiring severe interrupted cuts and in some machining operations on stainless steels, high-temperature alloys, and titanium. Micrograin carbides are also used as cutoff and form tools in some applications where they would otherwise chip or break if made from conventional carbide. One disadvantage of micrograin carbides is that they have a greater tendency to fail by cratering than other carbides.

Carbide Grade Classifications

The ISO and ANSI designations are key to selecting the appropriate carbide composition. Classification of cemented carbides for cutting tools is a controversial subject. They are available in a wide variety of compositions with different properties and from many suppliers. Frequent reference is made to cast iron and steel cutting grades, and to some extent, edge-wear- and crater-resistant grades. These terms are often misleading because the first two are limited to consideration of the workpiece material, while the latter two are related to the mode of tool failure.

The primary classification systems for cemented carbide cutting tools are:

- the unofficial C-classification system initiated by the U.S. automotive industry and commonly used in the United States. Table 49-1 illustrates the C-classification system.
- the International Organization for Standardization (ISO) system based on ISO Standard 513-1975(E), which is widely used in Europe. Table 49-2 illustrates the ISO classification system.

Coated Carbide

Carbide inserts coated with wear-resistant compounds for increased performance and longer tool life represent a growing segment of the cutting-tool material spectrum. The use of coated carbide inserts has permitted increases in machining rates up to five or more times over the machining rates possible with uncoated carbide tools. Coated carbides are best suited for use on free-machining steels, plain carbon and alloy steels, tool steels, many grades of stainless steel, and cast iron. The first coated insert consisted of a thin titanium carbide (TiC) layer on a conventional WC substrate. Since then,

Table 49-1. Classification of tungsten carbide per U. S. C system (Drozda and Wick 1983)

Classification Number	Materials to be Machined	Machining Operation	Type of Carbide	Characteristics of Cut	Characteristics of Carbide
C-1	Cast iron, nonferrous metals, and nonmetallic materials requiring abrasion resistance	Roughing cuts	Wear-resistant grades, generally straight WC-Co with varying grain sizes	Increasing cutting speed ↑↓ Increasing feed rate	Increasing hardness and wear resistance ↑↓ Increasing strength and binder content
C-2		General purpose			
C-3		Finishing			
C-4		Precision boring and fine finishing			
C-5	Steels and steel alloys requiring crater and deformation resistance	Roughing cuts	Crater-resistant grades, various WC-Co compositions with TiC and/or TaC alloys	Increasing cutting speed ↑↓ Increasing feed rate	Increasing hardness and wear resistance ↑↓ Increasing strength and binder content
C-6		General purpose			
C-7		Finishing			
C-8		Precision boring and fine finishing			

Table 49-2. Classification of tungsten carbide per ISO system

ISO Color Code	ISO Classification	Materials to be Machined	Machining Operation	Carbide Characteristics
Red	K01 K10-20 K30-40	Cast iron, austenitic stainless steel, high-temperature alloys, nonferrous materials, and nonmetallic materials	Finishing ↕ Roughing	Base grade with little or no titanium Good tolerance for abrasive wear Most commonly used for short-chipping materials
Yellow	M10 M20-30 M40	Plain carbon steel, alloy steel, pearlitic malleable iron, nodular iron, and martensitic stainless steel	Finishing ↕ Roughing	Medium grade with roughly 5–10% titanium Called "crater-resistant" grades, used to machine longer-chipping materials
Blue	P01 P10-20 P30-40	Steel and steel alloys	Finishing ↕ Roughing	Premium grade with up to 30% titanium and/or tantalum Used for chipping materials Recommended for high-speed finishing cuts Much more resistant to cratering Should not be used for the high abrasive materials such as those listed with the K grades

improved substrates better suited for coating have been developed to increase the range of applications for coated carbide inserts.

The base material, or substrate, provides the necessary toughness and thermal shock resistance required for cutting conditions. The coatings provide a barrier between the substrate and the workpiece. Therefore, the limitations of conventional substrate materials can be overcome. The high hardness of the coating materials allows for outstanding wear resistance. Typical coating materials include titanium carbide (TiC) for wear resistance, titanium nitride (TiN) for hot hardness and oxidation resistance, and aluminum oxide (Al_2O_3), which has excellent wear and thermal properties at high speeds. The titanium-based coated tools are typically operated in the medium cutting speed range, while the aluminum-oxide-coated carbides can be used at higher speeds. The coatings are applied in very small layers, as thin as 78–393 μin. (2–10 microns). Very few layers are required, as thicker coatings will be prone to adhesion and chipping problems. The coatings are applied using either the physical vapor deposition (PVD) or chemical vapor deposition (CVD) process at extremely high temperatures.

Increased productivity is the most important advantage of using coated carbide inserts. With no loss of tool life, they can be operated at higher cutting speeds than uncoated inserts. Longer tool life also can be obtained when the coated tools are operated at traditional speeds. However, higher-speed operation, rather than increased tool life, is generally the approach taken to improve productivity and reduce overall processing costs. The feed rate used is generally a function of the insert geometry, not the coating.

The increased versatility of coated carbide inserts is another major benefit. Fewer grades are required to cover a broader range of machining applications. This simplifies the selection process and reduces inventory requirements. Most producers of coated carbide inserts offer only three primary grades: one for machining cast iron and nonferrous metals and two for cutting steels.

Coated carbide inserts are not suitable for all applications. For example, they are generally not suitable for light finishing cuts, including precision boring and turning of thin-walled workpieces, two operations which usually require sharp cutting edges for satisfactory results. They also should not be used for machining workpieces containing surface sand or scale, inclusions, or other imperfections. Most heavy roughing operations and severely interrupted cuts are not recommended. Also, they are often not as suitable as uncoated carbide inserts or other tool materials for machining some nonferrous metals and nonmetallic materials.

CERMETS

Another important family of sintered cutting tool materials is *cermets*. These were originally developed in Germany during WWII but were too brittle at that time to gain widespread commercial use. Cermets reached commercial success in the 1960s. They bridge the gap between cemented carbides and ceramics in composition and function.

Cermets combine hard particles, such as titanium carbide (TiC), titanium nitride (TiN), and titanium carbonitride (TiCN), and a metallic binder, such as molybdenum, cobalt, and/or nickel. Cermets provide wear resistance (crater and flank wear) on the order of 20 times higher than cemented carbide. However, this is still less than the higher performing pure ceramic tools. Hardness characteristics, including hot hardness, of cermet tools are similar to cemented carbide. Unlike cemented carbide, the metal matrix of a cermet determines its hot hardness level. Cermet's strength, toughness, and thermal shock resistance is lower than cemented carbide.

Cermets work best with materials that produce a ductile chip, such as steels and ductile irons. Their increased speed capability enables them to machine carbon, stainless steels, and ductile irons at high speeds while producing excellent surface finishes. They work best in dry cutting applications; in fact, coolant tends to increase wear with this material. With strong geometry, cermets are also acceptable for deeper and intermittent cuts. Corrosion resistance is another positive characteristic of the cermet materials as well as their good abrasion and heat resistance.

Cermets are not recommended for rough or semi-finishing because of their lower strength at higher feed rates. They are inherently limited in their resistance to plastic deformation, primarily because of the metallic binder. Newer grades containing molybdenum carbides operate better on interrupted cuts, resisting plastic deformation and maintaining a sharper edge throughout the wear cycle.

49.2 INSERT GEOMETRY AND CLASSIFICATION

The use of inserts for cutting tools dominates the tooling industry. The wide variety of tool geometries available requires careful consideration of all tool materials, workpiece geometries, and machining conditions. The advantages of using inserts include the availability of multiple cutting edges on each insert and there is no need to resharpen them. The evolution to inserts has led to the availability of a proliferation of sizes, shapes, and configurations.

The selection of each parameter is critical to a successful machining process. Tool suppliers furnish extremely good documentation on each parameter, allowing the best selection of tooling, although some testing and development are often necessary. One key parameter is the tool's shape. The typical shapes are shown in Figure 49-2. While the shapes available vary greatly, special considerations about their actual application must be taken into account.

Typically, the macro geometry determines whether the insert is designed to produce light or rough cuts. The geometry dictates the parameters used for cutting, such as the use of higher rake angles, which are good for rougher cuts. The tool's nose radius may be large for strength, or sharp for fine radius turning. Since a sharp edge is weak and fractures easily, an insert's cutting edge is made in a particular shape to strengthen it. Those shapes include a honed radius, a chamfer, a land, or a combination of the three.

The insert size is designated by the largest circle that can be inscribed within the perimeter of the insert. This is called the *inscribed circle* (IC). For rectangular and parallelogram inserts, the size is noted by the length and width dimensions. The inscribed circle dimension is illustrated in Figure 49-3.

The International Organization for Standardization (ISO) and the American National Standards Institute (ANSI) provide standard insert classifications, which allow suppliers to offer a fairly standardized set of tool shapes. The ISO standard is 1832: 1991 and the ANSI standard is B212.4 2002. The classifications specify:

- insert shape,
- relief angle,
- tolerances,
- insert type,
- size (IC),
- thickness,
- corner radius,

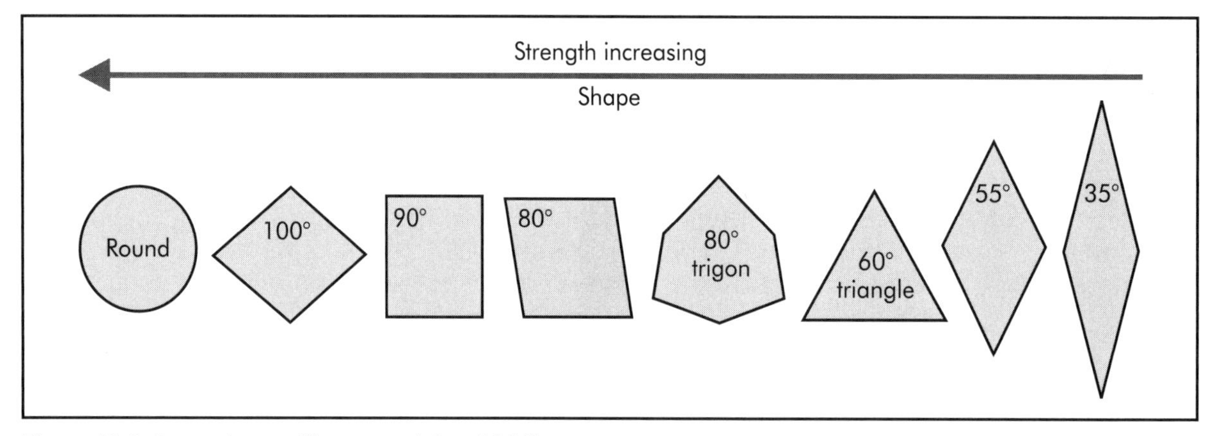

Figure 49-2. Insert shapes (Kennametal, Inc. 1996).

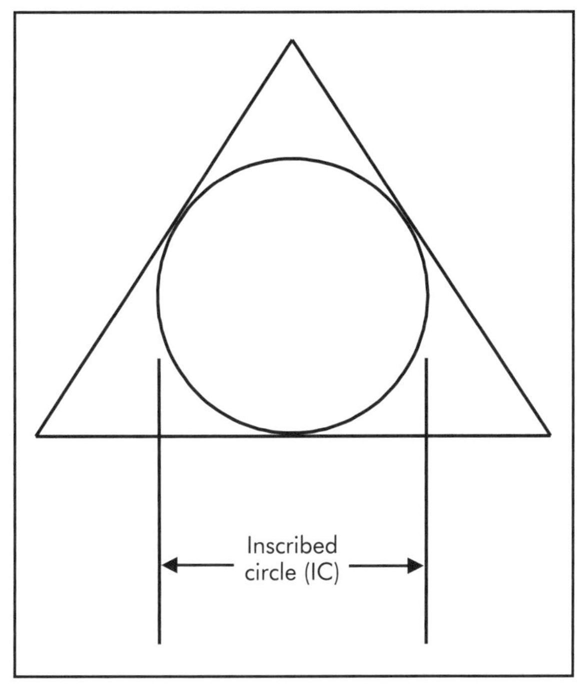

Figure 49-3. Inscribed circle (IC) on a triangular insert.

- hand of insert (left or right), and
- cutting edge condition.

More information about the insert classification system can be found in the literature provided by most insert manufacturers and suppliers.

49.3 TOOL SELECTION PROCESS

The tool selection process involves two key factors, cutting physics and economics. The quality of the cut is purely the result of the physics involved, which depend upon the cutting parameters selected. The goal is to choose a cutting tool shape and material that minimize cutting time, tool changes, tool cost, and setups, while maximizing tool life, accuracy, and surface finish.

In setting up the cutting process, there are several things that must be known about the cutting conditions:

- workpiece material;
- interrupted or continuous cut;
- machines available and their capacity for workpiece size, spindle geometry, speed, and power;
- maximum allowable cycle time, allowable number of tool changes, etc. (economic objectives);
- amount of material to be removed (rough or finish cut—tolerance and surface finish);
- workpiece contours (tool path and tool access to machining features);
- number of pieces desired per tool;
- cutting fluid availability; and
- tool shapes available (standard or custom ground).

49.4 HIGH-SPEED MACHINING

High-speed machining (HSM) achieves faster metal removal rates by running higher spindle speeds and feed rates but taking lighter cuts. To do this requires better tools, better machine control, and better machines. There are two main goals of high-speed machining. One is speed and productivity, and the other is better surface finish to eliminate or reduce the need for polishing, especially on molds and dies.

HSM is popular for milling aluminum at spindle speeds over 12,000 rpm with feeds over 600 in./min (15 m/min) (Aronson 2001). This approach provides a better finish, which has the potential of eliminating secondary operations such as grinding, hand rubbing, or electrical discharge machining (EDM). It can also reduce the warping that occurs with deeper cuts, especially in aerospace work. The ability to machine complex parts by contour milling is another reason HSM has been popular in the aerospace industry.

Basic HSM parameters are spindle speeds above 10,000 rpm, 400 in./min (10 m/min) or greater feed rates, and 2–5-axis machine control. By operating in multiple axes, the machine can keep the cutting tool at the optimal cutting angle, thereby reducing deflection and maintaining high feed rates. The complicated control of such a system requires advanced computer control. Improved spindles, speed multipliers, toolholder damping, and machine controls have all advanced the state of HSM.

Not all machines are versatile enough to run the spindle speeds necessary. One common approach is to add a geared spindle multiplier to speed up the cutting tool. Of course, higher spindle speeds result in the risk of vibration and, therefore, chatter on the workpiece and increased risk of tool breakage. This has been addressed through the addition of vibration damping systems in the toolholder.

Current high-speed spindles present a tradeoff between cutting force and cutting speed. First, the size of the motor is limited. High-speed spindles generally have direct-drive motors, meaning the motor must fit inside the spindle housing. The bearings are very critical and high-speed spindle bearings trade stiffness for speed. This is one more reason why high-speed machining generally employs light depths of cut. As spindle speed increases, the choice of toolholders has more impact on process effectiveness. HSK (German acronym for hollow taper shank) and collet-type holders are used to improve concentricity. Hydraulic and shrink-fit tool-holders are gaining acceptance in industry.

Other controls include feedback systems for tool deflection, tool breakage detection, and temperature compensation.

MACHINE CONTROL

CAD/CAM and computer control of the machine has been a critical enabler of HSM technology. Programming for pocket milling, for example, is a typical HSM routine. More advanced controls include look-ahead software routines to accelerate and decelerate the tool for more effective cutting, and tool-path optimization software, which can provide more effective setup and tool selection. CAD models allow for a complete digital description of the geometry of the workpiece. Post-processors are an often overlooked piece of the process. The post-processor is software that translates the output from the tool-path software into the specific machine's code.

While this all seems to indicate that computer programs and machines do it all, the computer programmer and machine operator's expertise are the final enablers and probably the most critical to integrating all of the pieces into a workable, competitive, and productive process. The programmer and machine operator are responsible for establishing the valid base parameters, selecting the tools, and then running the programs and machines.

TOOLING

Advances in cutting-tool materials and geometry are also enablers of HSM. Superior toughness is the key requirement for tool selection. Fine-grade carbides and coated carbides (coated to improve wear) are common choices for HSM applications. Because complex part geometry can sometimes result in interrupted cutting conditions, the more brittle ceramic tools are not recommended. Cermet materials can be used with the benefit of being able to cut dry, thereby reducing thermal shock to the tool and workpiece.

HSM continues to grow and evolve and has already replaced other more traditional machining methods and even some newer technologies. HSM can run faster than EDM, therefore many applications involving complex detail are now done with HSM. Even some superfinishing operations have been eliminated since the finish of an HSM operation can eliminate polishing requirements. Because of the requirements for CAD/CAM and advanced machining center capability, HSM is best applied on complex contours for molds and dies in batches of one. However, as part geometries become simpler, higher volumes and overall throughput must be considered. In many cases, traditional machining methods may still be advantageous.

49.5 BORING

Boring is a precision machining process for generating internal cylindrical forms by removing metal with single-point tools (Figure 49-4) or tools with multiple cutting edges. This process is most commonly performed with the workpiece held stationary and the cutting tool both rotating and advancing into the work. Boring is also done, however, with the cutting tool stationary and the workpiece rotating. Common applications for boring include enlarging or finishing cored, pierced, or drilled holes and contoured internal surfaces. In the past, jig boring equipment was commonly used for high precision and specialized milling or boring jobs. Except for situations where very large equipment is required, these applications are well within the capabilities of today's CNC milling machines.

In boring, there are many variations of tools, spindles, and machines available. The process can be performed with the workpiece stationary as in a milling machine, or with the tool stationary as in a lathe operation. Boring is generally done with machines such as large horizontal and vertical boring machines, vertical turret lathes, and boring, drilling, and milling machines. Motors can be directly mounted on the spindle shaft for higher-speed applications.

The key to a successful boring operation centers on the quality and selection of the spindle and boring bar. Spindle selection is influenced by many considerations, including (Drozda and Wick 1983):

- speed of rotation,
- direction and magnitude of loads,
- accuracy,
- surface finish,
- mounting requirements,
- drive requirements, and
- dimensional limitations, such as overhang and center distance.

Boring operations are usually harder on cutting tools than turning operations because of the confined machining area, which can cause chip removal problems, especially from deeper and smaller diameter bores. As a result, the size, strength, and stiffness of

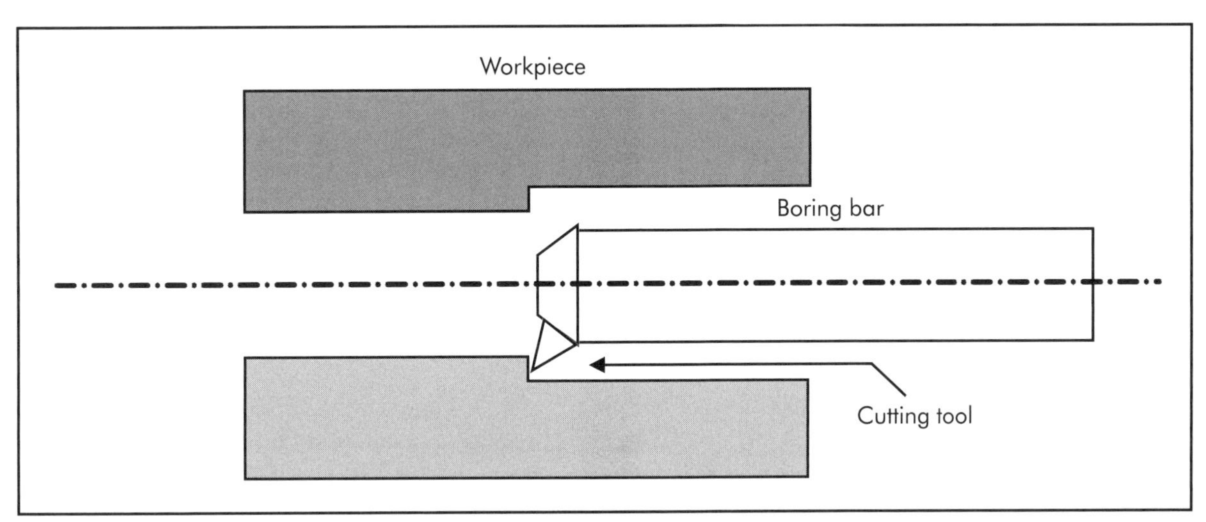

Figure 49-4. Boring.

boring tools are often limited by hole size and length of cut. Owing to the special nature of boring operations, however, some design considerations tend to be more critical. Boring tools are available as solid tools, with tips brazed to holders, and as indexable inserts.

Boring bars are made in a wide variety of styles and standard sizes. They are typically adjustable and allow for the use of adjustable inserts. A fine adjustment is included in increments up to 0.0001 in. (2.540 microns).

Boring bar stiffness is a critical issue and because of the concern for tool deflection, it is essential to use a tool that provides a high natural frequency for a stable cut. "Tuned" bars, containing internal damping fluids, are now available to reduce boring bar vibrations and workpiece chatter in specialized applications where a long reach or a superior surface finish is required. Internal cooling passages are also designed into certain boring bars, which can help in deep, horizontal applications.

When properly configured, a boring operation can yield a precisely sized, round hole with a high-quality finish. Using the appropriate spindle (as strong as possible) with the proper damping characteristics is important. When a deep boring operation is required, it is strongly recommended that the product design allow access for a support and clearance where possible for good chip removal.

49.6 BROACHING

Broaching is a traditional machining process used for special surface machining applications. It is a specialized cutting operation where the cutting tool is either pushed or pulled across the workpiece surface being machined. The tool is called a *broach* and it can produce straight, circular, or complex profiles. The broach has a set of cutting teeth where each tooth is at a different height to progressively remove material from the surface being machined as illustrated in Figure 49-5. The shape and spacing of those teeth is critical. It is important to give the chips the appropriate space as the next tooth makes its cut.

There are two major groups of broaches: internal and external. Internal broaches are pushed through the workpiece. Internal form broaches can finish all or part of the internal surface as illustrated in Figure 49-6. Keyways and splines are common applications for internal broaches. Also, round broaches can provide a finished size to a cast or rough-finished hole.

External broaches are pushed over the workpiece to machine-in specialized features or contours. They are mounted in a fixture and utilize a guide.

Tool selection is determined by the amount of material to be removed and the tool's ability to handle the required forces. The tools are made of typical materials such as high-speed steel or carbide.

Generally, any material that can be machined by other cutting processes can be broached. The material's ability to be broached is essentially the same as their machinability; however, because broaching is a high-force machining process, proper grain structure and hardness are of utmost importance. Softer steels tend to tear, and it is more difficult to produce smoother surface finishes when broaching them. Harder steels are more difficult to broach due to their hardness.

There are numerous advantages to using the broaching operation, including:

- The basic operation allows for unique and complex contours to be machined with a highly accurate finish.
- The versatility of broaching allows for several parts to be machined at the same time—even large or uniquely shaped workpieces—with relatively inexpensive tooling.

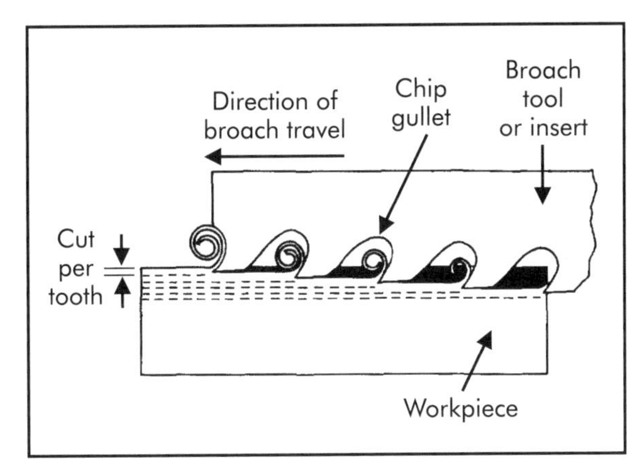

Figure 49-5. Broaching tool (Drozda and Wick 1983).

There are also limitations to broaching:

- Unobstructed access is required for tool travel.
- The workpiece or feature to be broached must be parallel to the direction of the tool.
- Complex contoured surfaces with curves in two or more planes cannot be formed in a single broaching operation.
- The typically high forces required in broaching require rigid machines and workholding devices.

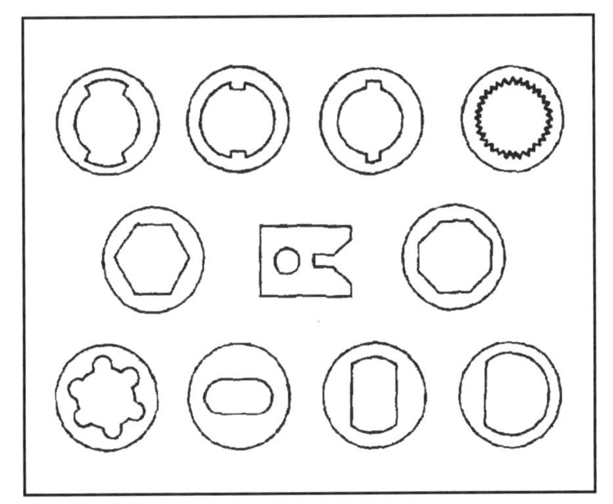

Figure 49-6. Internal broaching examples (Drozda and Wick 1983).

- Internal broaching requires a starting hole in the workpiece.

While high-speed machining (HSM) and other new processes have replaced broaching in certain applications, broaching continues to be a viable method of producing specialized, highly accurate shapes.

49.7 GRINDING

Basic grinding processes are capable of providing some of the most precise means of material finishing available. Grinding methods are typically selected to fulfill the following needs:

- a high level of accuracy,
- high-quality surface texture, and/or
- finishing a very hard workpiece material.

The variety of machines and tools available for grinding make it an extremely flexible process. However, there are several drawbacks to take into consideration. Grinding is typically time consuming and expensive compared to other methods of machining, so the feasible options should be given careful consideration. Also, tool selection and grinding parameters tend to be very sensitive and establishing the optimal process often requires development time.

As is true with most manufacturing processes, the primary factors that must be taken into consideration for the grinding process encompass the machine, the tool (grinding medium), and the workpiece. To develop the correct process, the following must be known:

- the amount of material to be removed,
- the workpiece's dimensions,
- the material's specifications, and
- the surface texture required.

The most critical parameter is the material's hardness and the basic grinding calculations vary depending on it. Soft steels are relatively easy to grind; tool steels and high-nickel alloys are difficult to grind.

Grinding can be done dry or wet. Like other cutting processes, coolant selection must be compatible with the tools being used. The chemical resistance of the bonding material in the grinding wheel must be considered. Generally, the requirement for parts feeding and recycling systems add cost to the operation. More specifically, water-soluble fluids can be a concern due to the possibility for rust, foaming, and bacteria growth. When using an oil-based fluid, its flash-point, chemical disposal, and environmental concerns (airborne inhalant/mist control) must be taken into consideration. Synthetic fluids resolve some of these shortcomings.

ABRASIVE MATERIALS

Conventional abrasives consist of aluminum oxide, silicon carbide, and zirconia alumina. Another important abrasive class is termed "superabrasives," which includes cubic boron nitride (CBN) and diamond materials.

For grinding ferrous materials, there are two options: aluminum oxide and CBN. Aluminum oxide is the most popular abrasive. It is used to grind steel, other ferrous alloys, and high-tensile, high-strength materials. Aluminum oxide is *friable*, that is, its abrasive grains shatter under pressure. It is best applied when there is a requirement to maintain sharp cutting points continuously through fracture of the individual grains, as in dry grinding of heat-sensitive steels.

CBN grinding wheels can be used on difficult-to-grind steel such as certain hard tool-and-die steels. CBN's abrasion resistance is on the order of four times higher than aluminum oxide.

Nonferrous, nonmetallic, and cast iron materials are generally ground with silicon carbide. Silicon carbide is also used for grinding very hard and dense metals, particularly when a smooth finish is required. A third alternative from the conventional materials is zirconia alumina, which has a unique self-sharpening characteristic that provides longer life when used for rugged stock removal of metals. Controlled fracture of the grains results in continuously sharp, new abrading points. Zirconia alumina is a manufactured abrasive mineral that is tougher than aluminum oxide, making it especially suitable for use when heavy grinding pressures are employed on nonferrous materials.

Manufactured diamonds are used for grinding ceramics and nonferrous metals. There is also a recent trend to use diamonds for difficult-to-grind steels because other abrasives break down too rapidly, thus generating added cost because of their inability to remove stock and hold geometry.

Grain size should be considered to obtain the correct surface finish. Typically, the smaller the grains, the finer the resulting surface finish. The density of the abrasive is also important. Higher-density grinding wheels should be used with lower-ductility workpiece materials.

GRINDING WHEEL BONDS

Just as important as selecting the correct abrasive for a grinding operation is selecting the correct bonding material. The function of the bonding material is to hold the abrasive during grinding. Once the abrasive starts to dull, which increases the forces on the grain, the bond must allow the grain to break away and expose the next layer of abrasive material. The typical bonding materials are vitrified, resinoid, rubber, shellac, and metallic. Material removal from the workpiece and grinding wheel must be taken into consideration when planning the process. Vitrified materials are strong and used for large wheels and higher grinding forces, but are typically limited to slower speeds. Where higher speeds or a slightly more flexible bond material is needed, resinoid is a better choice.

Rubber-bond materials are used in cutoff wheels and for centerless grinding. Shellac binder is used for wheels to produce high finishes on rolls and for cutlery grinding. The metallic bond is used for diamond abrasives commonly used in the grinding of ceramics. Metallic bonds are also used with aluminum oxide or diamond abrasive to provide conductive wheels for electrolytic grinding.

GRINDING PROCESS ANALYSIS

Wheel Wear

Wheel wear can by broken down into three categories: a) attrition—dulling of sharp grains from rubbing, b) grain fracture—primary cutting wear, and c) gross pullout of grains—excessive force and weak bond. Grinding is a sensitive process; if the force is too light, the result is burnish and buildup, resulting from rubbing instead of cutting, which can also burn the surface. If the force is too high, damage can occur to the work surface and grains are pulled out from the grinding wheel. Additionally, the pores of the grinding wheel can become filled with chips from the workpiece. This phenomenon is called *wheel loading*. Wheel loading decreases the cutting ability of the wheel and increases the heat generated during grinding.

Material Removal Rate

The material removal rate is a function of force intensity, rough or finish removal, effective diameter, and grinding ratio. *Force intensity* is the force across the width of the grinding wheel measured in lb/in. (N/mm). It is an indication of wheel wear and cutting depth. Consideration must be given to the amount of rough material that can be removed prior to a final operation to secure the correct dimension and desired surface quality on the workpiece. The *effective diameter* normalizes the conditions and allows calculation as if the workpiece were flat (required for internal and external radial grinding situations).

The *grinding ratio* is the volume of material removed from the work per unit volume of wheel wear. It is a useful measurement of the ease with which a material can be ground. The higher the ratio, the easier a workpiece material is to grind. The grinding ratio for a particular material, however, varies with different types of grinding operations and their specific conditions (speed, feed, grinding fluid, etc.).

TRUING AND DRESSING

Truing is the removal of abrasive material from the cutting face of the wheel so that the outer diameter will run concentric with the inner diameter. It also involves bringing the sides of the wheel parallel to one other and perpendicular to the spindle. *Dressing* removes the glaze from a dull wheel, including loaded material from the face, restoring a wheel to its original geometry and conditioning it to perform a specific job. Grinding wheels can be made to act harder and finer, or softer and coarser, by means of wheel conditioning.

49.8 THREAD CUTTING

Screw threads may be cut with single-point tools on a lathe or with multiple-tooth cutters that include dies, taps, and milling cutters on various types of machines. Threads may be formed on screws and bolts by rolling or pressing on thread-rolling machines as discussed in Chapter 50. They also may be ground or hobbed like gears and produced by die casting and plastic molding. This section will focus on thread cutting only.

Thread cutting on a lathe is called *chasing*. The process on a lathe is slow, requires skill, and is expensive. It is, however, very versatile and does not require special equipment. External and internal, right- and left-handed, straight and tapered, and practically all sizes and pitches of threads can be chased on engine lathes.

Threads also can be cut using a die for external threads and a tap for internal threads. A threading die, as illustrated in Figure 49-7, has an internal thread like a nut. Lengthwise grooves in the center hole expose the cutting edges of the die. The first few threads on a die are tapered so the die can be started on a circular workpiece. Dies can have a small slot extending from the outer body diameter into the thread diameter. This permits the die to be adjusted a small amount by means of a set screw to correct for wear (Drozda and Wick 1983).

Holes are usually threaded by taps. A tap has a shank and a round body with several radially placed cutting teeth as shown in Figure 49-8. Taps are made in many sizes and shapes to satisfy a number of purposes. They may be operated by hand or machine. A tap has two or more flutes that may be straight, helical or spiral, or spiral pointed. Those used in production tapping operations may be of carbide or high-speed steel. High-speed steel and carbide taps may be coated with abrasion-resistant materials to permit higher cutting speeds and improve performance.

DESIGN FOR TAPPING

The percentage of depth of thread being tapped is important to efficient and economical tapping. Too great a percentage strains the teeth of the tap and serves no useful purpose. The greater the percentage, the more power required to tap, the more difficult it is to hold size, and the greater the amount of tap breakage. Theoretically, a full (100%) thread is only 5% stronger than a 75% thread. Common practice is to use, and the Unified Thread Standard is based on, a 75% thread. However, in many cases, even less thread percentage is desirable.

The percentage of thread in a tapped hole should be governed by, in order of importance (Drozda and Wick 1983):

- the diameter and pitch of the tap,
- the hardness and toughness of the material being tapped,
- the depth of the tapped hole, and
- the type of hole, whether blind or through.

As an example, the factor of tap diameter and pitch would make it difficult to tap a No. 8-32 thread in a hole in tool steel with 75% depth of thread, since a No. 8-32 thread has a small diameter in proportion to its pitch. Yet, the same pitch on a 1/4-in. tap would be entirely practical, since a 1/4-32 tap has sufficient strength. Therefore, the smaller the diameter and coarser the pitch of the tap, the lower the percentage of thread that should be required.

The hardness and toughness of the material to a great extent governs the amount of material a tap is able to remove. Generally, the harder and tougher the material, the lower the percentage of thread that should be required. The minimum percentage should be adopted whenever possible.

The depth of the tapped hole is the third important factor governing the percentage of thread. Calculations should be based on the length of engagement, that is, the length of contact between a screw and a tapped hole measured axially. This length of engagement should equal the basic major diameter of the thread. It would be possible to tap greater percentages of threads if the depth of the tapped holes were less than their basic major diameter. Therefore, the percentage of thread should be reduced whenever the tapped hole exceeds the basic major diameter. This is particularly true in tapping blind holes, especially with the smaller taps and the coarser pitches where there is difficulty in finding room for the chips.

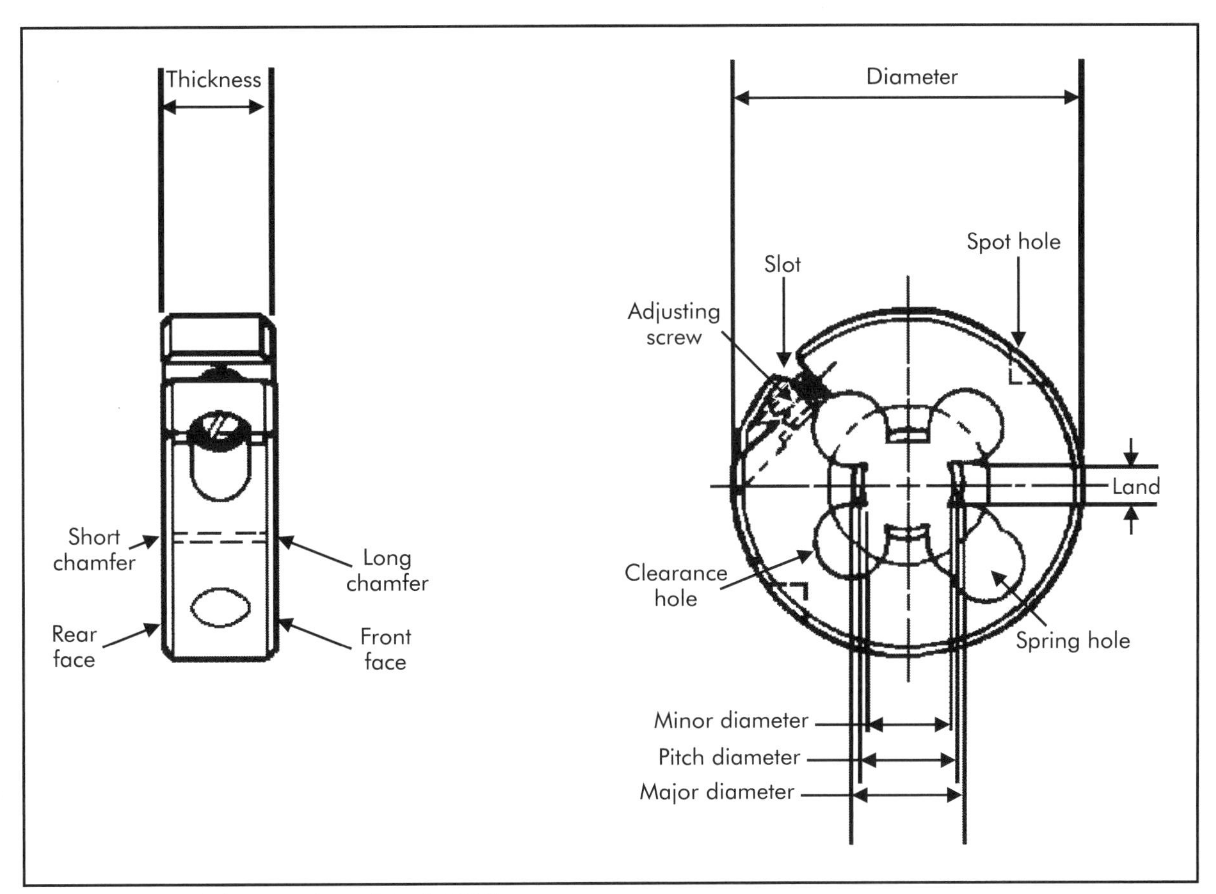

Figure 49-7. Solid die (Drozda and Wick 1983).

With the proper percentage of thread, very little pressure is required to start a tap. When too much pressure is applied in feeding or retracting, the tap cuts away each succeeding thread as the tap revolves.

A tap-drill chart with the most common sizes of taps can be found in *Machinery's Handbook* or a similar reference. For very accurate tapped holes with the required percentage of thread, reaming the hole before tapping is recommended.

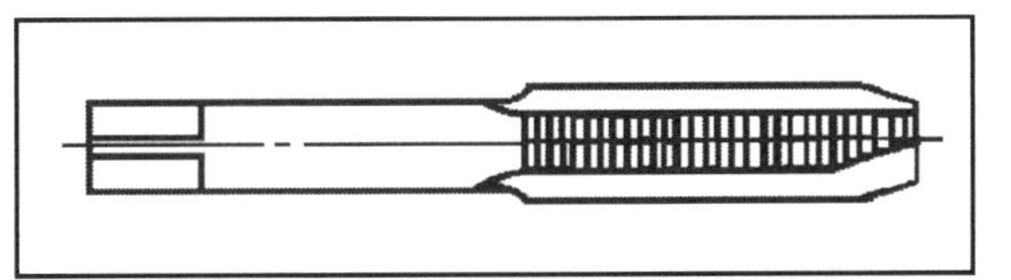

Figure 49-8. Solid tap (Drozda and Wick1983).

REVIEW QUESTIONS

49.1) Cermet cutting tools combine the properties of which two cutting-tool materials?

49.2) What is used to designate the size of a triangular carbide insert?

49.3) What is the minimum spindle speed required for a machining process to be considered high speed?

49.4) In broaching, can the feature to be broached be perpendicular to the direction of the tool?

49.5) What type of bond is used for cutoff wheels?

49.6) What is a common measure of how easily a workpiece can be ground?

49.7) How does the hardness of the workpiece influence the thread percentage required?

REFERENCES

Aronson, Robert B., ed. 2001. "The Changing World of HSM—Making HSM Work." *Manufacturing Engineering*, October.

Drozda, T. 1982. "*Wheels and Discs, Grinding Technology*." Dearborn MI: Society of Manufacturing Engineers.

Drozda, Thomas J. and Wick, Charles, eds. 1983. *Tool and Manufacturing Engineers Handbook*, Fourth Edition. Volume 1, *Machining*. Dearborn, MI: Society of Manufacturing Engineers.

Kennametal, Inc. 1996. "Lathe Tooling," Catalog 6000. Latrobe, PA: Kennametal, Inc.

Schrader, George F., and Elshennawy, Ahmad K. 2000. *Manufacturing Processes and Materials*, Fourth Edition. Dearborn, MI: Society of Manufacturing Engineers.

Chapter 50
Forming Processes Analysis

50.1 EXTRUSION

There are several variations of the extrusion process that allow for a variety of ductile materials and shapes to be produced. The three primary extrusion methods are direct, indirect, and hydrostatic.

In a *direct extrusion* process, the material, in billet form, is loaded into a thick-walled chamber (container) and then pushed through a stationary die to form the desired shape as illustrated in Figure 50-1. The direct method is by far the most common method of extrusion, because the tooling is not complicated and the process, while not applied to high-precision applications, provides an acceptable product.

There are, however, a few disadvantages mainly related to the friction of moving the entire length of the billet along the sides of the container wall. The extrusion material and presence or absence of a lubricating film are key to the amount of friction and force required.

The extrusion process can produce flat shapes, profiles (beam type) of reasonable complexity, and pipe shapes. But the friction issue limits the length of billet used. Example extrusions are illustrated in Figure 50-2.

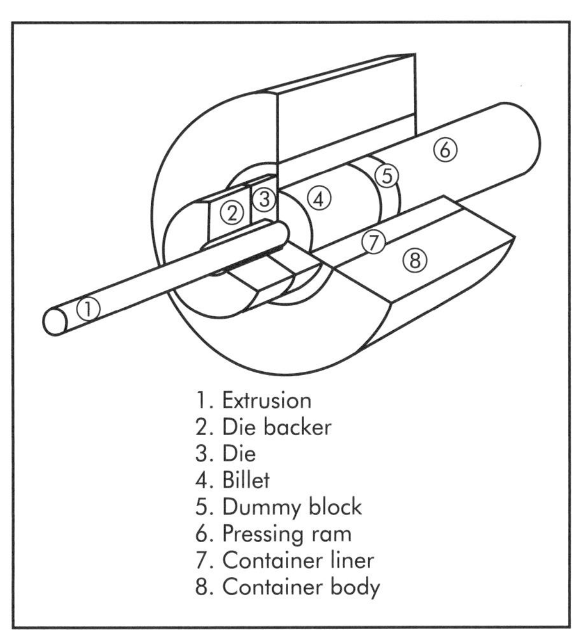

Figure 50-1. Direct (forward) hot extrustion.

Figure 50-2. Example extrusions (Wick, Benedict, and Veilleux 1984).

Another variation is *indirect extrusion*, where the billet is held stationary and the die moves toward the billet, as illustrated in Figure 50-3. This is sometimes called *backward extrusion* since the extruded part moves back along the ram or punch. This method requires less force since the ram does not have to overcome the weight and friction of a moving billet. The lower force also results in less chance of fracturing the billet, more overall uniformity in the final product, and longer die life. Indirect extrusion is not capable of producing shapes as large as those of the direct process, since the die shape is constrained by the structure required for the ram's function.

The third process, *hydrostatic extrusion*, reduces the friction effects even further. A hydraulic fluid is used in the ram cavity and, when pressurized, force is exerted not only to push the material through the die, but also away from the walls of the cavity. This allows extrusion of materials normally too difficult for other methods and the ability to extrude longer billets. Since the hydraulic pressure is controllable during the cycle, hydrostatic extrusion is similar to mechanical impact extrusion, where the ram pressure is higher at the start of the cycle. Wire made from less ductile materials is a common application for hydrostatic extrusion. The process is fast and can accommodate a high extrusion ratio. The *extrusion ratio* is defined as:

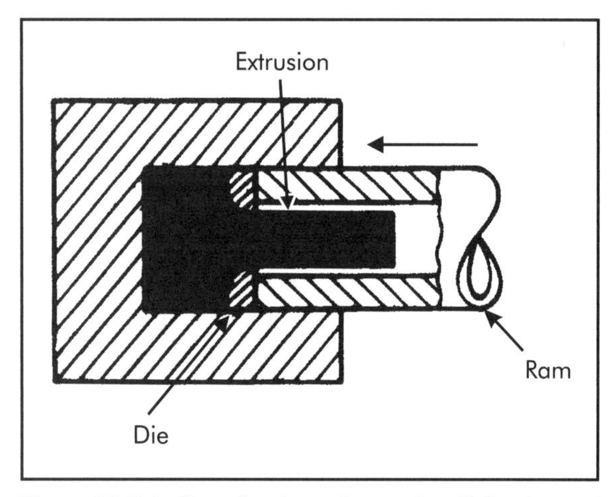

Figure 50-3. Indirect (backward) extrusion (Schrader and Elshennawy 2000).

$$\frac{A_o}{A_f} \tag{50-1}$$

where:

A_o = original cross-sectional area of the starting billet (in.^{2} [mm^2])
A_f = final cross-sectional area of the extruded part (in.^{2} [mm^2])

Hydrostatic extrusion requires high-pressure seals and more accurate billets that seal at the die face at the start of the extrusion cycle. Figure 50-4 illustrates the hydrostatic extrusion process.

There are a number of key parameters to consider in setting up an extrusion operation, such as:

- material composition—alloy type (ductility and toughness are key properties);
- shape complexity/shape factor (typically a factor based on the surface area of the part);
- the billet's length-to-diameter ratio;
- press equipment condition;
- ram pressure;
- tool geometry;
- material lubrication;
- die angle; and
- extrusion ratio.

In all cases, extrusion can be carried out under cold, warm, or hot conditions. Cold and warm forming refers to work done below the critical recrystallization temperature of the metal. At the recrystallization temperature, primarily new, strain-free grains are formed.

Important advantages of cold and warm extrusion include substantial cost savings for many applications, fast production rates, improved physical properties, accommodation of close tolerances (0.005 in. [0.13 mm]), energy conservation, and elimination of pollution problems. Cold extrusion is limited by the complexity of the shape, as well as the length-to-diameter ratio, and maximum size to which the part can be produced. Warm extrusion can help in some cases by raising the billet temperature.

Typical parts produced by cold extrusion include bearing races, a variety of fasteners, piston pins, spark-plug shells, socket wrenches, transmission and axle shafts, switch housings, and steering and suspension components.

Hot extrusion is done at a temperature above the metal's recrystallization temperature, thereby reducing the forces required. This makes it more appropriate for higher volume production, more complex shapes, and less ductile materials. Hollow shapes are especially suited for hot extrusion as illustrated in Figure 50-5. Even though the forces are lower, the friction of a hot material increases, which also increases die wear. Surface finish and tolerances are also degraded in this process.

Defects sometimes occur in extrusions as the result of non-uniform flow of metal in the billet container. Such defects include funnel-shaped, hollow forms in the center of the billet (usually restricted to the unextruded portions of the billet); piping (hollow core); and surface defects, such as cracks, tears, scale, blisters, and die lines.

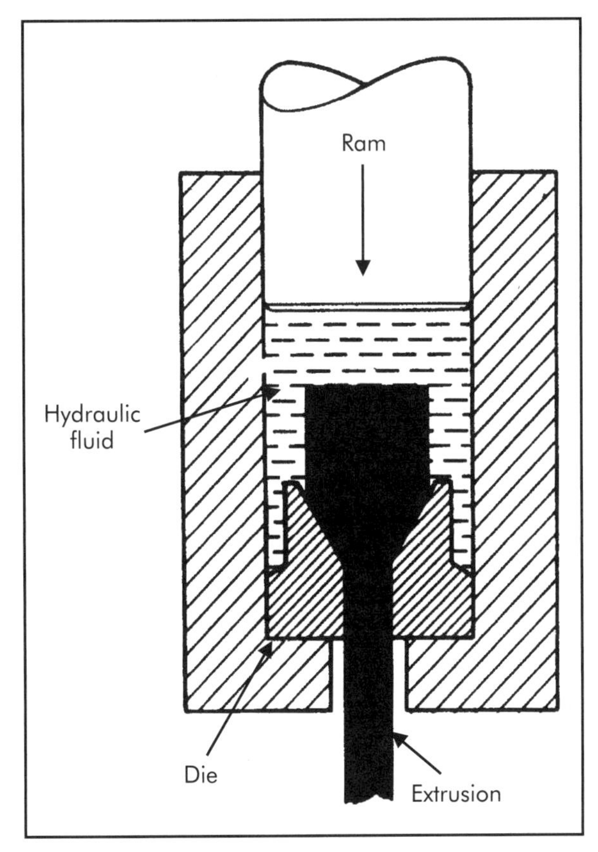

Figure 50-4 Hydrostatic extrusion (Schrader and Elshennawy 2000).

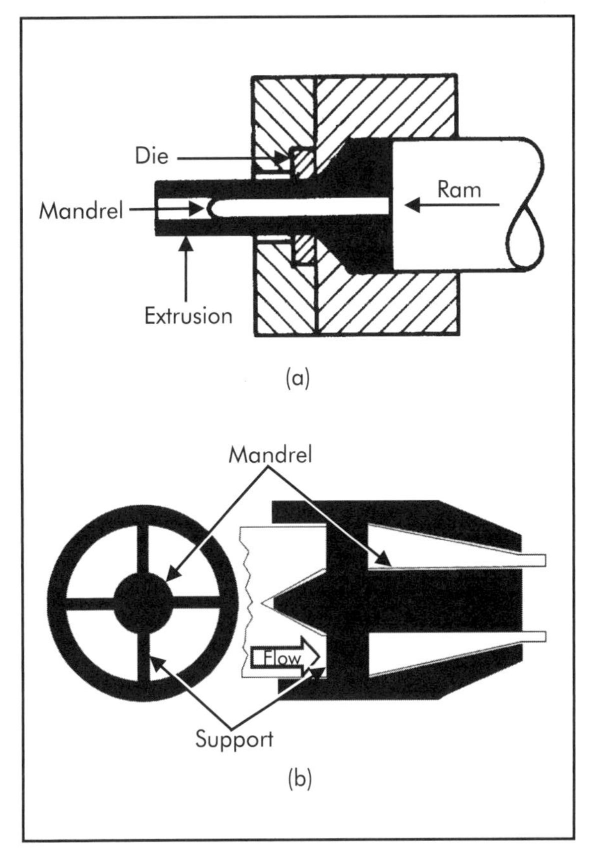

Figure 50-5. Hollow extruding process: (a) mandrel (Schrader and Elshennawy 2000), and (b) spider die.

50.2 ROLLING

Rolling is a highly productive, continuous forming process that can be done hot or cold. The primary shapes produced are plates, bars, rods, and structural shapes. However, other shapes are also possible as described in Chapter 22 of *Fundamentals of Manufacturing*, Second Edition.

As illustrated in Figure 50-6, when metal is rolled, the crystals or grains are elongated in the direction of rolling and the material emerges at a faster rate than it enters. In *hot rolling* (above the recrystallization temperature) the crystals or grains start to reform after leaving the reduction zone. However, in *cold rolling* (below the recrystallization temperature) they retain substantially the same shape created by the action of the rollers.

Material can be displaced at higher rates and lower forces if done hot. However, there is a loss of dimensional accuracy at elevated temperatures. Cold rolling can provide better surface finishes, but requires higher forces. Forces and product requirements also drive the rolling mill requirements. Several configurations are used in industry, ranging from two-high cogging mills for roughing work, to three-high, and four-high as shown in Figure 50-7.

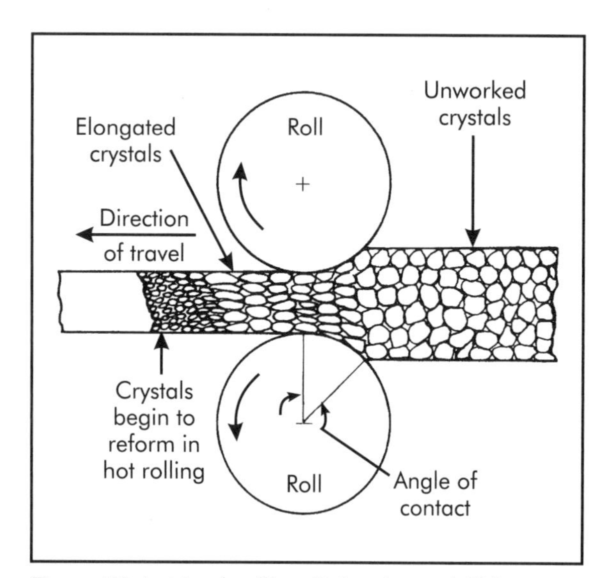

Figure 50-6. Metal rolling (Schrader and Elshennawy 2000).

Rolling defects consist of:

- scabs in the ingots;
- porosity;
- impurities, inclusions, etc.;
- heating variation and inconsistencies across the workpiece or rolling mills;
- wavy edges—buckling, indicative of severe camber;
- zipper cracks—internal fractures;
- edge cracks—external fractures;
- alligatoring—horizontal fracture of the bar or sheet, caused by non-uniform rolling force; and
- crown and camber variation across the rolled area, which can be the result of deflections in the rollers (crown, thickest in the middle, and camber, thicker at the sides).

Another application of rolling is *thread rolling*, a simple cold-rolling process for producing threads on cylindrical or conical workpieces. The helical threads are produced by displacing or rearranging the blank material as illustrated in Figure 50-8, rather than by removing material as in thread cutting or grinding.

Production rates for rolling are generally higher than those for cutting or grinding. The threads produced have improved strength and fatigue properties; the surface finish is good; and the work-hardened surface often provides additional advantages. Most rolling is performed with the blanks at room temperature, although heat may be applied to facilitate metal displacement, most often in the case of high hardness materials.

There are many advantages to thread rolling. There are no chips formed as in a conventional cutting process and rolling tends to generate a stronger thread because of the compressive forces on the grain structure.

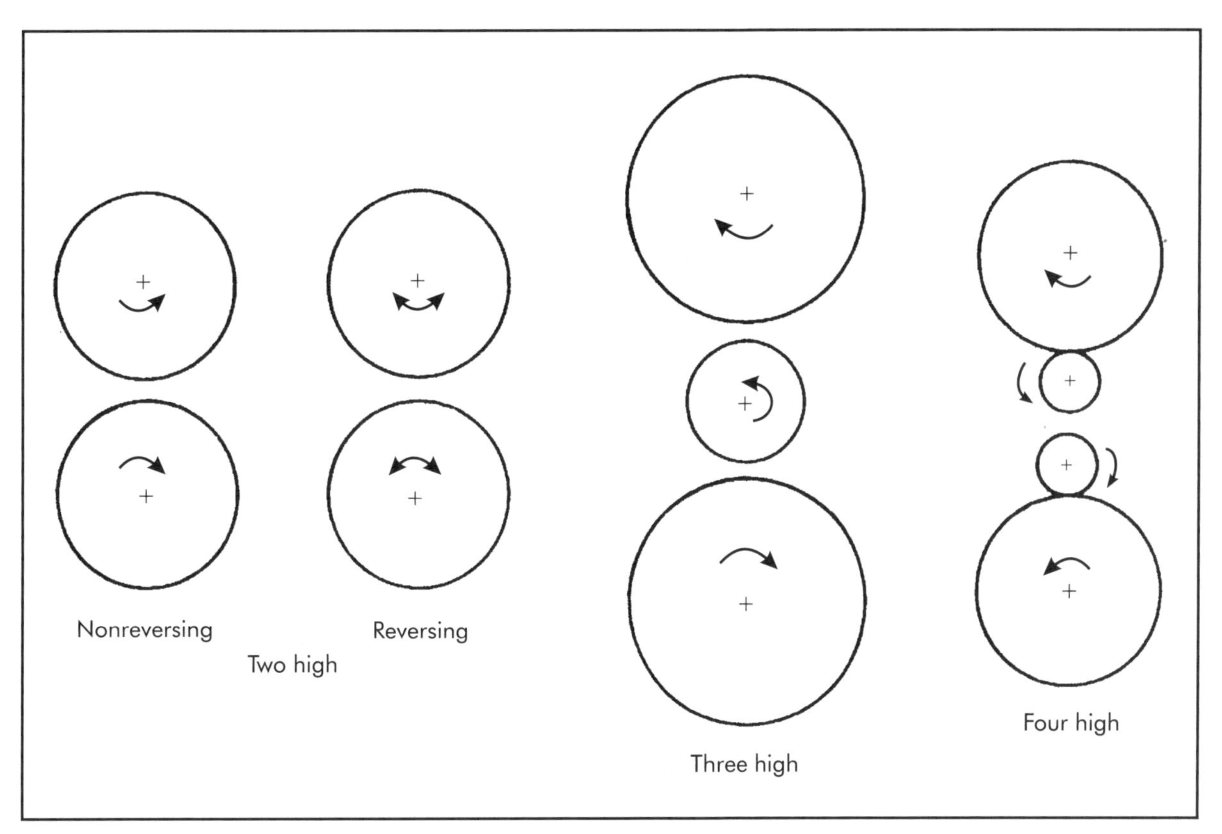

Figure 50-7. Types of rolling mills (Schrader and Elshennawy 2000).

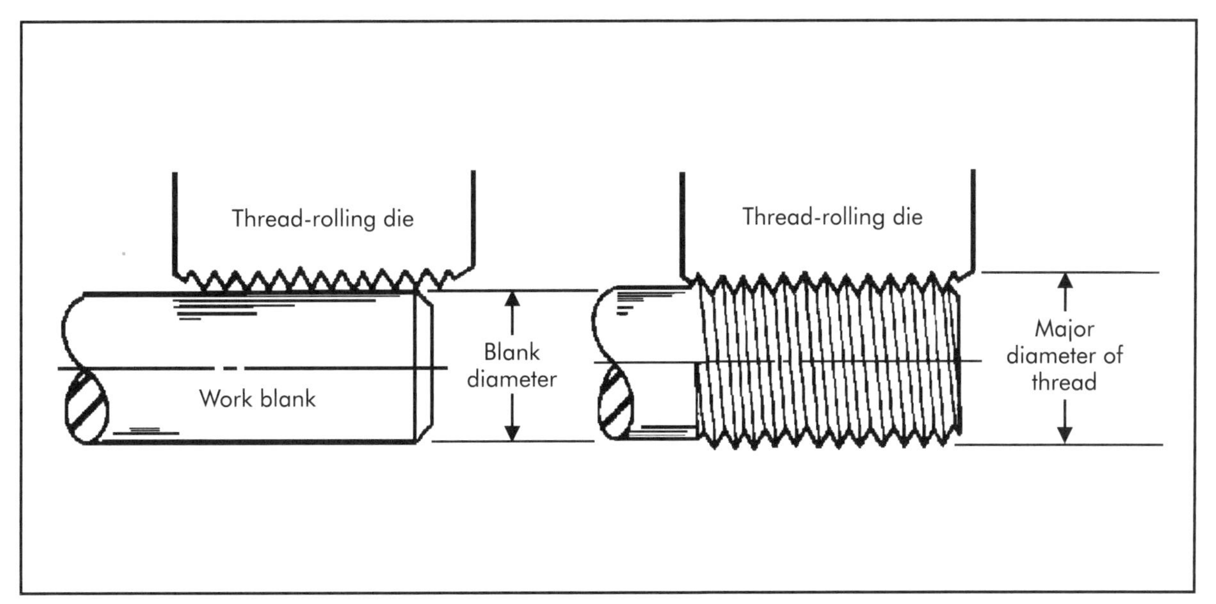

Figure 50-8. Material displacement in thread rolling (Drozda and Wick 1983).

Figure 50-9a illustrates how the grain follows the contour of the rolled threads, thereby increasing their strength as compared to the machined threads shown in Figure 50-9b.

50.3 FORGING

Forging is another way to form and reshape material to produce parts and tools. *Hot forging* is the controlled plastic deformation or working of metals into predetermined shapes by means of pressure or impact blows, or a combination of both. In hot forging, plastic deformation is conducted above the recrystallization temperature to prevent strain hardening of the metal. During the deformation process, the crystalline structure of the base metal is refined and any nonmetallic or alloy segregation is properly oriented. Hot forging develops a directional alignment of the grains or fibers, which follows the part's shape as illustrated in Figure 50-10, helping to increase ductility, resistance to impact, and fatigue strength of the metal. Chapter 22 of *Fundamentals of Manufacturing*, Second Edition discusses the various types of forging in more detail.

There are several product design related aspects that must be taken into account when applying forging techniques. A part designed for forging can have enhanced function over those produced by other processes, and reduced cost of the basic manufacturing process and tooling. Ductile materials are well suited for forging. They provide good formability during the forging process, allowing the part's strength to be set up such that the grain flow of the material is in the same direction as the required maximum strength of the part.

Shape complexity is an important consideration in forging as it has implications on how much the material will have to be displaced. As the material flows, and the complexity of the part shape increases, features such as fillets and corner radii should be increased. Extreme or abrupt variations in the local areas and volumes should be avoided to ensure smooth flow of the material. Also, in consideration of the use of dies in this process, draft is required on all vertical surfaces. Generally, 5° draft is necessary, but the amount can be less if the material has good ductility. Typically, internal surfaces should have slightly more draft than external surfaces.

There are also several process considerations. The workpiece temperature is very critical to achieving the proper forces and correct material flow. The material selected and the alloying elements will, of course, dic-

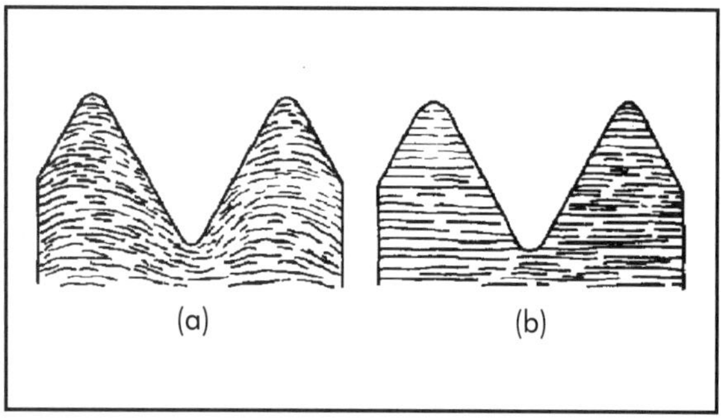

(a) (b)

Figure 50-9. Typical grain flow: (a) in a rolled thread and (b) in a machined thread (Drozda and Wick 1983).

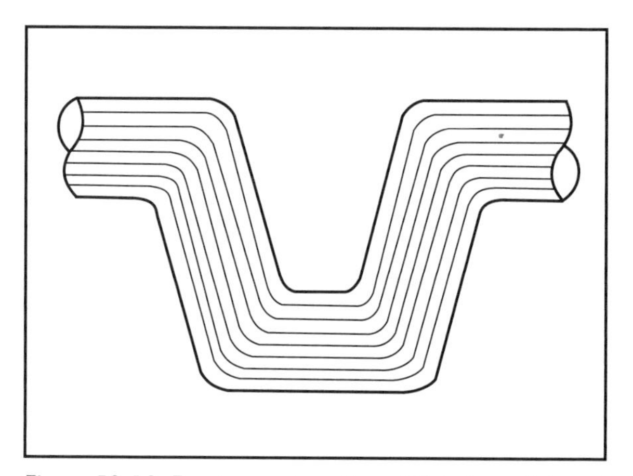

Figure 50-10. Forging grain pattern (Wick, Benedict, and Veilleux 1984).

tate the temperatures. However, a safe window of operation must allow for the time to remove the workpiece from the furnace and complete the forging operation.

In setting up the process, decisions must be made on several key process parameters such as the number of preliminary dies, the need for a preform or specific billet size, and the number of strikes in each die. Cracking and buckling during the forging process can be a concern if the force or the material flow is not properly addressed. This is clearly evident in upset forging as mentioned in Chapter 22 of *Fundamentals of Manufacturing*, Second Edition.

Upsetting, also known as *heading*, consists of applying lengthwise pressure to a bar to enlarge its diameter by sacrificing some of its length. The basic rules pertaining to the gathering of stock are as follows.

- The limit of length of unsupported stock that can be gathered or upset in one pass without injurious buckling is not more than three times the diameter.
- Lengths of stock more than three times the diameter of the bar can be successfully upset in one blow by displacing the material in a die cavity no greater than 1.5 times the diameter of the bar, provided the stock extending beyond the die face is no greater than half of the diameter.
- An upset requiring more than three diameters of stock in length and extending up to 2.5 bar diameters beyond the die face can be made if the material is confined by a conical recess in the punch that does not exceed 1.5 bar diameters at the mouth and 1.125 bar diameters at the bottom. This is provided the heading tool recess is not less than 2/3 the length of the working stock or not less than the length of the working stock minus 2.5 times its diameter.

Figure 50-11 illustrates the application and violation of these rules (Nee 1998).

50.4 METAL SPINNING

Metal spinning is a very specialized metal-forming process used for hollow products that can be turned to produce the desired shape. Products made by spinning include urns, orifices, nose cones, dishes, swaged tube, and more. Metal spinning can be used to form a part from a plain sheet or tube, or to add a flange to a partially formed part.

Spinning is typically a cold forming process and the basic physics are similar to sheet bending. The process also can provide additional strength through work hardening as the part is formed. Metal spinning requires minimal tooling so it is especially appropriate for small quantities of parts.

The spinning process can be classified into four basic types: manual (hand) spinning, power spinning, shear forming, and tube spinning. Smooth surface finishes and superior mechanical properties are possible with spinning. The primary limitations to spinning are the shapes that it is confined to and its dependence on the skill of the operator. The process must be refined to avoid wrinkling and buckling as the material is displaced. There is no appreciable thinning of the material, so the starting blank diameters must be considerably larger than the diameters of the finished workpieces. Higher volumes of parts are not recommended due to the length of time required for processing a part. While metal spinning can be faster than machining, it is not as fast as the typical stamping operation.

The size and tolerance limits are dependent on the size of the machine and the skill of the operator. Typically, blanks can be up to 0.250 in. (6.35 mm) thick in steel and 0.375 in. (9.53 mm) in aluminum. The tolerance capability of the typical spinning process is in the range of 0.005–0.015 in. (0.13–0.38 mm).

Manual (hand) spinning can be performed by having the operator apply muscle power and pressure to force the tip of the tool back

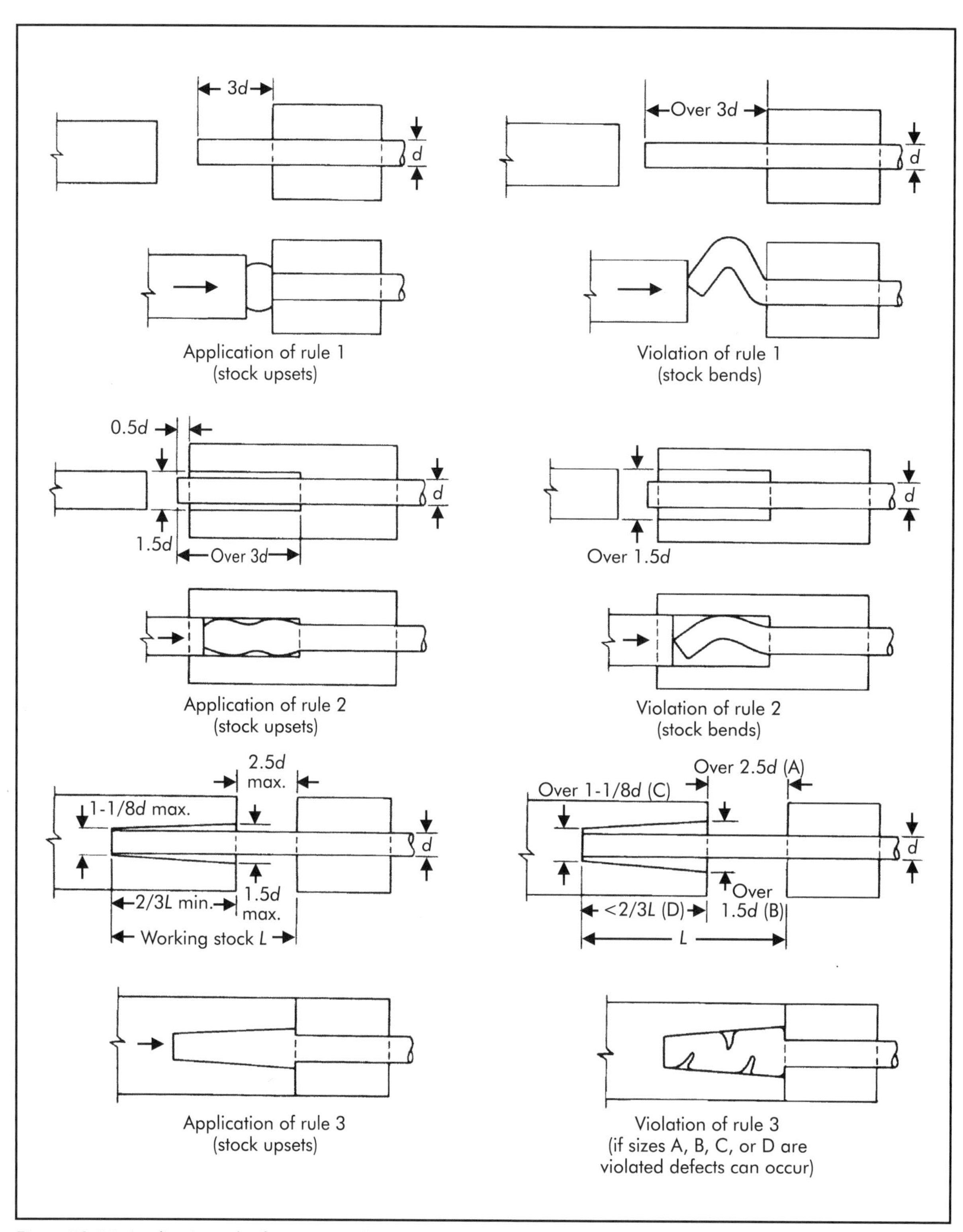

Figure 50-11. Application rules for upsetting (Nee 1998).

and forth across the rotating disc. Another method is to use a scissor-like tool that provides a fulcrum and lever assist to reduce the manual force required. The operator works from the center axis outward, gradually forcing the flow of the metal to conform to the shape of the pattern as illustrated in Figure 50-12.

Power spinning and power-assisted spinning processes provide the same results as manual spinning but utilize various automated feeding devices to apply the forming forces. These devices include toolholder carriages or compounds powered by mechanical, air, hydraulic, or mechanical/hydraulic means; and hydraulic or electronic tracing and copying systems using single, multiple, or swivel templates. CNC controlled machines are also used.

50.5 BENDING

Bending refers to the basic process of folding over the edge of flat material. It is achieved by stressing the metal beyond its yield strength, but not exceeding its maximum tensile strength. A unique distortion takes place in bending as the external surface is stretched and the internal surface compressed. As illustrated in Figure 50-13, the neutral axis of the bend is the location inside the material where there is neither tension nor compression. For thin materials, the neutral axis is assumed to be in the middle of the bend, but for thicker materials, it is located approximately 30% of the material thickness from the inside of the bend.

Bending is primarily done on press brakes, though stamping dies can be used. This equipment may be mechanically or hydraulically operated similarly to other metal-forming operations. There are two types of press brake operations: air bending and bottom bending as shown in Figure 50-14. There are trade offs between the two methods with respect to springback and bent flange accuracy.

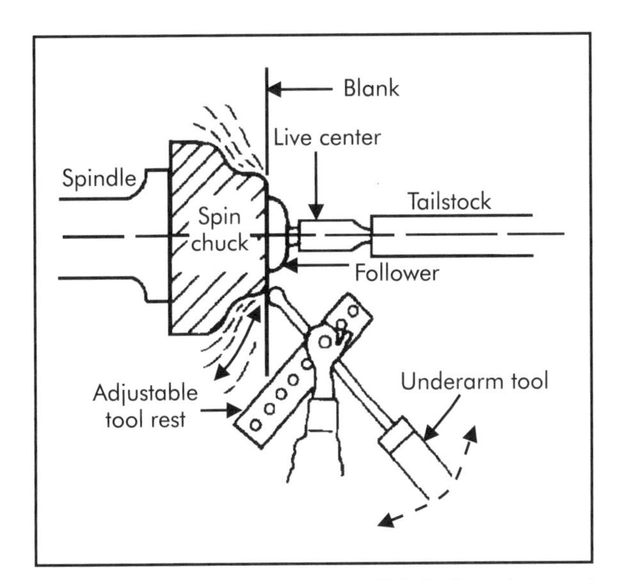

Figure 50-12. Manual spinning (Wick, Benedict, and Veilleux 1984).

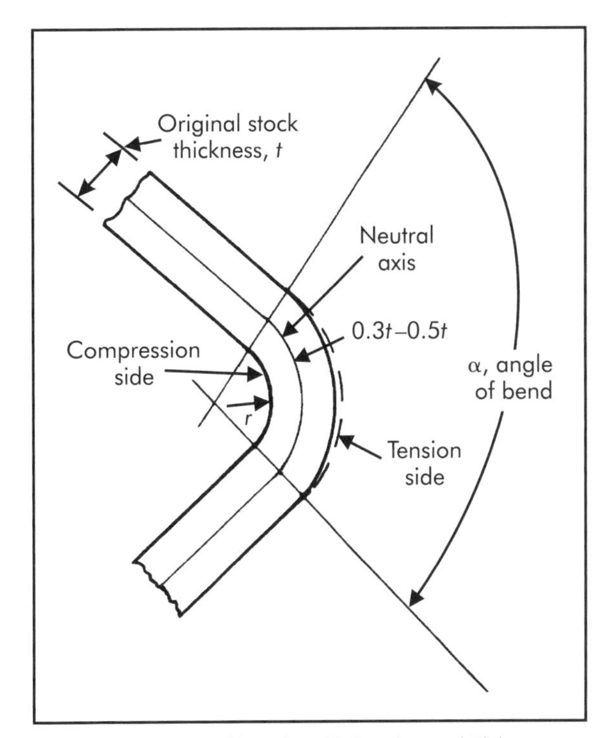

Figure 50-13. Metal bending (Schrader and Elshennawy 2000).

Air bending refers to bending operations performed in V-dies in which the punch does not bottom, resulting in low force requirements. In *bottom bending*, the work is completely pressed into the female die and the internal radius is accurately formed by the male die. Thus consistently accurate flange sizes are possible. Due to the higher force required, bottom bending has a limitation with respect to maximum work thickness, which usually can be no greater than 0.125 in. (3.18 mm) for steel.

Some common parameters of the process include: the workpiece material and thickness, bend radius, and springback. The minimum bend radius is the smallest radius that can be formed without part cracking. It is a parameter that is specific to material ductility and material thickness. In general, the minimum ratio of the bend radius to material thickness, Equation 50-2, increases as ductility decreases.

$$\frac{r}{t} \tag{50-2}$$

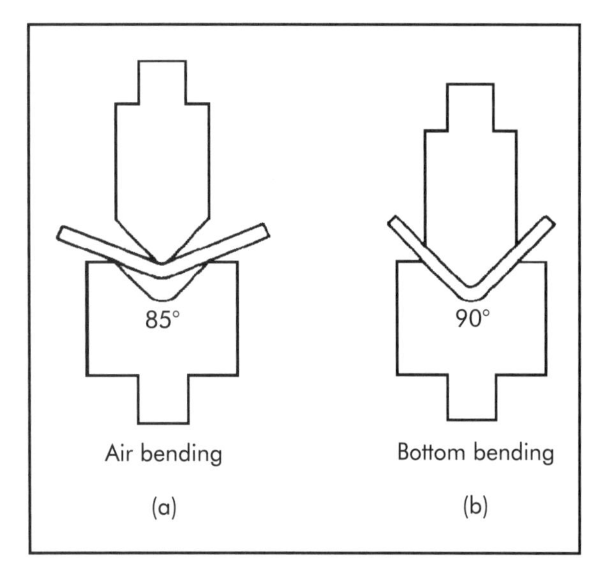

Figure 50-14. Basic methods of bending: (a) air bending and (b) bottom bending (Wick, Benedict, and Veilleux 1984).

where:

r = bend radius (°)
t = material thickness (in. [mm])

In other words, less ductile materials require a larger bend radius or thinner material. Another factor, stress concentrations, can be very sensitive in less ductile materials. Where such concerns exist, experiments and detailed stress analysis studies should be performed.

One additional factor that must be considered in bending relates to the elastic behavior of the material or *springback*. While the material is generally undergoing plastic deformation, it is not occurring throughout the material in the bend. Therefore, the material will attempt to return or spring back to its original form. Springback can be predicted to a limited extent, which in typical sheet materials is in the range of 1–4°.

Springback can be compensated for by over-bending and bottoming (setting). Overbending beyond the part's desired shape allows it to spring back to the desired shape. Bottoming involves plastic deformation at the root of the bend during the bending process as shown in Figure 50-15. Plastic deformation at the root prevents springback.

STAMPING

Sheet-metal stamping is another press-forming process. It is considered the most common sheet-metal forming method. In this process, a flat blank is formed into a finished shape between a pair of matched dies as shown in Figure 50-16. Stamping incorporates two principal types of deformation, drawing and stretching.

In an idealized forming operation in which *drawing* is the only deformation process that occurs, the clamping force of the hold-down dies (pressure pads) is just sufficient to permit the material to flow radially into the die cavity without wrinkling. Deformation of the sheet takes place in the flange and over the

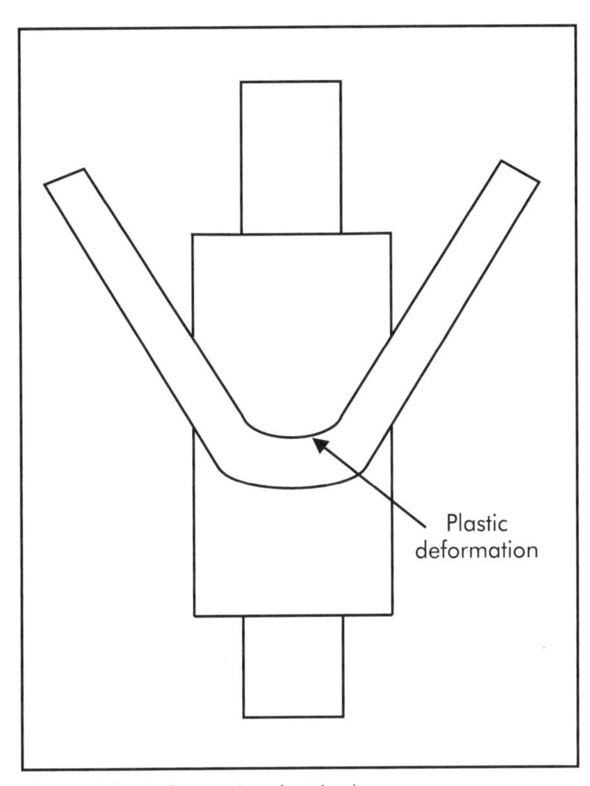

Figure 50-15. Bottoming (setting).

lip of the die; no deformation occurs over the nose of the punch. This is analogous to wire drawing in that a large cross-section is drawn into a smaller cross-section of greater length. Consequently, this type of forming process is called drawing to distinguish it from stretching.

A blank of sheet metal is clamped firmly around the periphery or flange to prevent the material in the flange from moving into the die cavity as the punch descends in an ideal *stretch-forming* operation. In this case, hold-down dies prevent radial flow of the flange. All deformation occurs over the punch, at which time the sheet deforms by elongating and thinning. As in tensile testing, if the deformation exceeds the ability of the material to undergo uniform straining, deformation is localized and fracture is imminent. This stage is similar to the elongation at maximum load in a tensile test.

Equipment

The word *die* is a generic term used to describe the tooling used to produce stamped parts. A die-set assembly consisting of a male and female component is the actual tool that produces the shaped stamping. The male and female components work in opposition to form and punch holes in the stock. The upper half of the die set, which may be either male or female, is mounted on the press ram and delivers the stroke action. The lower half is attached to an intermediate bolster plate, which in turn is secured to the press bed. Guide pins are used to ensure alignment between the upper and lower halves of the die set.

Multiple stamping operations may be performed within a single die, or at a number of die stations within a die set and with a single stroke of the press. Single-station dies can be either compound dies or combination dies. A *compound die* performs basic cutting operations, such as blanking and hole punching, to produce parts. *Combination dies* com-

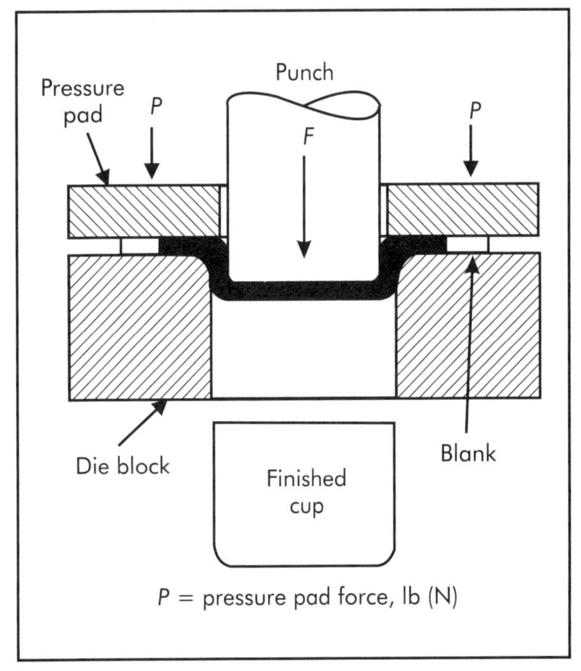

Figure 50-16. Basic drawing process.

bine shaping and forming with cutting to manufacture parts.

Multiple-station dies are arranged so a series of sequential operations are accomplished with each press stroke. Two die types are used: progressive and transfer. With *progressive dies*, coil stock is fed into the press. Individual stampings are connected with a carrier strip as they progress through the various die operations and are ultimately separated and then discharged from the press. In *transfer die* operations individual stock blanks are mechanically moved from die station to die station within a single die set. Large stampings are done on tandem press lines where the stock is moved from press to press, on which specific operations, such as drawing or trimming, are performed.

Die making is as much of an art as it is a science. Finite element analysis based software programs can be used for die analysis. However, when all the dynamics of stamping are taken into account, the resulting part may not meet all expectations. To help fine-tune the stamping process and finalize die design, die makers use an analytical tool called *circle grid analysis* (CGA). The application of CGA involves the etching of a pattern of small circles on the blank's surface. This pattern deforms along with the blank as it is formed, providing point-to-point calculations of the deformation that occurred. Analyzing this stamped grid pattern suggests the location and type of rework that must be performed on the die to produce easily manufactured parts. The CGA process is repeated on the die until an acceptable part is produced.

50.6 HYDROFORMING

Hydroforming uses hydraulic pressure to move the workpiece material (sheet or tube) so it will conform to the shape of a metal die. This is one process that evolved from several others to overcome the significant manufacturing limitations of sheet-metal stamping and tube forming. Historically, deep drawing and forming tight corners in shallow draws have been major problems with regular steel die forming methods, even if the stamping is done in several stages. Hydroforming allows shapes to be made that are beyond the capabilities of the other sheet-forming processes.

To hydroform a part, a sheet or tube must be clamped into a die, similar to any stamping operation. With the cavity sealed, hydraulic pressure is applied to fill the cavity and stretch the workpiece uniformly. Advanced computer control is a key enabler, since the pressure cycle and die motion have to be well coordinated throughout the process.

There are three key stages to a hydroforming process. They are expansion, calibration, and displacement. In the expansion stage, once the die cavity is sealed, initial fluid pressure is applied to uniformly expand the material. At this point, the sheet or tubing material makes initial contact with the die surfaces. Also during this phase, the die is moving to compensate for the material displacements. In the case of tube forming, the ends of the die sealing the fluid cavity have to move inward as the tube shortens in length while the diameter increases to contact the walls of the die cavity.

The critical phase is the calibration phase. At this point, the pressure is increased and the difficult part details completed. Features such as tight corners are now formed. This part of the process requires very close control over the pressure cycle as the die cavity now remains stationary and the material is locally displaced.

The displacement phase follows as dictated by the design of the workpiece. Any special protrusions are now added by moving segments of the die cavity and then repressurizing to complete the part. This phase is particularly unique to hydroforming since it allows a new level of design freedom. Features that previously required pins and movable die sections in inaccessible areas can now be formed. In some cases, the movement of the die can be coordinated with the pressurization cycle to ensure uniform displacement of the material. Figure 50-17 illustrates the hydroforming process.

Specialized tubing products are becoming a primary niche for the hydroforming industry. Hydroformed parts are worthy substitutes for weldments and castings. They offer product performance benefits over regular tubular products given that there is almost unlimited capacity for shape variation. Other benefits include part consolidation, reduced secondary operations, structural improvements, and weight reduction.

Hydroforming is a relatively new process, which can be very difficult to implement, especially due to the current limited knowledge base. Getting the process to run correctly and consistently is very important. The tooling costs for hydroforming can be higher than other processes, so production volumes must justify it. Because of the complexity, cycle times are rather long. However, this is compensated for by the reduction in the need for secondary operations.

REVIEW QUESTIONS

50.1) Define the extrusion ratio.

50.2) Is cold rolling performed above or below the recrystallization temperature?

50.3) Define what the neutral axis is with respect to bending.

50.4) How is springback compensated for?

50.5) What analytical tool is used for finalizing stamping die designs?

REFERENCES

Drozda, Thomas J. and Wick, Charles, eds. 1983. *Tool and Manufacturing Engineers Handbook*, Fourth Edition. Volume 1: *Machining*. Dearborn, MI: Society of Manufacturing Engineers.

Nee, John G., ed. 1998. *Fundamentals of Tool Design*, Fourth Edition. Dearborn, MI: Society of Manufacturing Engineers.

Schrader, George F., and Elshennawy, Ahmad K. 2000. *Manufacturing Processes and Materials*, Fourth Edition. Dearborn, MI: Society of Manufacturing Engineers.

Wick, Charles, Benedict, John, and Veilleux, Raymond, eds. 1984. *Tool and Manufacturing Engineers Handbook*, Fourth Edition. Volume 2: *Forming*. Dearborn, MI: Society of Manufacturing Engineers.

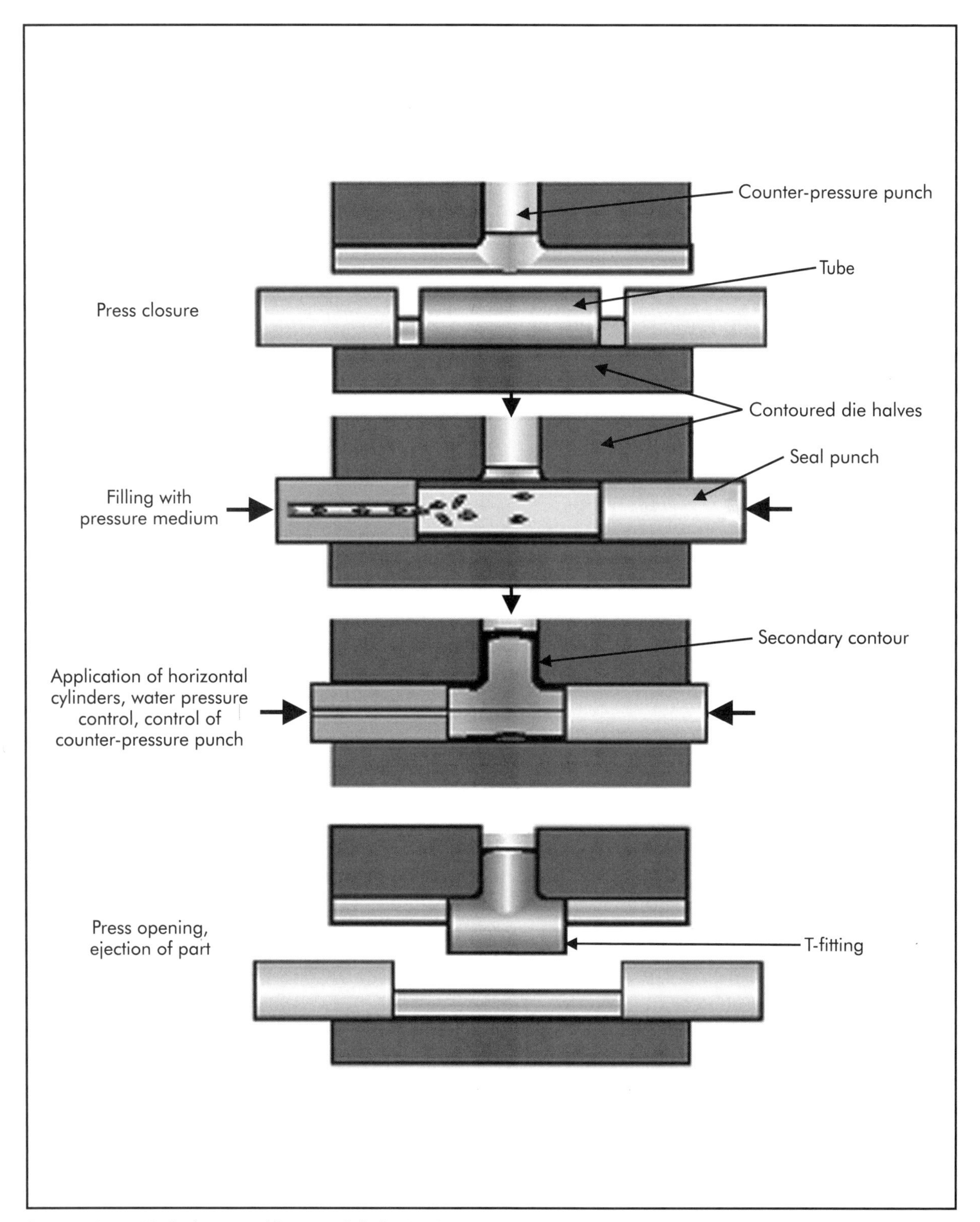

Figure 50-17. Hydroforming. (Courtesy Schuler, Inc.)

Chapter 51

Joining and Fastening Analysis

51.1 WELDING PROCESS SELECTION

There are more than 50 different welding processes, some of which are discussed in Chapter 26 of *Fundamentals of Manufacturing,* Second Edition. Welding processes can be classified as either fusion or solid-state (nonfusion).

In *fusion welding*, workpieces are melted together at their faying surfaces. The fusion welding processes discussed in this chapter are electric arc, weld bonding, electron-beam welding (EBW), and thermit welding. Filler metals often used with fusion welding methods have melting points about the same as or just below those of the metals being joined.

In *solid-state welding*, the workpieces are joined by the application of heat and usually pressure, or by the application of pressure only. However, in solid-state welding processes, the welding temperature is essentially below the melting point of the materials being joined, or if any liquid metal is present, it is squeezed out of the joint. No filler metal is added during solid-state welding. Friction and inertia welding are the solid-state welding processes discussed in this chapter.

Every welding process has advantages and disadvantages with respect to specific applications. Factors that should be considered in selecting the optimum process include the following:

- materials to be joined;
- joint design, including location and orientation, as well as the thickness and configuration of the parts being joined;
- access to the joint from one or both sides;
- production requirements (rate and total);
- available equipment;
- tooling requirements;
- edge preparation;
- welder or machine-operator skills;
- environmental condition requirements;
- effects of the process on the properties of the weldment;
- weld quality;
- service conditions to be satisfied, including loading and operating temperature for the finished product or structure;
- cost—the economics of the process, including labor and overhead rates, and the cost of power and consumable materials; and
- safety considerations.

51.2 TYPES OF WELDED JOINTS

Most common fusion welds (except resistance welds) can be designed for either complete or partial joint penetration. The selection of the type of joint and weld to use in a particular application depends on many factors, including the following:

- the load magnitude and type—tension, compression, bending, or shear;
- the manner in which the load is applied—steady, variable (possible fatigue), or sudden (impact);
- distribution of the load or stresses in the weld joint; and

- the cost of the joint preparation, welding, and inspection.

Various types of joints are illustrated in Figure 51-1.

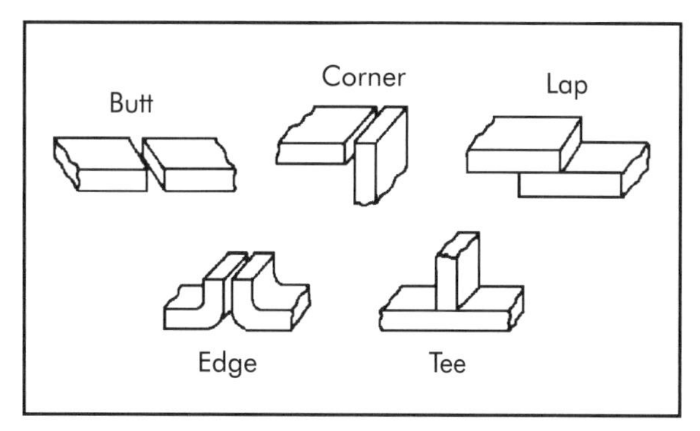

Figure 51-1. Welding joints (Wick and Veilleux 1987).

51.3 SHIELDED-METAL-ARC WELDING (SMAW)

Shielded-metal-arc welding, also known as *stick welding*, is one of the oldest and most widely used of the various arc welding processes. It is basically a manual process used by small welding shops, home mechanics, and farmers for equipment repair. The process also has extensive application in industrial fabrication, structural steel erection, weldment manufacture, and other commercial metal-joining operations. Figure 51-2 illustrates the process.

Major advantages of shielded-metal-arc welding include application versatility, flexibility, and the simplicity, portability, and low cost of the equipment required. The process is capable of welding thin and thick steels and some nonferrous metals in all positions. Power-supply leads can be provided over long distances, and no hoses are required for shielding gas.

Required periodic changing of the electrode is one of the major disadvantages of shielded-metal-arc welding for production applications. This decreases the percentage of time actually spent welding. Another disadvantage is the limitation placed on the current that can be used. Welding current is limited by the resistance heating of the electrode. The electrode temperature must not exceed the breakdown temperature of the flux covering. If the temperature is too high, the covering chemicals react with each other or with the air and, therefore, do not function properly at the arc. In addition, the heating of the core wire increases the melt-off rate and changes the arc's characteristics.

Essential requirements for optimum arc welding include using the correct electrode type and diameter, welding current, arc length (arc voltage), travel speed, and electrode position. Selecting the correct type and size of electrode requires careful consideration of many factors, including the following:

- the type, position, and preparation of the joint to be welded;
- the ability of the electrode to carry the required welding current without damage to the weld metal or loss of deposition efficiency;
- the base metal type, mass, and its ability to maintain acceptable properties after welding;
- the characteristics of the assembly with reference to the effect of stresses set up by heating;

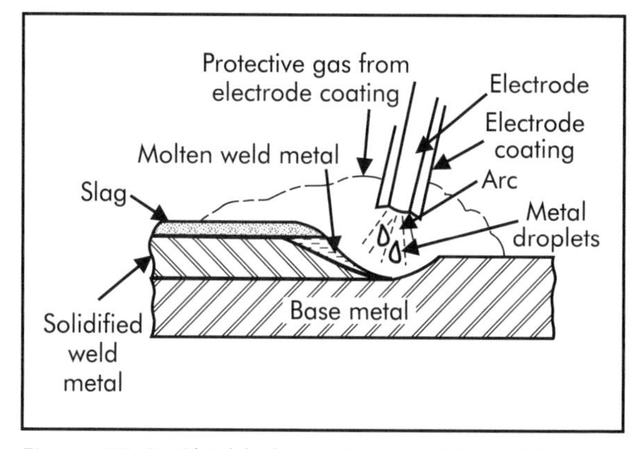

Figure 51-2. Shielded-metal-arc welding. (Courtesy Hobart Institute of Welding)

- specific weld quality requirements; and
- electrode cost.

A standard classification for electrodes, known as the AWS-ASTM (American Welding Society-American Society for Testing and Materials) classification, designates the characteristics of steel electrodes by a letter and series of numbers, such as E6010 and E6013, for example. The prefix "E" indicates the filler material is a conducting electrode. The first two digits specify the minimum tensile strength times 1,000 psi (6.9 MPa) of the weld metal in the "as-welded" condition. For example, a 6010 electrode would provide weld metal with 60,000 psi (414 MPa) tensile strength in the as-welded condition. The third digit specifies the welding position for which the electrode is suited: (1) for all positions, (2) horizontal and flat, and (3) flat positions only. The fourth digit indicates the type of flux coating surrounding the electrode, and thereby the conditions for which the electrode is intended, including the type of current and polarity (Schrader and Elshennawy 2000).

The flux covering the electrode generally provides these functions:

- Part of the flux covering burns off during welding, which forms a protective gas shield surrounding the weld pool.
- Part of the flux melts, which forms a protective slag layer on the weld surface that must be chipped off after welding.
- The remaining flux melts and provides a method of adding scavengers, deoxidizers, and alloying elements to the weld metal.

Shielded-metal-arc welding is done with either alternating current (AC) or direct current (DC). AC offers the advantage of eliminating arc blow, thus permitting the use of larger electrodes and higher currents. *Arc blow* is the deflection of the arc due to magnetic forces created in the workpiece by the electrical current. The power source for AC is also lower in cost than that for DC. However, direct current provides a steadier arc and smoother metal transfer than AC. Most covered electrodes operate better on direct current/reverse polarity (DCRP), which produces deeper penetration. With DCRP, the electrode is positive and the workpiece is negative. Some covered electrodes are designed for direct current/straight polarity (DCSP), which produces a higher electrode melting rate. With DCSP, the electrode is negative and the workpiece is positive. Direct current is generally preferred for welding thin sections, vertical and overhead welding, and welding with a short arc. However, arc blow may be a problem.

If the *arc length* is excessive (the arc voltage drop is too high), filler metal melts from the electrode in large globules that wobble from side to side as the arc wavers. This results in a wide, spattered, and irregular bead, with poor fusion between the base metal and deposited metal. A long arc is also more susceptible to air entrapment and possible porosity in the weld. If the arc length is too short (the arc voltage drop is too low), the heat is insufficient to melt the base metal properly and the electrode often sticks to the work. High, uneven beads with irregular ripples and poor fusion are the result of a short arc length. Maintaining the proper arc length concentrates the welding current in the joint and minimizes spatter.

When the *welding speed* is too fast, the pool does not remain molten for a sufficiently long time, the bead produced is narrow, penetration is reduced, and impurities and gases may be locked into the weld. When the speed is too slow, excessive filler metal is added to the joint, and the resultant weld bead is high and wide.

51.4 GAS-METAL-ARC WELDING (GMAW)

In *gas-metal-arc welding,* coalescence (joining) of metals is produced by heating with an arc between the work and a continuous, solid (consumable) electrode as the filler metal. Shielding is provided by an externally supplied gas or gas mixture. Gas-metal-arc welding is also known as *metal-inert-gas* (MIG) *welding*, *dip-transfer welding*, and *wire welding*. Figure 51-3 illustrates the process.

Gas-metal-arc welding is performed using either a handheld gun or mechanical welding head or torch to which the electrode is fed automatically. The process is used extensively for high-production welding operations.

The major features of gas-metal-arc welding are:

- the capability of obtaining high-quality welds in almost any metal,
- its all-position capability,
- its relatively high speed and economy, and
- its elimination of slag entrapment in the weld.

A limitation of GMAW is that it requires an externally supplied inert shielding gas or gas mixture, which adds to welding costs. Flux-cored wire is an alternative to using a shielding gas. However, the flux-cored wire is more expensive than plain wire. The welding gun must be kept close to the work to ensure adequate shielding, making it difficult to use on hard-to-reach joints.

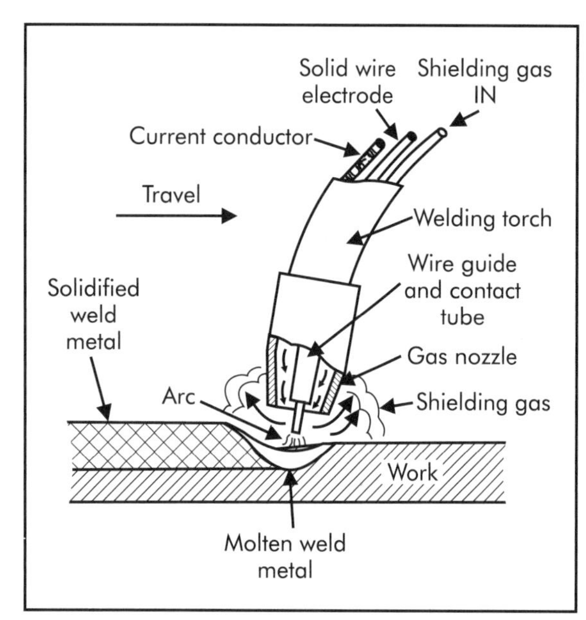

Figure 51-3. Gas-metal-arc welding. (Courtesy Lincoln Electric Co.)

The shielding gases used in the gas-metal-arc process, helium, argon, carbon dioxide, or mixtures thereof, protect the molten metal from reacting with the constituents of the atmosphere. For example, carbon dioxide-argon mixtures are suitable when welding low-carbon and low-alloy steels. Pure inert gas (argon), however, may be essential when welding highly alloyed steels. Although the gas shield is effective in shielding the molten metal from the air, deoxidizers are usually added as alloys in the solid electrode. Sometimes light coatings are applied to the electrode for arc stabilizing or other purposes. Lubricating films also may be applied to increase the electrode's feeding efficiency in semi-automatic welding equipment.

Gas-metal-arc welding is essentially a direct-current process. Alternating current is rarely used. DCRP (electrode positive) is used most extensively because it produces a stable arc, smooth metal transfer, and deeper penetration than DCSP (electrode negative). DCSP tends to produce arc instability, poor transfer of molten metal, and spatter, making it generally undesirable. However, it is sometimes used when minimum penetration is desired. Amperage is determined by wire feed speed.

51.5 GAS-TUNGSTEN-ARC WELDING (GTAW)

Gas-tungsten-arc welding produces coalescence (joining) of metals by heating them with an arc between a tungsten (nonconsumable) electrode and the work. Shielding is obtained by an envelope of an inert gas or gas mixture. Gas-tungsten-arc welding is also known as *tungsten-inert-gas* (TIG) *welding*. Figure 51-4 illustrates the process.

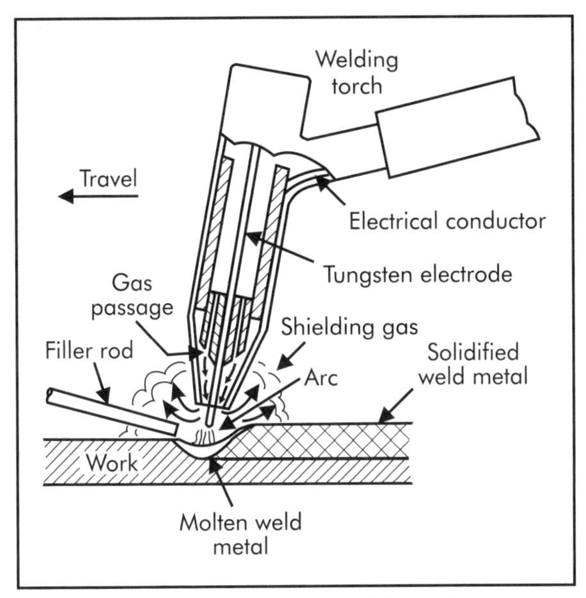

Figure 51-4. Gas-tungsten-arc welding. (Courtesy Lincoln Electric Co.)

While a wide range of metal thicknesses can be welded, the gas-tungsten-arc method is especially adapted for welding thin metals where the requirements for quality and finish are exacting. The process is performed manually, semi-automatically, or automatically.

Important advantages of the gas-tungsten-arc process are its suitability for welding most metals, both ferrous and nonferrous, and its ability to produce high-quality joints. Another advantage is the process does not produce weld spatter because no filler metal crosses the arc. Also, because fluxing agents are not used, cleaning after welding is seldom required.

A possible limitation to the use of the gas-tungsten-arc process is that it is slower than consumable-electrode, arc-welding processes. Also, gas-tungsten-arc welding requires an externally supplied inert shielding gas or gas mixture, adding to the cost of welding. Any transfer of tungsten particles from the electrode to the weld can cause hard, brittle contamination.

Essentially, the nonconsumable tungsten electrode is a torch—a heating device. Under the protective gas shield, metals to be joined are heated above their melting points, without melting the electrode, so that material from one part coalesces with material from the other part. Upon solidification of the molten area, unification occurs. Pressure may be used when the edges to be joined are approaching the molten state to assist coalescence. Welding in this manner requires no filler metal.

If the work is too thick for the mere fusing of abutting edges and if groove joints or reinforcements, such as fillets, are required, filler metal must be added. This is supplied by a filler rod, manually or mechanically fed into the weld puddle. The tip of the nonconsumable tungsten electrode and the tip of the filler rod are kept under the protective gas shield as welding progresses. Compositions of the filler metals should match the base metals being joined. In automatic welding, filler wire is fed mechanically through a guide into the weld puddle, generally at the leading edge of the weld puddle. All of the standard types of joints can be welded with the gas-tungsten-arc process and filler metals.

Tungsten is used as the nonconsumable electrode because of its high melting temperature and ability to generate a stable arc. The tungsten electrodes used may be pure tungsten, which is the least costly composition. Pure tungsten is used for aluminum and magnesium AC welding. Zirconiated tungsten, which provides less contamination, is also used for aluminum. With DC, 1–2% thoriated tungsten is used to weld carbon and stainless steels and nickel alloys. The radioactivity of thorium, however, has given rise to ceriated and lanthanated tungsten electrodes, which can be substituted for the thoriated tungsten. The shielding gas may be argon, helium, or a mixture of both. Argon is used to provide better shielding at low flow rates.

Gas-tungsten-arc welding is done with either direct or alternating current. However, pulsed current is used for some applications. DCSP (electrode negative) is used most extensively for the gas-tungsten-arc process. It is satisfactory for welding most metals except aluminum and magnesium, which typically require AC. The capability for removing surface oxides plus deep penetration often makes AC preferable for welding aluminum and magnesium.

DCRP (electrode positive) is the least used current for the gas-tungsten-arc process. It produces a wide bead with shallow penetration and requires the use of large-sized electrodes with comparatively low currents to dissipate the heat produced.

Pulsed current offers the advantages of minimizing the heat-affected zone (HAZ) and increasing the depth-to-width ratios of weld beads. It is used to join precision parts and for the automatic welding of pipe. The HAZ is the portion of the base metal that has not been melted, but whose mechanical properties or microstructure has been altered by the heat of welding (Wick and Veilleux 1987).

A supply of high-frequency current is generally provided from a separate source for gas-tungsten-arc welding. When welding with DC, the high-frequency current is used to initiate the arc (instead of touch or scratch starting) and is generally turned off after ignition. When welding with AC, the high-frequency current is on continuously to initiate the arc and ensure reignition at each half cycle (when the voltage is 0) during welding.

51.6 PLASMA-ARC WELDING (PAW)

Plasma-arc welding is similar to gas-tungsten-arc welding in that both processes use a nonconsumable tungsten electrode. In gas-tungsten-arc welding, the electrode extends beyond the torch, whereas in plasma-arc welding, the electrode is recessed in the torch as illustrated in Figure 51-5. In plasma-arc welding, an electric arc is formed between the tungsten electrode and the welding gun or workpiece and contained or restricted in a small-diameter nozzle (orifice). A gas, such as argon, is forced through the restricted arc, thereby forming a high-velocity ionized or high-energy plasma gas. The plasma stream is surrounded by an inert shielding gas to protect the weld pool.

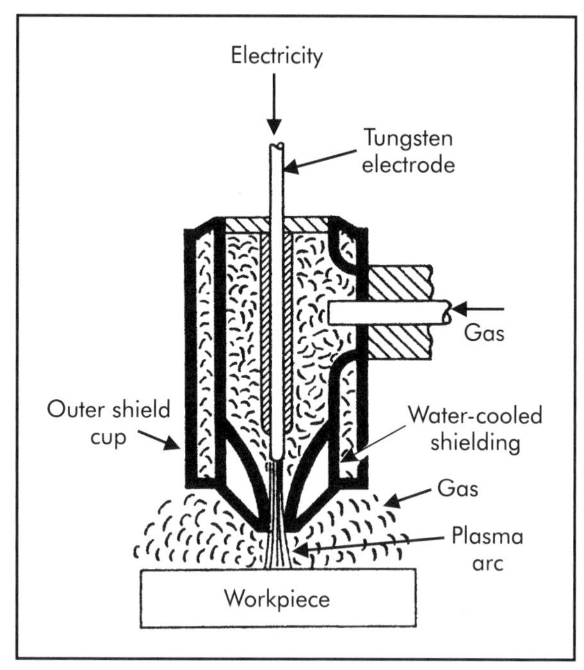

Figure 51-5. Plasma-arc welding (PAW) (Schrader and Elshennawy 2000).

The intensity of the narrow plasma stream creates a greater depth-to-width ratio weld than gas-tungsten-arc welding. PAW uses a pure tungsten electrode and typically welds the same metals as gas-tungsten-arc welding.

51.7 ARC-WELDING DEFECTS

Defects commonly occurring as the result of arc welding include spatter, undercut, incomplete fusion, cracks, porosity, slag inclusions, inadequate joint penetration, overlap, and distortion.

Weld spatter may be objectionable from an appearance standpoint, but is of no consequence to the structural function of the weld; however, it may interfere with service if entrapped in an inaccessible cavity. Excessive spatter is not necessary, and its appearance on a weldment is an indication of improper welding. It may be caused by:

- excessive welding current,
- wrong electrode,
- wrong electrode polarity,
- too large an electrode,
- improper electrode position,
- arc blow,
- excessive arc length,
- low gas flow,
- contaminated weld joint surfaces, or
- a combination of these factors.

Unless serious, *undercut* is more of an appearance discontinuity than a structural detriment. However, some inspection agencies will not accept undercut welds, particularly where cyclic loading is involved, and will demand that it be repaired. For this reason, undercut should be avoided. It may be caused by high welding current and fast travel speed, improper electrode position or manipulation, excessive arc length, or too large an electrode. A uniform weave of the electrode will tend to prevent undercutting when making groove welds. Excessive weaving will cause undercut and slag inclusions and should be avoided.

Incomplete fusion is sometimes associated with inadequate joint penetration or insufficient current and is a structural fault. Complete fusion is essential for full-strength welds. Incomplete fusion may be caused by improper current setting, welding technique, joint preparation, welding electrode size, electrode manipulation, or weldment positioning. All contact surfaces must be molten to achieve complete fusion by arc welding. Thick plates require higher welding currents for a given electrode than thin plates. Therefore, sufficient welding current should be used to ensure correct deposition of filler metal and good depth of fusion in the base metal. The electrode size for the first passes should be small enough to reach the root of the groove and permit good fusion as shown in Figure 51-6.

There are different kinds of *cracks* in welds, all of which are serious. All cracks should be examined to determine what corrective measures are needed. Common cracks in welded joints include crater cracks, under-bead cracks, transverse cracks, longitudinal cracks, and toe cracks as illustrated in Figure 51-7. Cracks can be detected using the methods discussed in Chapter 54.

Martensite formed in the welding steel is a leading cause of cracking. It is brittle and does not yield but breaks when the stresses in the weld become high enough (Schrader and Elshennawy 2000). Some alloys weaken and crack at welding temperatures and during solidification. Grain boundaries become brittle when elements, such as sulfur, segregate to the grain boundaries, thereby causing cracking when the weld metal solidifies. Hydrogen embrittlement (cold cracking) is also a major cause of cracking in steel. Hydrogen is highly soluble in hot steel but is largely expelled on cooling. If the gas can not escape, it exerts pressure to crack the metal. This is likely to happen under the same conditions that form brittle martensite (Schrader and Elshennawy 2000). Cracks due to hydrogen embrittlement generally occur after weld solidification.

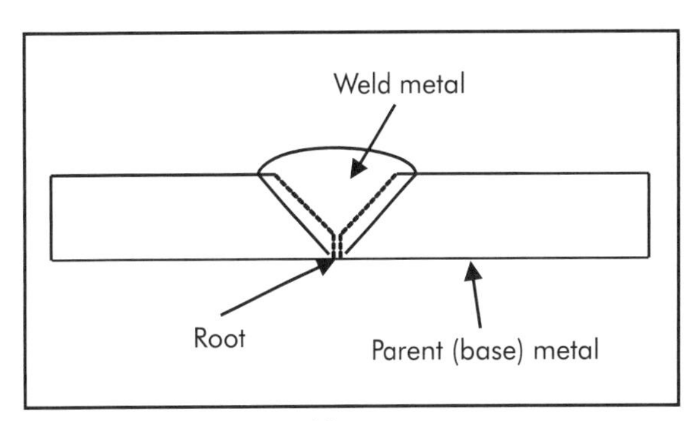

Figure 51-6. Groove weld.

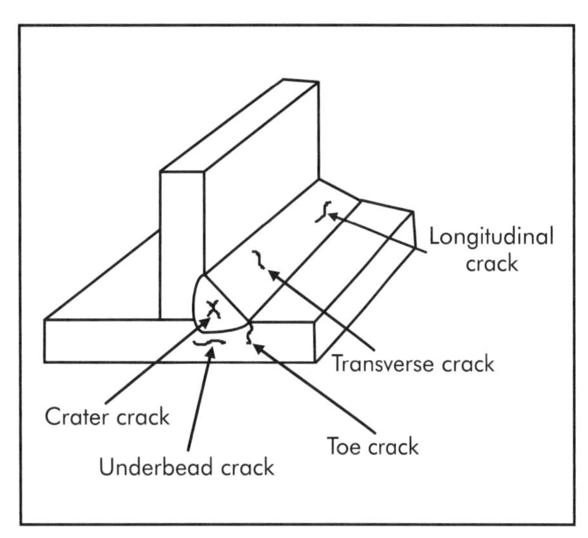

Figure 51-7. Fillet weld cracks.

To reduce the likelihood of cracking, preheating and post-heating are typically required. Preheating and post-heating are used for the following reasons.

- Especially with highly restrained joints, preheating and post-heating reduce shrinkage stresses in the weld and adjacent base metal.
- Preheating and post-heating provide a slower rate of cooling through the critical temperature range (about 1,600–1,330° F [870–720° C] for carbon steel). This reduces hardness or susceptibility to cracking of both the weld and heat-affected area of the base plate.
- Preheating and post-heating provide a slower rate of cooling to the temperature of about 400° F (204° C) for carbon steel, allowing more time for any hydrogen to diffuse away from the weld and adjacent plate to avoid underbead cracking.

Porosity in welds does not seriously affect weld strength unless the welds are extremely porous. Surface holes in the weld bead are undesirable from appearance and stress standpoints, and can affect fatigue life. Another common form of porosity, aside from surface holes commonly referred to as blowholes, is gas pockets. One of the major causes of porosity is incomplete cleaning prior to welding.

The residual stresses in a piece of steel resulting from the heat cycle of the welding process can cause distortion. The forces that cause distortion are present in every weld made and, unless proper techniques are used, the weldment may distort enough so that considerable time and money must be spent to correct it.

Simple guidelines can be followed that will aid materially in prevention and control of distortion. In many cases, the application of a single guideline will be sufficient. In others, a combination of the following may be required.

- Change the workpiece design.
- Use the most suitable welding process.
- Reduce the effective shrinkage force.
- Make shrinkage forces work to reduce distortion.
- Balance shrinkage forces with other forces.

51.8 WELD BONDING

Weld bonding is a combination of resistance spot or seam welding and adhesive bonding. The procedure used most commonly is to apply a structural adhesive to the area to be joined, followed by spot welding through the adhesive. Another method of weld bonding consists of applying tape or film adhesive, with holes cut in the adhesive where welds are required.

The advantages of weld bonding over spot welding or adhesive bonding alone include improved fatigue life and durability, and better resistance to peeling and cleavage as illustrated in Figure 51-8. Also, the adhesive acts as a seal in the joints to provide better corrosion protection and generally tighter construction.

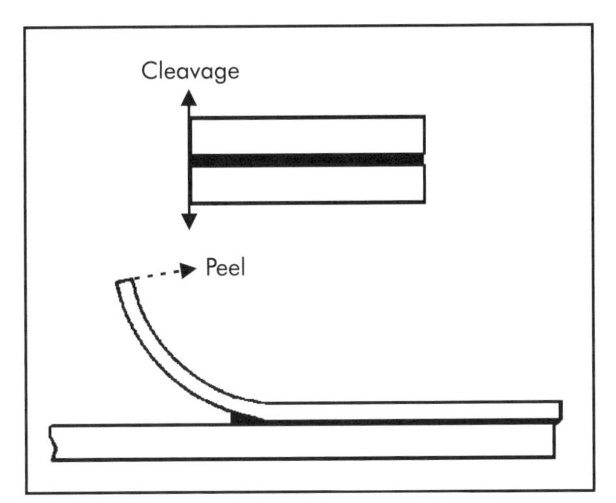

Figure 51-8. Common joint stresses (Wick and Veilleux 1987).

51.9 ELECTRON-BEAM WELDING (EBW)

Electron-beam welding (EBW) is a fusion joining process accomplished by impinging a high-intensity beam of electrons on the joint. This results in precise melting and coalescence of the joint interface surfaces. EBW produces a weld with a high depth-to-width ratio and a small heat-affected zone, thereby reducing workpiece distortion.

Electron-beam welding is employed in a variety of precision and production applications. In the automotive industry, semi-automated and fully automated partial vacuum and nonvacuum EBW systems are used for the welding of a large number of transmission components of varying materials. Other products joined by electron-beam welding include solenoid valves, transducers, sealed bearings, and medical implants.

When using the vacuum mode of EBW, production rates are limited by the time required to pump down the work chamber. With nonvacuum EBW, the distance that workpieces can be placed from the electron gun is limited. Electron-beam welding also requires safety precautions for protection from x-ray and other radiation (Wick and Veilleux 1987).

51.10 THERMIT WELDING

In *thermit welding*, coalescence (joining) of metals is accomplished by heating with superheated molten metal produced by a reaction between a metal oxide and aluminum. Although termed a welding process, thermit welding actually more closely resembles metal casting. The process offers advantages for certain specialized applications, especially for joining heavy and/or complex cross-sections that often are not weldable with conventional gas- or electric-arc processes. The most common application is welding rail sections into continuous lengths. Other applications include welding and splicing concrete-reinforcing steel bars together, and repairing large components.

51.11 FRICTION WELDING AND INERTIA WELDING

Friction welding is a solid-state joining process that produces coalescence of metals or nonmetals using the heat developed between two surfaces by a combination of mechanically induced rubbing motion and applied load. Mechanical energy is directly converted to thermal energy at the joint interface. Under normal conditions, the faying surfaces do not melt. Filler metal, flux, and shielding gas are not required with this process, but shielding gas is sometimes used for welding reactive metals. Typical applications include engine valves, drive shafts, steering shafts, transmission shafts, and hydraulic piston rods.

Inertia welding is a modification of friction welding where the rotating workpiece is attached to a flywheel. The flywheel is brought to a predetermined speed and then disengaged from the driving mechanism. The stationary workpiece and rotation workpiece are pressed together and the friction and heat produce the weld. When the weld is complete the flywheel stops. Inertia welding is shown in Figure 51-9.

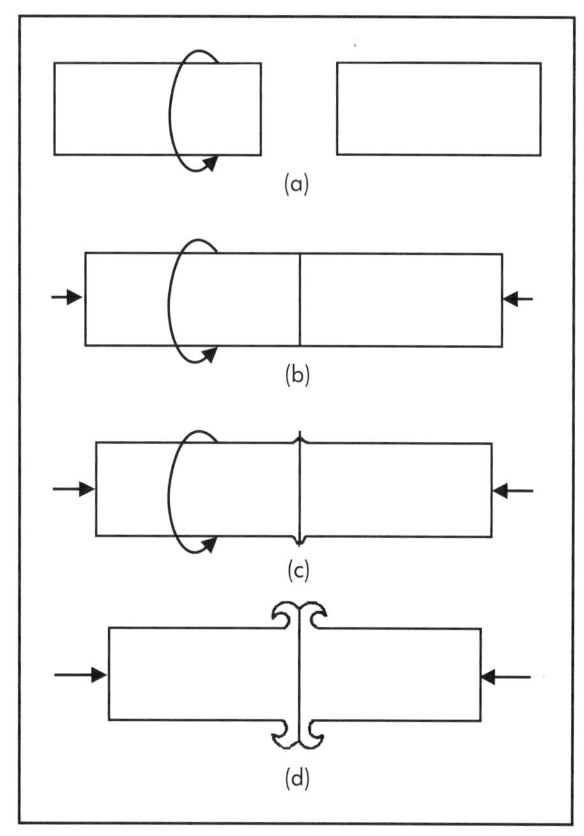

Figure 51-9. Inertia welding (Wick and Veilleux 1987).

51.12 MECHANICAL FASTENERS

Mechanical fasteners can be divided into two categories, integral and discrete. *Integral fasteners* are formed areas of a component that function by interfering or interlocking with other areas of the assembly. This type of fastening is commonly applied to formed sheet-metal products. It is also commonly used on plastic assemblies made from injection-molded parts.

Discrete fasteners comprise threaded fasteners such as bolts, nuts, screws, and other fasteners such as rivets, pins, and retaining rings.

THREADED FASTENERS

A primary application of threaded fasteners is joining and holding parts together for load-carrying requirements, especially when disassembly and reassembly may be required (Nee 1998). Various threaded fastener applications are shown in Figure 51-10.

Bolts and Studs

Bolts are externally threaded fasteners generally assembled with nuts. Bolts with hexagonal heads, frequently called hex heads, are the most commonly used. The heads have a flat or indented top surface, six flat sides, and a flat bearing surface. Hex heads are often used on high-strength bolts and are easier to tighten than bolts with square heads. They are generally available in standard strength grades (classified by the Society of Automotive Engineers [SAE], American Society for Testing and Materials [ASTM], and the International Organization for Standardization [ISO]) and special strength grades for specific applications.

Round-head bolts have thin circular heads with rounded or flat top surfaces and flat bearing surfaces. When provided with an underhead configuration that locks into the joint material, round-head bolts resist rotation and are tightened by turning their mating nuts.

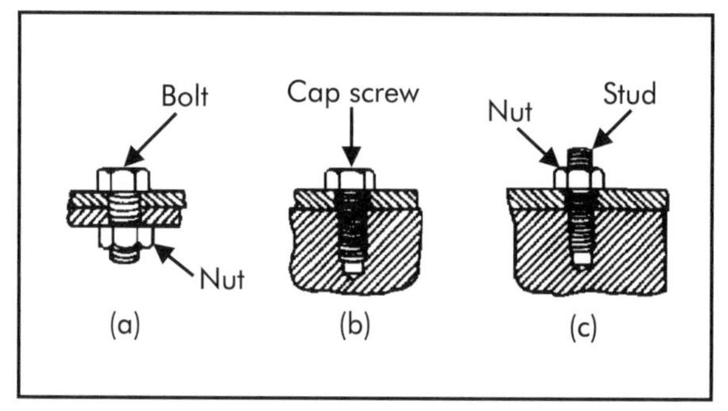

Figure 51-10. Typical assemblies using threaded fasteners: (a) bolt and nut; (b) cap screw; (c) stud (Wick and Veilleux 1987).

Studs are unheaded, externally threaded fasteners. They are available with threads on one or both ends or continuously threaded. Heat-treated and/or plated studs are available to suit specific requirements.

An advantage of studs for some applications, such as the assembly of large and heavy components, is their usefulness as pilots to facilitate mating of the components, which expedites automatic assembly. For many applications, studs provide fixed external threads and nuts are the only components that must be assembled (Nee 1998).

Nuts

Nuts are internally threaded fasteners that fit on bolts, studs, screws, or other externally threaded fasteners for mechanically joining parts. They also serve as an adjustment means in some applications.

Hex and square nuts, sometimes referred to as full nuts, are the most common. Hex nuts are used for most general-purpose applications. Square machine screw nuts are usually limited to use on light-duty and special assemblies. Regular and heavy square nuts are often used for bolted flange connections.

Single-thread nuts, sometimes called spring nuts, are formed by stamping a thread-engaging impression in a flat piece of metal as illustrated in Figure 51-11. These nuts are generally made from high-carbon spring steel; however, other metals can be used.

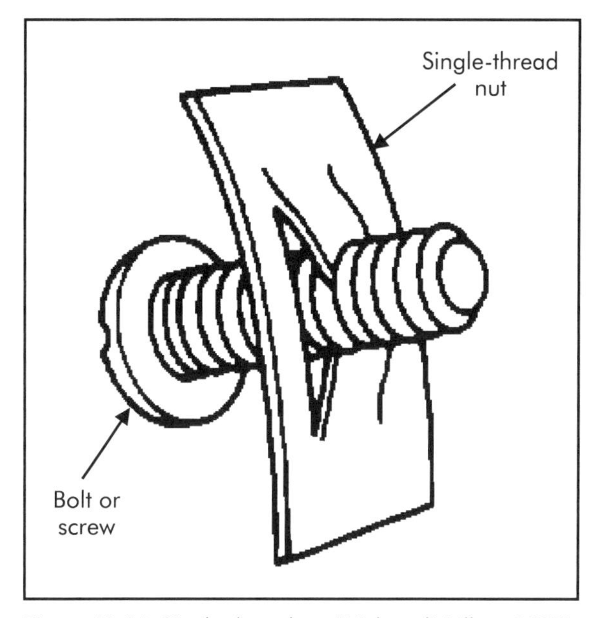

Figure 51-11. Single-thread nut (Wick and Veilleux 1987).

Screws

Screws are externally threaded fasteners capable of insertion into holes in assembled parts, mating with preformed internal threads, or cutting or forming their own threads. Because of their basic design, it is possible to use some screws, which are sometimes called bolts, in combination with nuts (Nee 1998).

Screws are available in a wide variety of types and sizes to suit specific requirements for different applications. Major types discussed in this section include machine screws, cap screws, sems (screw and washer assemblies), and tapping screws.

Machine screws are usually inserted into tapped holes, but are sometimes used with nuts. They are generally supplied with plain (as sheared) flat points, but for some special applications they are made with various types of points. Machine screws are usually made from steel, stainless steel, brass, or aluminum, and have a variety of head styles. Many machine screws are made from unhardened materials, but hardened screws are available (Nee 1998).

Cap screws are manufactured to close dimensional tolerances and designed for applications requiring high tensile strengths. The shanks of cap screws are generally not fully threaded to their heads. They are made with hex, socket, or fillister-slotted heads as illustrated in Figure 51-12. Most cap screws are made from steel, stainless steel, brass, bronze, or aluminum alloy.

Set screws are hardened fasteners generally used to hold pulleys, gears, and other

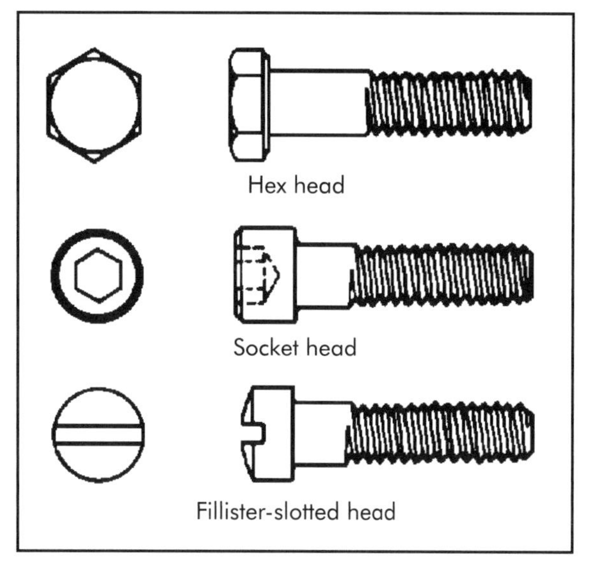

Figure 51-12. Cap screws with various heads (Wick and Veilleux 1987).

components on shafts. They are available in several styles with several point configurations as shown in Figure 51-13. Holding power is provided by compressive forces, with some set screws providing additional holding resistance to rotation by penetration of their points into the shaft material (Nee 1998).

Sems (screw and washer assemblies) is a generic term for pre-assembled screw and washer fasteners. The washer is placed on the screw blank prior to roll threading and becomes a permanent part of the assembly after roll threading, but is free to rotate. Sems are available in various combinations of head styles and washer types. They are used extensively in the manufacture of automobiles and appliances. These fasteners permit convenient and rapid assembly by

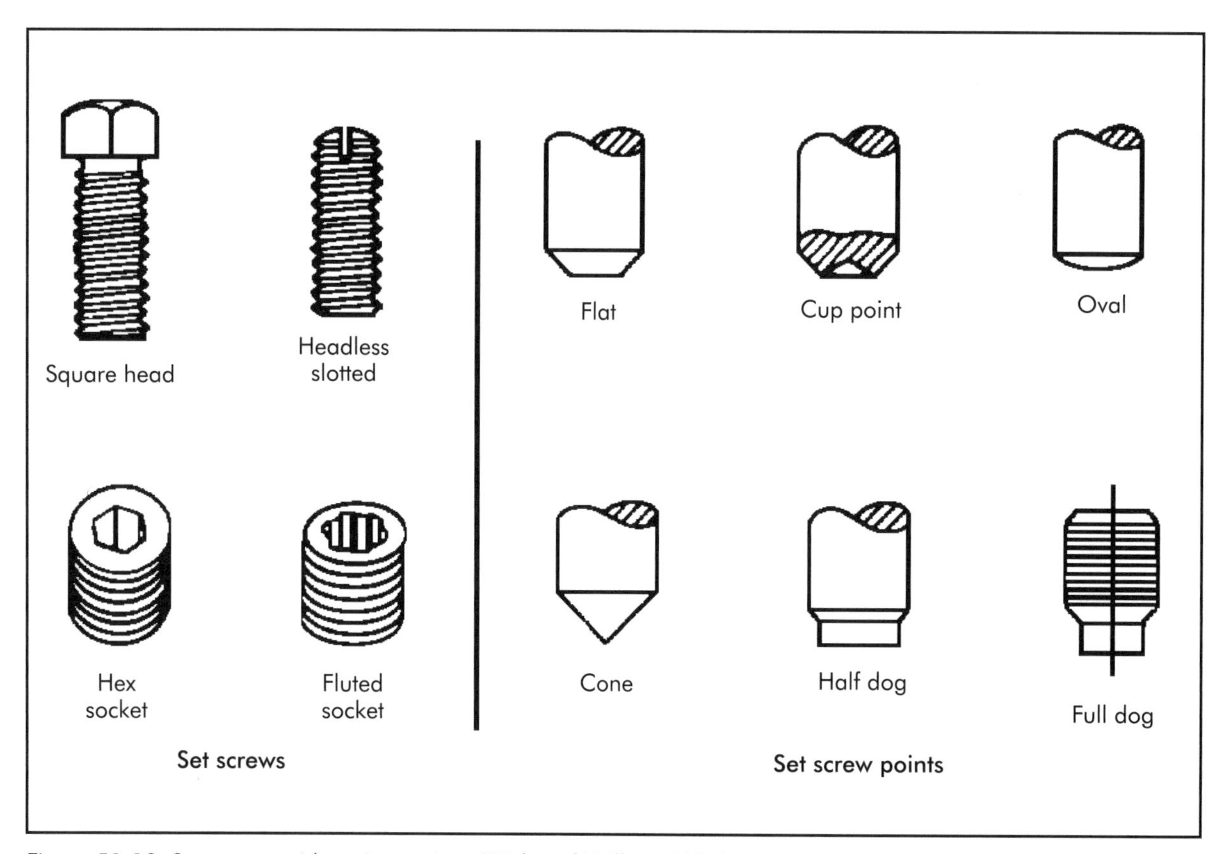

Figure 51-13. Set screws with various points (Wick and Veilleux 1987).

eliminating the need for a separate washer assembly operation.

Tapping screws will cut or form mating threads when driven into holes. Self-drilling, self-piercing, and special tapping screws are also available. They are made with slotted, recessed, or wrenching heads in various head styles. Tapping screws are generally used on thin materials. The advantages of tapping screws include rapid installation because nuts are not needed and access is required from only one side. Mating threads fit the screw threads closely with no clearance necessary.

51.13 SCREW THREAD TERMINOLOGY

The chief features of an external thread are illustrated in Figure 51-14. Internal threads have corresponding features. They determine the size and shape of a thread.

The *pitch* is the distance parallel to the axis from any point on a screw thread to a corresponding point on the next thread. The pitch is the reciprocal of the number of threads in a unit of length. Thus if a screw has a pitch of 2 mm, it has 0.500 threads per millimeter. Another that has eight threads per inch has a pitch of 0.125 in.

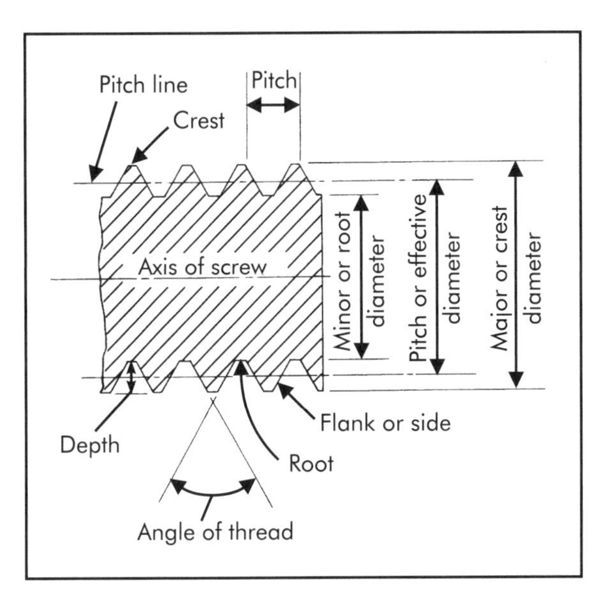

Figure 51-14. Features of an external screw thread (Schrader and Elshennawy 2000).

The *lead* is the distance a screw advances axially in one full turn. The lead is the reciprocal of the number of turns required to advance the screw axially a unit of length. Thus a drive screw that requires 20 turns to move forward 1 in. (25.4 mm) has a lead of 0.050 in. (1.27 mm). A single-thread screw has only one continuous thread on its surface as found on most commercial screws, bolts, and nuts. The lead of a single-thread screw is equal to its pitch.

The *major diameter* of a straight thread is the diameter of a cylinder on which the crest of an external thread or the root of an internal thread lies. The *minor diameter* applies to the root of an external thread or the crest of an internal thread. The *pitch diameter* of a straight thread is the diameter of a cylinder that cuts the threads where the width of the threads is equal to the width of the space between the threads (Schrader and Elshennawy 2000).

At one time, screw threads lacked uniformity in size and shape. Standardization of sizes, shapes, and pitches was undertaken to create a condition of order. A number of standards have prevailed over the years. An accord of the United States, Great Britain, and Canada in 1948 set up the Unified Screw Thread Form. The standard has been revised periodically since then. The purpose of the uniform thread system was to promote the interchangeability of products. The standard designates a coarse thread series by unified national coarse (UNC) and a fine thread series by unified national fine (UNF). There is a less often used extra-fine thread series called unified national extra-fine (UNEF), among others such as the American National Standard Taper Pipe Thread (NPT) for piping.

Each series specifies the threads per inch (tpi) and basic dimensions and tolerances for certain nominal inch size diameters as illustrated in Figure 51-15.

The last digit in the screw thread form designation shown in Figure 51-15 is the thread class. Screw threads are divided into classes to designate the fit between internal and external mating threads. For some applications, a nut may fit loosely on a screw; in other cases, the two must fit together snuggly. Different fits are obtained by assigning appropriate tolerances on the pitch, major and minor diameters, and allowances to the threads for each class.

The Unified Screw Thread Form Standard recognizes several classes of threads. Those classified as "A" are screws, "B" for nuts. Classes 1A and 1B are loose fitting, where quick assembly and rapid production are important and shake or play is not objectionable. Classes 2A and 2B cover most commercial screws, bolts, and nuts. Classes 3–5 provide for closer fits and interference fits. Screws from one class may be used with nuts from another class for more fits.

In addition to the Unified Screw Thread Standard there exists the ISO metric thread standard. Figure 51-16 illustrates the standard used for specifying a metric screw. The metric thread designation begins with "M," indicating it is a metric thread. Following the "M" is the nominal size in millimeters. After the times sign is the pitch. At the end of the designation is the tolerance class. The first digit and letter specify the pitch diameter tolerance grade and position, respectively. The second digit and letter refer to the major diameter tolerance grade and position, respectively. The digits refer to the grade, where the lower the number assigned, the closer the fit. Grade 6 is approximately equivalent to 2A or 2B in the Unified Screw Thread system. The letters refer to the tolerance positions for the pitch diameter and major diameter, respectively. The tolerance position specifies the allowable deviation between the location of the tolerance with respect to the basic pitch and major diameter sizes. Lower case "e," "f," and "g" refer to external threads. Upper case "G" and "H" refer to internal threads. A large allowance is indicated by "e" and a small one by "g" or "G." No allowance is indicated by "h" or "H."

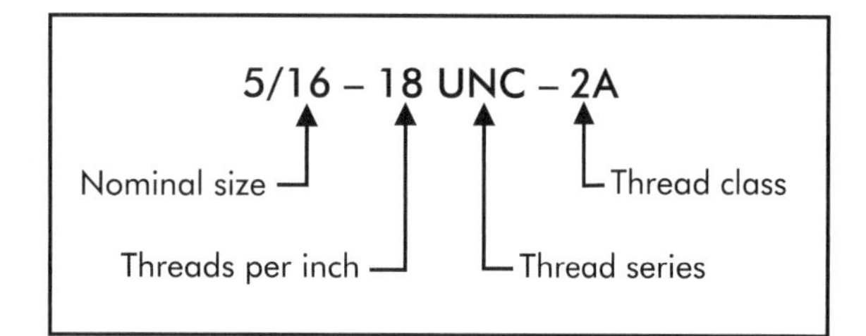

Figure 51-15. Unified Screw Thread Form designation.

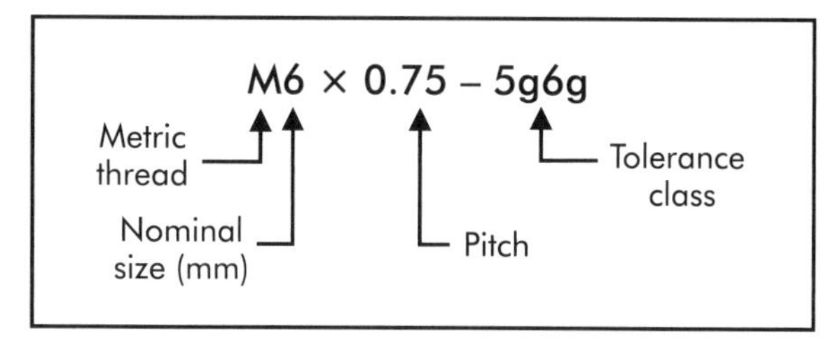

Figure 51-16. ISO thread designation.

51.14 RIVETS

A *rivet* is a one-piece, unthreaded, permanent fastener consisting of a head and a body. It is used for fastening two or more pieces together by passing the body through a hole in each piece and then clinching or forming a second head on the body end as illustrated in Figure 51-17. Once set in place, a rivet cannot be removed except by chipping off the head or clinched end, or by drilling it out.

A major advantage of rivets is that they can be installed economically and rapidly, so they are suitable for automatic assembly operations. Other advantages include their low cost and the fact that rivets are good for joining dissimilar materials of different hardnesses or thicknesses. A possible limitation is that the impact required for clinching can deform thin sheets. Rivets are

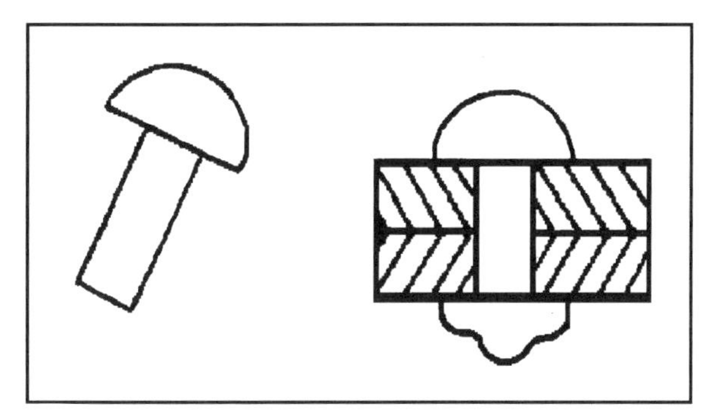

Figure 51-17. Solid rivet (Wick and Veilleux 1987).

usually less expensive than threaded fasteners, but their strength in shear or tension may be lower, especially when compared with heat-treated bolts.

Rivets are made with various head styles. Selection of a specific type and head style depends primarily on the job to be done, the joint location, strength requirements, and the appearance of the joint. For example, when a flush surface is desired, the countersunk-head rivet may be used.

Any metal that can be cold worked is suitable for making rivets. The metals most commonly used for rivets include aluminum alloys; brass, bronze, and other copper alloys; low, medium, and high-carbon steels; alloy steels; and corrosion-resistant steels. Rivets are often supplied with a natural (as-processed) finish, with no plating or other coating. However, they can be provided with various plated finishes, including zinc, cadmium, nickel, tin, copper, and brass.

Clinching, setting, or driving of rivets, commonly called riveting, is done in several ways with a variety of equipment. Riveting can be done hot and cold; however, cold riveting is the most common process. Typical riveting is done by impact or squeezing.

When selecting solid rivets, a variety of factors should be considered. Use of the correct length of rivet shank is important to form the desired upset head. If too long a length is used, buckling occurs; if too short a length is used, an incomplete head will be formed. Hole diameter also must be correct. Too large a hole can cause buckling and/or underfilled holes, resulting in loose joints. Punched holes generally improve the quality of the joint because the holes are perpendicular to the surface and have no heavy burrs. Other factors that can result in poor clinching include improper hole diameters; inadequate rivet support; worn, misaligned, or improperly designed tools; improper rivet material or size; and flaws in the rivets.

Blind rivets (*pop rivets*) are fasteners with self-contained mechanical, chemical, or other features that form upsets on their blind ends and expand their shanks to join parts of assemblies. This design permits the fasteners to be installed in the holes of joints that are accessible from only one side.

51.15 PINS

Pins provide a simple and low-cost method of mechanical fastening. They are available in straight-cylindrical or tapered designs, with or without heads, and generally rely on elastic compression for their gripping power. Applications of pins are common in industrial machines and commercial products. Pins are used as locking devices, locating elements, pivots, and bearing faces. They often secure the positions of two or more parts relative to each other. Applications are primarily for shear loading where there is not a high amount of axial loading.

Most pins are hardened for maximum strength and permanent assembly, but some are used soft so they will shear before the assembly or mechanism is damaged.

An advantage of pins, in addition to their low cost and effectiveness, is that many of them can be inserted in simple drilled or cored holes. However, some press-fit pins require varying degrees of hole preparation for proper insertion.

A wide variety of standard types and sizes of pins is available, and special designs are made for specific applications. The types of pins discussed in this section are machine pins including straight pins, dowel pins, taper pins, clevis pins, cotter pins, and spring pins.

Solid, straight, cylindrical pins are usually cut from wire or barstock and have unground surfaces. They are available in chamfered or square-end designs. Square-end straight pins can have the corners on both ends broken with small radii. Straight pins are generally made from cold-drawn, low- or high-carbon steel wire, rod, or bar. High-carbon steels are generally heat-treated. Pins are also available in stainless steels, brasses, and other metals.

Dowel pins are precision, straight, cylindrical pins available in hardened and ground, and unhardened and ground types. They are used extensively in the production of machines, tools, dies, jigs, and fixtures to retain parts in fixed positions or preserve alignments.

Tapered pins, commonly called *taper pins*, inserted by a drive fit, are often used to position parts or transmit low torque forces. Tapered pins are available in commercial and precision classes, with the precision pins having closer tolerances.

Clevis pins are solid pins with cylindrical heads at one end and a drilled hole for a cotter pin at the other end as illustrated in Figure 51-18a. They are commonly used as pivots in many mechanisms and with clevises and rod-end eyes in industrial applications.

Cotter (split) pins are double-bodied pins formed from half-round wire. A loop at one end of each pin provides a head as illustrated in Figure 51-18b. Cotter pins are used in clevis pins and other pinned assemblies.

Spring pins are made in slotted (split) tubular and coiled (spirally wrapped) designs as illustrated in Figure 51-19. Pins of both designs have smaller cross-sectional areas than solid pins of the same diameter, sometimes resulting in lower shear strengths.

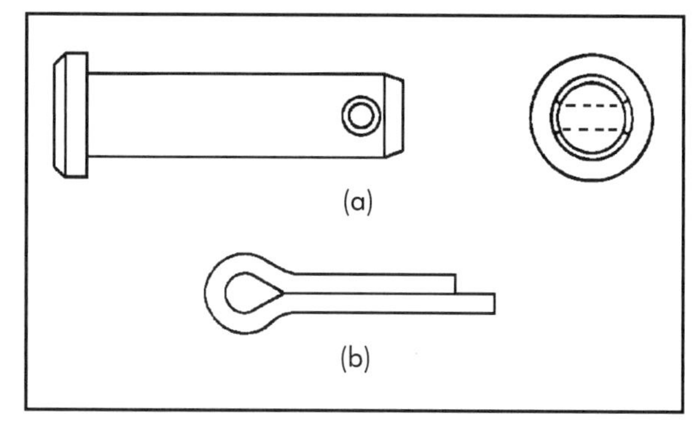

Figure 51-18. (a) Clevis pin and (b) cotter pin (Wick and Veilleux 1987).

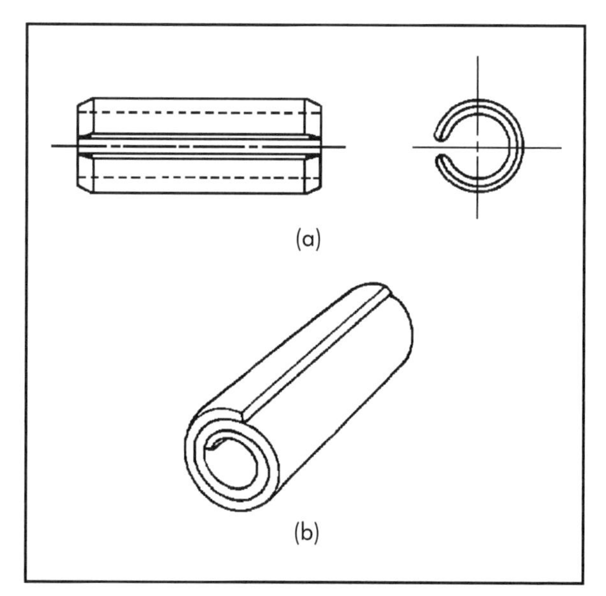

Figure 51-19. (a) Slotted tubular spring pin and (b) spirally coiled spring pin (Wick and Veilleux 1987).

However, some heat-treated spring pins have higher shear strengths than low-carbon solid pins. Spring pins provide good shock and vibration absorption, and stresses are distributed equally. Their inherent springiness makes insertion in holes easier. When manufactured, spring pins are made oversized with respect to the diameters of the holes in which they will be inserted. When inserted, the pins are compressed, resulting in radial forces against the hole walls to retain the pins in the desired positions.

51.16 RETAINING RINGS

Retaining rings, sometimes called *snap rings*, are fastening devices that provide shoulders and/or bearing surfaces for locating or limiting the movement of parts on shafts or inside holes. They are designed to exert a radial clamping force. For most applications, the rings provide a removable means of fastening. Some are designed to take up end play caused by accumulated tolerances or wear of the parts being retained (Wick and Veilleux 1987).

Retaining rings are usually made from spring steel or other materials with good spring properties to allow deformation during assembly and a return to original ring shape. Figure 51-20 illustrates some common snap rings.

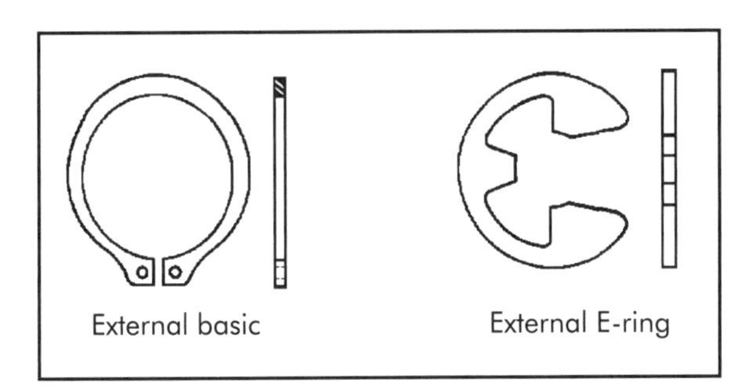

Figure 51-20. Retaining rings (Wick and Veilleux 1987).

REVIEW QUESTIONS

51.1) Describe the functions of flux in SMAW.

51.2) In GMAW, how is adjusting the amperage accomplished?

51.3) In GTAW using DC, what is the function of the high-frequency current?

51.4) How can hydrogen embrittlement be reduced?

51.5) Define thread pitch.

51.6) What type of rivet is used when access is only from one side of the hole?

REFERENCES

Nee, John G., ed. 1998. *Fundamentals of Tool Design*, Fourth Edition. Dearborn, MI: Society of Manufacturing Engineers.

Schrader, George F., and Elshennawy, Ahmad K. 2000. *Manufacturing Processes and Materials*, Fourth Edition. Dearborn, MI: Society of Manufacturing Engineers.

Wick, Charles and Veilleux, Raymond, eds. 1987. *Tool and Manufacturing Engineers Handbook*, Fourth Edition. Volume 4: *Quality Control and Assembly*. Dearborn, MI: Society of Manufacturing Engineers.

Chapter 52

Deburring and Finishing Analysis

This chapter introduces surface preparation and treatment techniques not discussed in Chapter 27 of *Fundamentals of Manufacturing,* Second Edition. Finishing is a very important part of manufacturing. It is performed regularly and, like other processes, has evolved to include new technology. This chapter progresses from very fundamental to more sophisticated methods. It covers deburring, honing, lapping, shot peening, electropolishing, anodizing, and polymer coatings.

52.1 DEBURRING

Deburring processes discussed in this section consist of hand deburring, mass finishing, tumbling, vibratory finishing, abrasive flow machining, thermal energy method, electrochemical deburring, and wire brushing. A combination of burr accessibility and size, edge and surface requirements, available equipment, and cost usually dictates the selection of a specific deburring process.

HAND DEBURRING

Hand deburring is a viable deburring process, even though it is slow, labor intensive, costly, and often provides less consistent results than desired. Advantages of hand deburring include the versatility of the process and minimal capital investment. Other advantages of hand deburring include:

- increased flexibility,
- the ability to access hard-to-reach burrs,
- convenient when only a small number of parts are needed and/or a short time is required, and
- the ability to handle delicate parts.

MASS FINISHING

Mass finishing processes include cleaning, deburring, deflashing, edge and corner radiusing, and surface finishing. In general, mass finishing infers handling workpieces in bulk quantities. All mass finishing processes are based on loading parts into a container, which usually holds abrasive or nonabrasive media, water, and/or a compound. Action of the container causes the media to rub against the workpieces or the workpieces rub against one another, thus producing the desired results.

Mass finishing is a simple, versatile, and low-cost means of conditioning the edges and surfaces of various components. Normally, individual handling or fixturing of work-pieces is not required, thus eliminating the costs associated with manual and most other mechanized finishing processes. With proper control, consistent results can be attained from workpiece to workpiece and batch to batch.

A limitation of mass finishing is that its action is generally only effective on surfaces, edges, and corners of workpieces that contact the media. It is not normally possible to give preferential treatment to specific areas; however, masking specific areas has been successful for some applications. The action is greater on the edges of workpieces than on equally exposed surfaces. The action in holes and recesses is significantly less than on exposed areas; and in small, deep recesses, it is unusual to be able to do any finishing unless the workpiece is fixtured. Long cycles may result in critical dimensions becoming out of tolerance. The media may become lodged in workpieces with different sized openings, but this problem generally can be avoided by selecting a different size or shape of media or by sealing the openings. Popular mass finishing processes include tumbling (rotary barrel) and vibratory finishing.

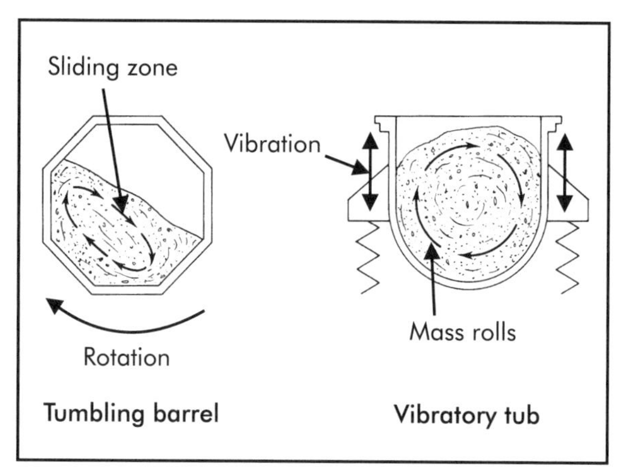

Figure 52-1. Principles of tumbling and vibratory finishing (Schrader and Elshennawy 2000).

Tumbling

The operation called *tumbling, rolling,* or *barrel finishing* consists of loading workpieces in a barrel, which is about 60% full of abrasive grains, sand, or other scouring agents, depending on the work and action desired. Water can be added and is usually mixed with an acid, rust preventive, detergent, and/or lubricant. The barrel is closed or tilted and rotated at a slow speed for a time period according to the treatment required. As the barrel rotates, its load moves upward to a turnover point. The force of gravity overcomes the tendency of the mass to stick together. Then the upper layer slides toward the bottom of the barrel as illustrated in Figure 52-1. Although abrading action may occur as the workload rises in the barrel, about 90% of the rubbing action occurs during the slide (Schrader and Elshennawy 2000).

Vibratory Finishing

Vibratory finishing does the same work as barrel (tumbling) finishing, but is done in an open tub or bowl filled with workpieces and media, and vibrated at 1,000–1,200 Hz as shown in Figure 52-1. The vibration action makes the entire load rotate slowly in a helical path, but the whole mass is agitated. Scouring, trimming, and burnishing take place throughout the mixture. Vibratory finishing is faster than tumbling. Under proper conditions it can deburr the insides of workpieces and in recesses where tumbling is limited to the outsides of workpieces (Schrader and Elshennawy 2000).

ABRASIVE-FLOW MACHINING (AFM)

Abrasive-flow machining (AFM) is a process in which a semisolid abrasive media is forced or extruded through a workpiece passage. The AFM process can be thought of as the use of a self-forming abrasive tool to precisely remove workpiece material from surfaces, edges, holes, slots, cavities, and any restricted places through which the media can be forced to flow, including complex internal passages. Several to hundreds of holes, slots, or edges can be deburred, radiused, and/or polished in one operation. Depending upon workpiece and machine sizes, several or even dozens of parts can be processed in one fixture load, resulting in production rates up to hundreds of parts per hour. After processing, however, the abrasive media normally remains within the part's interior and surrounds its exterior, making fully automated processing and handling difficult (Wick and Veilleux 1985).

THERMAL ENERGY METHOD (TEM)

The *thermal energy method* (TEM) of deburring uses instantaneous and intense heat energy to burn away or oxidize burrs. Parts are placed inside a chamber, which is sealed and then pressurized with a mixture of combustible gas and oxygen that completely encloses the parts, reaching into internal passages and cavities. When the gaseous mixture is ignited, an intense combustion takes place, generating intense heat lasting several milliseconds. The intense heat burns, or oxidizes, the burrs. Only the burrs are removed because of their high ratio of surface area to mass. The part is then cleaned to remove the oxides created during the process. TEM can remove burrs and flash without affecting the parent metal surfaces.

ELECTROCHEMICAL DEBURRING (ECD)

In *electrochemical deburring* (ECD), burrs are dissolved from metallic workpieces electrochemically and flushed away by pressurized electrolyte (Wick and Veilleux 1985). With the ECD process, electrolyte flows through a gap between a tool (cathode) and the workpiece (anode), thus completing the electrical circuit needed for the DC power to dissolve the burrs. The tool and workpiece do not contact each other, and the workpiece is not exposed to any mechanical or thermal stresses. As a result, there are no changes in the physical or chemical properties of the metal. Special workholding fixtures and tools are used to ensure the conductive surfaces of the tools conform to the areas or edges of the workpieces to be deburred. Proper insulation of the tools and protective shielding or masking of the workpieces limit the electrochemical action to the desired surfaces. Electrochemical deburring is ideal for selective (controlled) deburring and the removal of burrs that are inaccessible with most other processes. For parts with surfaces that cannot be altered or scratched, ECD is often the only possible method of deburring. The absence of mechanical forces permits the deburring of thin sections and fragile parts without distortion or damage. One obvious limitation to ECD is that parts made from nonconductive materials cannot be deburred by this method.

WIRE BRUSHING

Wire brushing is a very prominent method used in industry for scale removal, deburring, and deflashing. Hand brushing tools and techniques are well known and continue to be effectively used. Power-driven, rotary industrial brushes are widely used for deburring, cleaning, and finishing because they result in time and cost savings for many applications. Applications include edge blending, deburring, controlled surface roughening or refinement, cleaning, and finishing. Stock removal with brushes varies from minimum to substantial amounts.

Several workpiece parameters drive the process of wire brush finishing. Key geometric parameters such as part shape, size, access, and criticality of neighboring surfaces must be considered to select the proper brush size, shape, and tool path. The workpiece material type and the amount of material removal are important in the selection of the brush material and density. Depending upon the brush's filament size and speed, the brushing operation can affect workpiece dimensions. However, due to their flexibility, brushes are not suitable tools for achieving final dimensional size.

The availability of a wide range of brush types and the flexibility of the process allows wire brushing to be used in numerous applications. Standard brush shapes are wheel, cup, end, wide face, and tube as shown in Figure 52-2.

Wheel brushes are appropriate when higher, localized forces are required. Cup and wide-faced brushes are preferred in applications where a single plane or surface of the workpiece must be deburred or descaled.

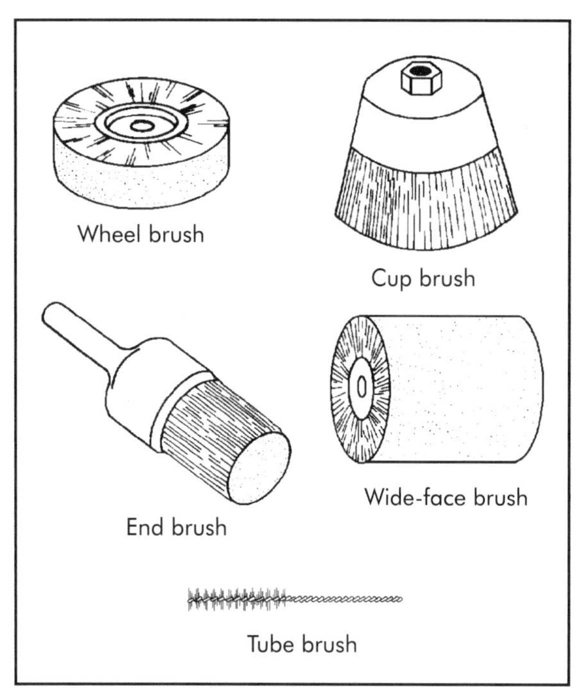

Figure 52-2. Primary types of power brushes (Wick and Veilleux 1985).

End and tube brush applications are more geometry specific; end brushes are good where there is limited access and tube brushes are used for bores and internal geometry.

The following are considerations for selecting the correct filament for the brush.

- Material type—strength, elasticity, and abrasiveness; typically steel filaments are used for material removal applications and nylon filaments are used to remove burrs or improve surface finish.
- Length—shorter filaments are more rigid and can transmit higher forces.
- Fill density—a more densely packed brush will be stronger and therefore transmit higher forces.
- Filament configuration—straight, crimped, or braided; which respectively increase in stiffness and strength.

The power equipment used to drive wheel brushes includes bench-mounted motors, portable tools, and CNC-based machine tools. Speeds, feed rates, and the depth of interference between the brush and workpiece are the critical process parameters. Potential limitations to the use of power brushing include the possibility of contaminating the workpieces, changing the color or surface finish of the workpieces, the generation or turning over of burrs, and hardening of the workpiece surfaces. When improperly used, wire brushing can displace material rather than remove it.

52.2 HONING

Honing is an abrasive finishing process used to achieve final sizing, correct contour errors, and provide desired surface finishes. Stock removal is usually 0.010 in. (0.25 mm) or less. The surface finish is typically 4–32 μin. (0.10–0.81 μm) R_a. The applications are typically internal bores that can range from diameters as small as 0.08 in. (2.0 mm) up to as large as 48 in. (1.2 m). To a much lesser extent, honing also can be used for flat surfaces and other geometries. The honing process typically follows a rough boring operation, which sets the feature location and initial rough size. Figure 52-3 illustrates common boring errors corrected by honing.

Honing is well suited for applications where accurate fits are required, or where sliding friction takes place and boundary lubrication is desired, such as in automotive engine cylinders. A honed, cross-hatched surface finish works well to retain oil.

There are several variations to honing. The honing method and speed, in addition to the honing stones, play a critical role. To control contours, stones are long and the cutting action occurs by rotating and reciprocating the tool. Either the workpiece or the tool location will float.

The bonded abrasive stones used for honing are contoured to provide face contact with the workpiece. The high stone contact area and slow speeds make honing quite different from a grinding operation where the grinding wheel geometry provides linear

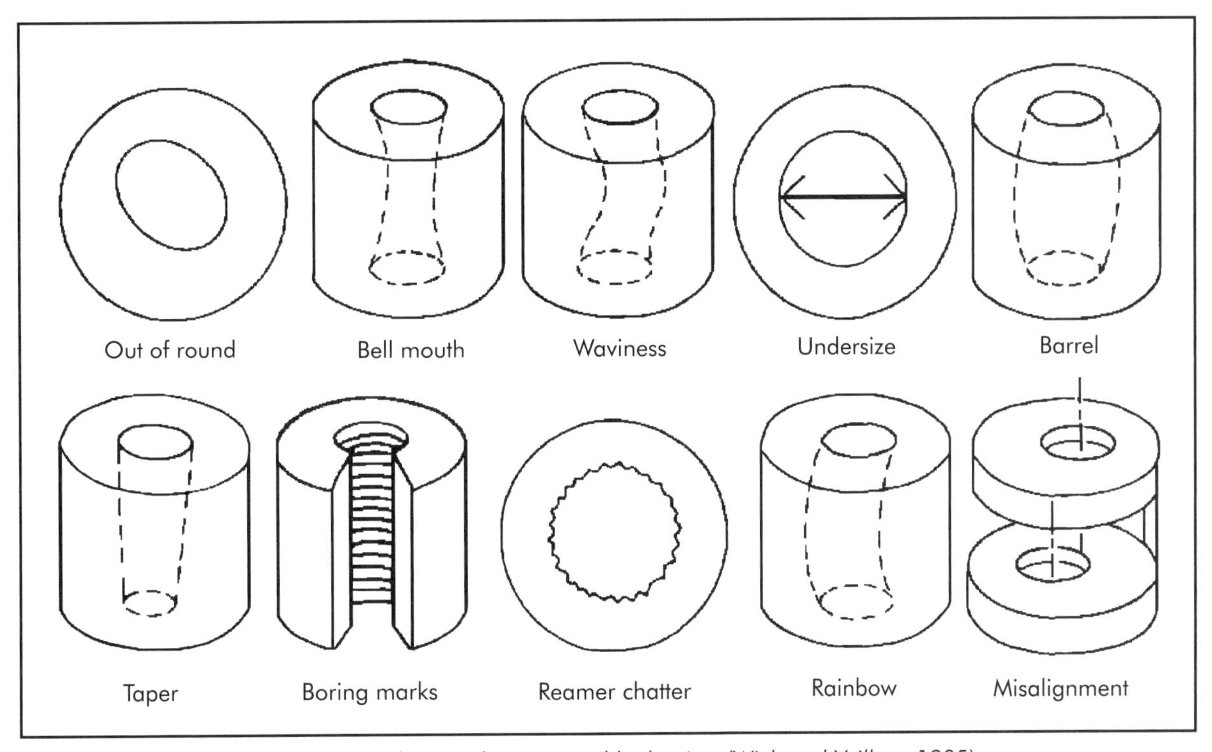

Figure 52-3. Common boring errors that can be corrected by honing (Wick and Veilleux 1985).

contact with the workpiece to enable high speed and high material removal rates. The abrasive material must be selected based on the material of the workpiece. General selections are shown in Table 52-1. The grit will be selected based on the surface finish required.

The honing tool holds one to eight stones. Generally, the larger the bore the more stones recommended. The tool also has guide surfaces that contact the workpiece. The stone expands outward to apply pressure to the work surface, which is critical to ensuring proper results. Figure 52-4 illustrates a typical honing tool.

To keep honing stones cutting at all times and prevent glazing of their cutting surfaces, a steady, continuous, and consistent breakdown of the stone must be ensured. The critical forces come from three sources: 1) pressure against the workpiece, 2) rotation, and 3) reciprocating action of the tool. Setting the contact pressures for the stones is critical for the material removal rate, but the structural capability of the workpiece must also be taken into consideration. To ensure a favorable geometry result, it is recommended that the stone length be at least 50% of the bore diameter. However, in most cases, the length-to-diameter ratio should be greater than one and set such that the stroke for the tool is 1/3 of the length of the stone.

A lubricant aids in maintaining consistent pressure on the bore, functions as a coolant, and carries the microchips away from the area being honed. The fluid must be compatible with the stones. Oil-based and water-soluble fluids are used, depending on the conditions.

Table 52-1. Honing abrasive applications

Abrasive	Application
Silicon carbide	Cast iron and nonferrous materials
Aluminum oxide	Steel
Cubic boron nitride (CBN)	All steels (soft and hard), nickel and cobalt-based superalloys, stainless steel, and a variety of other metals, including beryllium copper and zirconium
Diamond	Chromium plate, carbides, ceramics, glass, cast irons, brass, bronze, surfaces nitrided to depths greater than 0.001 in. (0.03 mm), blind holes (where little overstroke is available), intermittent and interrupted bores, and automatic honing in mass production

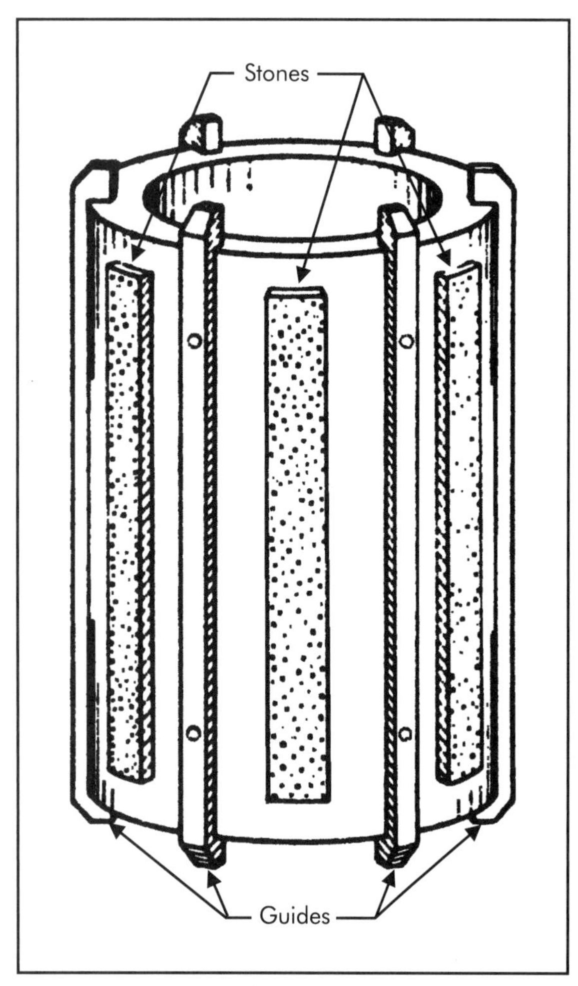

Figure 52-4. Sketch of a honing tool (Schrader and Elshennawy 2000).

52.3 LAPPING

Lapping, like honing, is an abrasive machining process used for highly accurate stock removal under 0.001 in. (25.4 μm) and in applications where finishes better than 2–16 μin. (0.05–0.40 μm) R_a are required. Lapping is a highly process-sensitive manufacturing operation that can generate extremely precise geometries for applications such as gaging, the microfinishing of high-tolerance metering devices, and valves. The process utilizes a lapping tool moving in rotation and reciprocating at low pressure. The abrasive material is a compound that is placed between the tool, called a lap, and the workpiece. Lapping is very unique because it can be done by using a tool or the mating workpiece. It is generally a final finishing operation that results in four major refinements to the workpiece:

- extreme accuracy,
- correction of minor imperfections of shape,
- final surface finish, and
- close fit between mating surfaces.

Given that lapping pressures are low, workholding issues such as clamping and chucking that cause distortion are minimized. Also, less heat is generated due to the low material removal rates so heat distortion is not a factor. A flat workpiece can be lapped on both sides simultaneously, improving the accuracy of flatness and parallelism. Key factors impacting the lapping process are workpiece material, lap tool material, abrasive medium, speed and motion, and pressure applied.

Hand lapping operations are common, in which the compound is brushed onto a lap. The finish produced will vary according to the grit of the abrasive selected and the operator's skill. Power-driven equipment includes driven flat laps, cylindrical laps, and internal laps. Automated lapping equipment allows the operator to fix the speed and motion of the lap while varying the pressure applied to the workpiece. An automated feed system is used to continuously apply the compound.

Abrasive compounds are classified by material and grit size. Common lapping materials include aluminum oxide, silicon carbide, boron carbide, and diamond. Aluminum oxide is the softest material used. It is not suitable for material removal, but can be used to improve finishes on hardened steel. Silicon carbide is a harder material with sharp grains that make it appropriate for fast material removal; however, it will not generate a fine surface finish. Boron carbide can be used to lap very hard workpiece materials, such as tool steel, carbides, and ceramics, and to improve finish, but it lacks the necessary properties for fast material removal. Diamond is becoming the dominant lapping material in most applications due to its ability to cut any material fast. Diamonds can be furnished in a wide selection of grits and they do not deteriorate.

The lapping compound is formed by mixing loose grains of the abrasive into an oil or water solution. The compounding agent holds the abrasive particles in suspension. It is typically furnished in various viscosities. Compounds are commercially available in grits ranging from 90–600, where the higher number designates the finest or smallest grit.

Lapping tools are available in a wide range of materials. The lap material should be softer than the workpiece and slightly porous to retain the abrasive compound and ensure even distribution on the workpiece. Cast iron is one of the most desirable materials due to its strength, reasonable hardness, long life, and ability to hold the abrasive compound. Other lapping tool materials used include copper, brass, nylon, and other plastics.

The lapping speed and pressure is dependent on the workpiece material. When improving finish, high speeds and low pressures are used. For material removal, slower speeds and higher pressures will cut faster, but the technique must be perfected to minimize heat build up and avoid breakdown of the lapping compound (Wick and Veilleux 1985).

52.4 SHOT PEENING

Shot peening is the cold working of a metal surface by a stream of spherical shot particles applied to the surface at high velocity under carefully controlled conditions.

Shot peening is most effective in reducing fatigue failures in parts subject to cyclic loading. It prevents these failures by creating a compressive stress layer in the surfaces of the parts. Additionally, as peening cold works the part surface, it blends surface imperfections and effectively eliminates them as stress concentration points. These surface imperfections or flaws may be localized areas of tensile stresses or phase transformations from machining or grinding, as well as pits, scratches, and other surface defects (Wick and Veilleux 1985).

52.5 ELECTROPOLISHING

Electropolishing is an electrochemical process, originally developed for sample preparation in metallography. It is used to enhance the surface finish and appearance of a workpiece. Whereas electroplating adds material to the workpiece, electropolishing removes it. The process is best suited for applications where the complete workpiece must be treated. In electropolishing, an electrical charge is passed through a chemical bath containing the workpiece. The current is applied to form a polarized film on the workpiece surface, which causes a leveling or fine smearing action, as shown in Figure 52-5. Projections, such as burrs, detach themselves from the work surface.

The electropolishing process can be used on most metals, including aluminum, copper, steel, and stainless steels. The process not only improves surface finish, but can provide corrosion resistance and leave surfaces clean for additional coating processes. Since electropolishing involves submerging the workpiece in a solution, it can be used for large parts and those that have internal features and passages that require cleaning and deburring.

Mechanical preparation of the workpiece can depend on the workpiece material. In most cases, milled surfaces are acceptable. Solvent cleaning or low-voltage electrolytic cleaning is necessary to remove grease and oil.

The equipment used is relatively flexible and includes a plastic-lined tank, electrolyte, and rods to mount the parts and act as electrical conductors. The parts are mounted on anode rods; and cathode rods are mounted parallel to the anode rods as illustrated in Figure 52-6. The rods are submerged in liquid electrolyte and the current flow into the workpiece is determined by the material composition, size, and shape of the workpiece, and its proximity to the cathode rods. The electrolyte or polishing solutions vary and, in many cases, they are heated to just below their boiling points.

After the electropolishing process is complete, several water-rinsing steps are done with light agitation. The preferred solution is distilled water with a wetting agent. Upon rinsing, the parts can be plated, anodized, or coated as necessary. Extremely large parts can be spray rinsed.

Electropolishing has been used for appearance enhancement in applications such as cookware, fountain pen caps, and some tub-

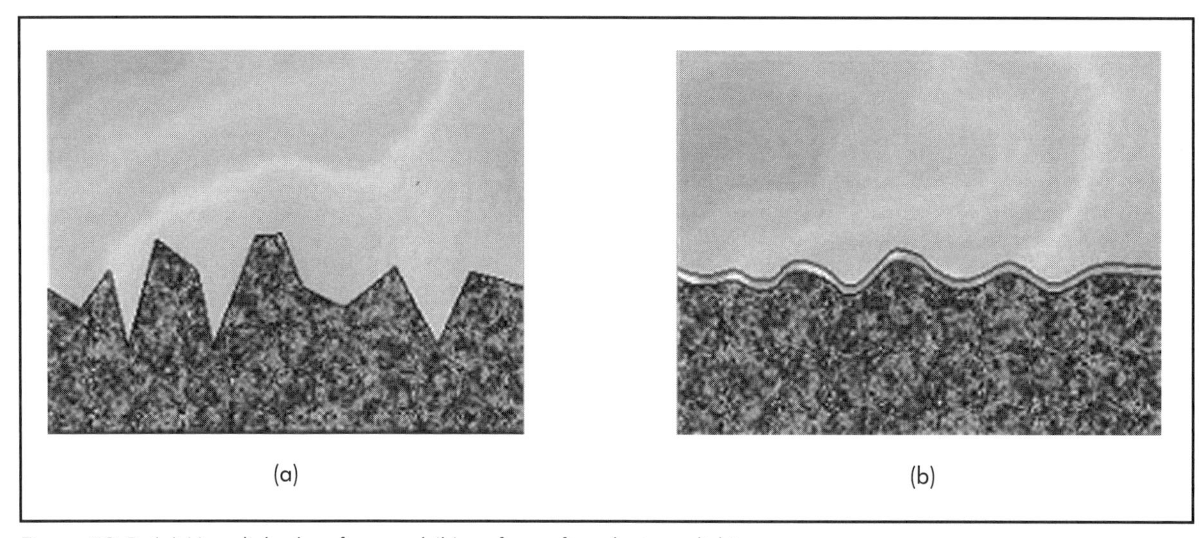

Figure 52-5. (a) Unpolished surface and (b) surface after electropolishing.

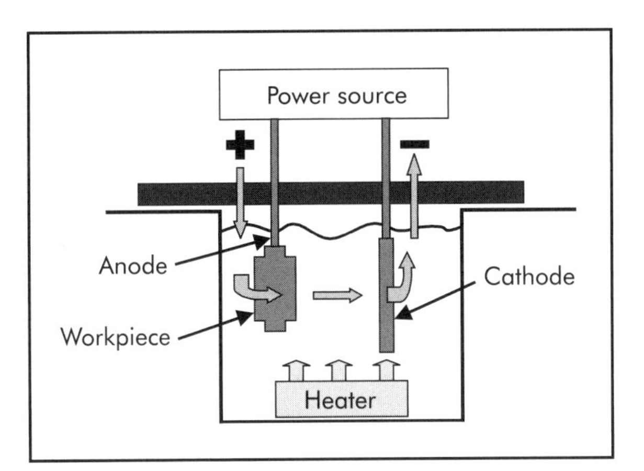

Figure 52-6. Electropolishing equipment.

ing. In addition, its ability to improve surface quality and clean parts has increased its use in the food and medical industries.

52.6 ANODIZING

The trend toward lighter-weight alloys has increased the need to refine processes that enhance particular material properties. Used for nonferrous parts, there are several variations of the anodizing process.

Anodizing is the common commercial term for the electrolytic treatment of metals, which forms stable films or coatings on the metal's surface. Aluminum and magnesium are anodized to the greatest extent on a commercial basis, but other metals, such as zinc, beryllium, titanium, zirconium, and thorium also respond to anodic treatment to form films of varying thicknesses. These coatings have historically been used for corrosion protection and appearance improvement. Dyes can be added in the anodizing process to give products a decorative colored surface.

Anodizing differs from electroplating in two ways. In electroplating, the work is made the cathode, and the metallic coatings are deposited on the work. In anodizing, the work is made the anode, and its surface is converted to a form of its oxide, which is integral with the metal substrate. The metallic portion of the electrolytic container is made the cathode.

Similar in nature to many other electrochemical methods, in anodizing the workpiece is electrically charged while submerged in an acid-based, electrolyte solution. The workpiece acts as the anode and oxides build on the surface. In the case of aluminum, aluminum oxide is formed, which is hard, inert, electrically insulating, transparent, and porous, so it accepts dyes. Traditionally, thicknesses of 0.0001–0.0030 in. (2.540–76.200 μm) are achieved. The coating thickness has a direct correlation to corrosion resistance.

Hard anodizing is a relatively new process that provides a hard case surface on nonferrous parts. In the hard anodizing process, two layers are built up on the workpiece surfaces, whereas in regular anodizing one layer is built primarily for corrosion protection. The initial layer is called the barrier layer. This is a nonporous layer that adheres to the surface, penetrating roughly 0.0001–0.0003 in. (2.540 –7.620 μm) deep. The second layer is a hard and porous protective layer. The thickness of these layers depends on the formation voltage for penetration, the electrolyte used, and processing time. The thickness of a hard anodized surface can be built up to 0.012 in. (0.31 mm).

The method of anodizing depends upon the type of electrolyte being used. Generally, the process is relatively straightforward and consists of cleaning the workpiece, etching or pickling, anodizing, coloring, and then sealing. *Pickling* refers to dipping a workpiece in an acidic bath to remove oxidation, rust, and surface scale. A thorough rinsing between each step is extremely important.

While anodizing makes the surface stronger than other coatings, it does not fill in voids or defects in the surface of the workpiece. Further, hard anodizing will degrade the surface finish, so additional honing or polishing may be required. The harder, more brittle surface can also reduce the fatigue strength of the material. Anodizing works best on external surfaces rather than internal features or complex geometries. One other factor to consider is that the buildup of the coating will be perpendicular to the surface. As a result, sharp edges and corners result in gaps in the coating, making these areas prone to chipping, especially with thicker coating levels. Hard anodic coatings are used to lower the coefficient of friction or to produce good release properties.

Commercially, standard anodized parts have long been used in numerous applications such as automotive trim, cosmetic cases, aerosol caps, keys, picture frames, name plates, and sporting goods. Hard coating continues to increase in use mainly because of its sliding wear and corrosion-resistant properties. It must be remembered that since this is a relatively thin layer, it does not add to the bearing strength of the underlying aluminum. Typical applications for hard anodizing include hydraulic cylinder bores, gun parts, aluminum gears, and ordnance and missile components.

52.7 POLYMER COATINGS

Polymer surface coating is one of the most cost effective and easiest methods of providing corrosion protection and improved appearance in applications where the duty cycle is not severe and there is accommodation for the coating thickness. Polymer coatings are available in the more traditional liquid form, but also can be applied in powder form. The material selection and application methods vary significantly between these two alternatives. Coatings are also available in both thermoplastic and thermoset resin forms.

Coatings may seem very straightforward, simple to select, and easy to apply, but their selection and application are actually complex. To achieve the correct results, the correct coating must be selected for the application, the work surface must be properly prepared, and application and curing processes must all be precisely followed (Wick and Veilleux 1985).

LIQUID ORGANIC COATINGS

Liquid organic coatings are complex polymer mixtures provided in several forms, such as water reducible, powder, and high-solid content, and typically applied by spraying or dipping.

The makeup of an organic coating includes a polymer binder, pigments, solvents, and other additives to enhance the processing. The binder is primarily polymer based and critical to adhesion to the work surface because it holds the pigment and other additives together in solution.

Options for the binder are:

- natural oils—cure in air, however not consistent, predictable, or easy to control;
- alkyds—fast air-curing resin, but not very hard;
- epoxy—two-part, room-temperature cure;
- silicone—high temperature, non-yellowing, strong chemical resistance;
- novolac—phenol-formaldehyde—air/heat-cured with solubility limitations;
- acrylic—thermoset and thermoplastic formulations, high-emulsion applications—latex previously used extensively in the automotive industry; and
- fluorocarbons—high chemical resistance, but lubricity and anti-stick properties require special techniques for application.

The appearance is a function of the addition of pigment to the coating. The chemistry behind the pigment materials and their characteristics is extremely complex. Organic, colored pigments are used, as well as numerous variations of metallic and inorganic types. Some of the key characteristics in pigment selection are tinting strength, light-fastness, bleed characteristics (metallic oxidation), hiding power (opaqueness), and chemical and thermal resistance. The concentration level of the pigment can affect the mechanical and optical properties of the coating, so this must be taken into consideration when selecting the binder/pigment combination. Other additives, such as solvents, stabilizers, defoamers, and catalysts, are critical to the coating process and ensure satisfactory results. Solvent selection is critical due to environmental considerations. Solvents are classified by their content of volatile organic compounds (VOCs).

The typical application process involves cleaning and preparing the surface, applying a primer coat if necessary, curing that coat, applying a base coating (with pigment) by spray or dip, and then either air drying or baking. An additional protective coating can be applied to improve the strength of the coated work surface.

Coatings are used in virtually all manufacturing industries for parts that are large, small, weak, flexible, or rigid. The coating selection criteria must take into account the temperature and chemical exposure limits of the product to be coated. Traditional liquid-base coatings are typically easy to apply and control. The equipment is flexible and can be adapted to many different product configurations. It is typically desirable to use a controlled environment since some binders are more sensitive to environmental conditions than others.

Coating defects consist of several types such as orange peel, runs and sags, holidays, and fish eyes. *Orange peel* is a term descriptive of a surface finish where the coating has not flowed out or leveled to a perfectly smooth finish, leaving the coating with the pebbled appearance of the skin of an orange. *Runs* and *sags* are apparent where excess coating has been applied too quickly and the liquid has run down in sheets. *Holidays* are small bare spots left by oily or greasy surface spots. *Fish eyes* are small (under 0.10 in. [2.5 mm] in diameter) raised or crater-like areas around a central thin spot caused by a tiny oil drop or surface contamination from silicones.

POWDER COATINGS

In powder-based coating, a finely pulverized polymer is applied to the work surface at ambient conditions and then cured at an elevated temperature. This technology uses similar binders and pigments as in the liquid process, but utilizes a much different carrier to distribute and cure the coating material. A major advantage of powder processes and a key reason for the minimal environmental effects is that the coating material can be reclaimed and reused, resulting in extremely low VOC levels.

Powder coatings are available in thermoplastic and thermoset forms. However, the thermoplastic materials are more difficult to pulverize into a fine powder, so these are restricted to applications where a thicker, more pliable coating is needed. Common thermoplastic powders include nylon, polyvinyl chloride (PVC), polypropylene, and polyethylene. Thermoset powders are based on lower-molecular-weight resins, which are cross-linked upon application, providing a dense, heat-stable coating. Thermoset powders include epoxy, acrylic, and polyester.

The primary powder-coating application methods are electrostatic spraying and fluidized bed dipping. In *electrostatic spraying* parts are cleaned and dried prior to spraying. The fluidized powder is electrostatically charged and then sprayed onto the grounded workpiece as illustrated in Figure 52-7. After spraying, post curing and cool down are necessary. Electrostatic spraying allows for coverage of the total surface area of the workpiece.

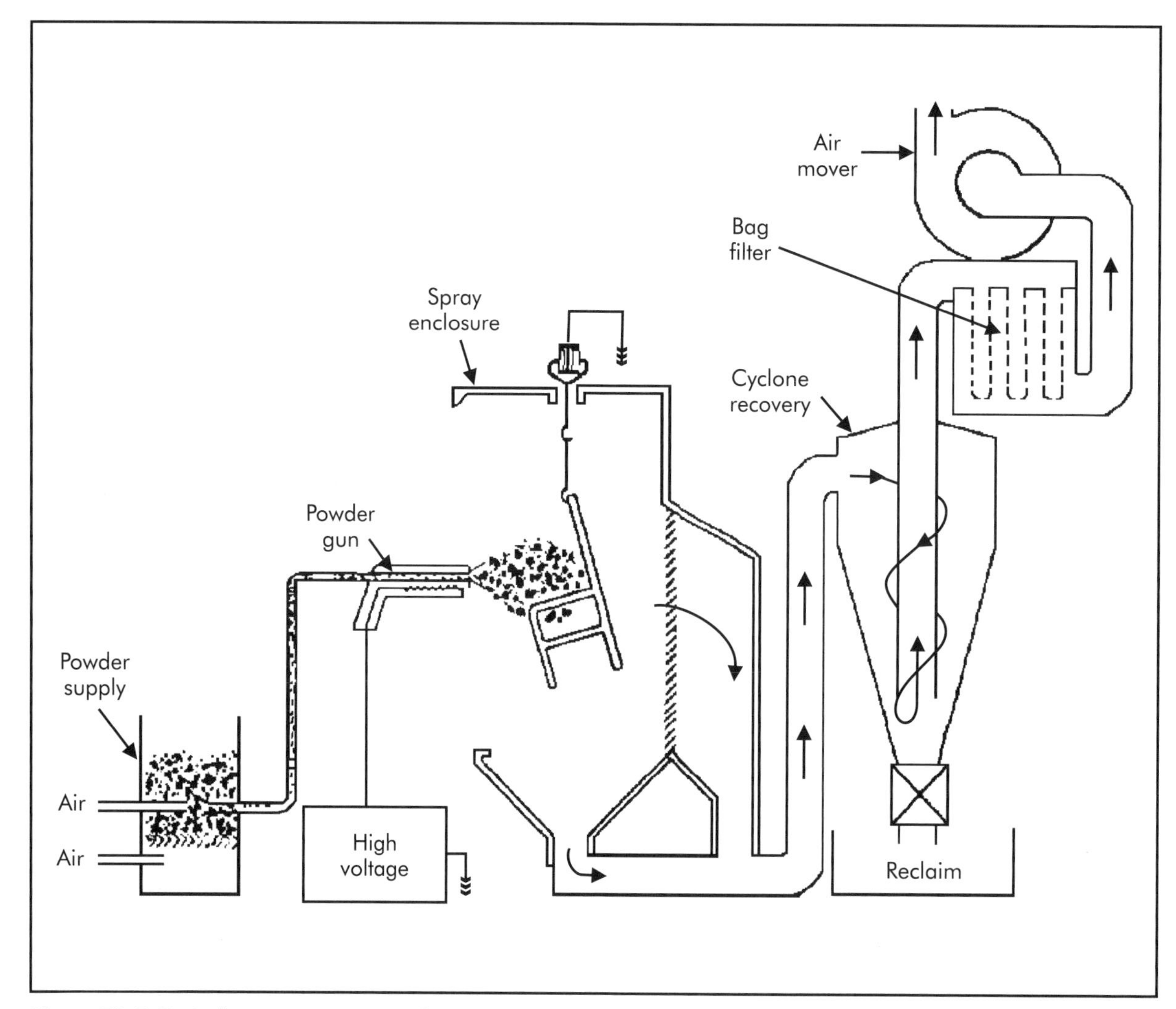

Figure 52-7. Typical components in an electrostatic spray system (Wick and Veilleux 1985).

Fluidized bed dipping is an application alternative with two variations. One variation includes a heated workpiece that attracts the powder as illustrated in Figure 52-8. The parts are first cleaned and dried. During the coating process, parts are heated to temperatures ranging from 300–500° F (149–260° C) and then immersed in the fluidized bed. The powder adheres to the heated surface, melts and flows over the surface, and then cools and hardens forming a smooth layer. The other variation is electrostatic, whereby charged powder is attracted to the grounded workpiece as illustrated in Figure 52-9. Compressed air is used to keep the powder in suspension and a post-curing operation is typically used. This process is more appropriate for coating smaller parts.

A newer method of powder application utilizes a flame spray to heat the powder in a spray gun, applying it as a liquid to the work surface. This process has the advantage of being able to apply a powder coating to a nonmetallic part.

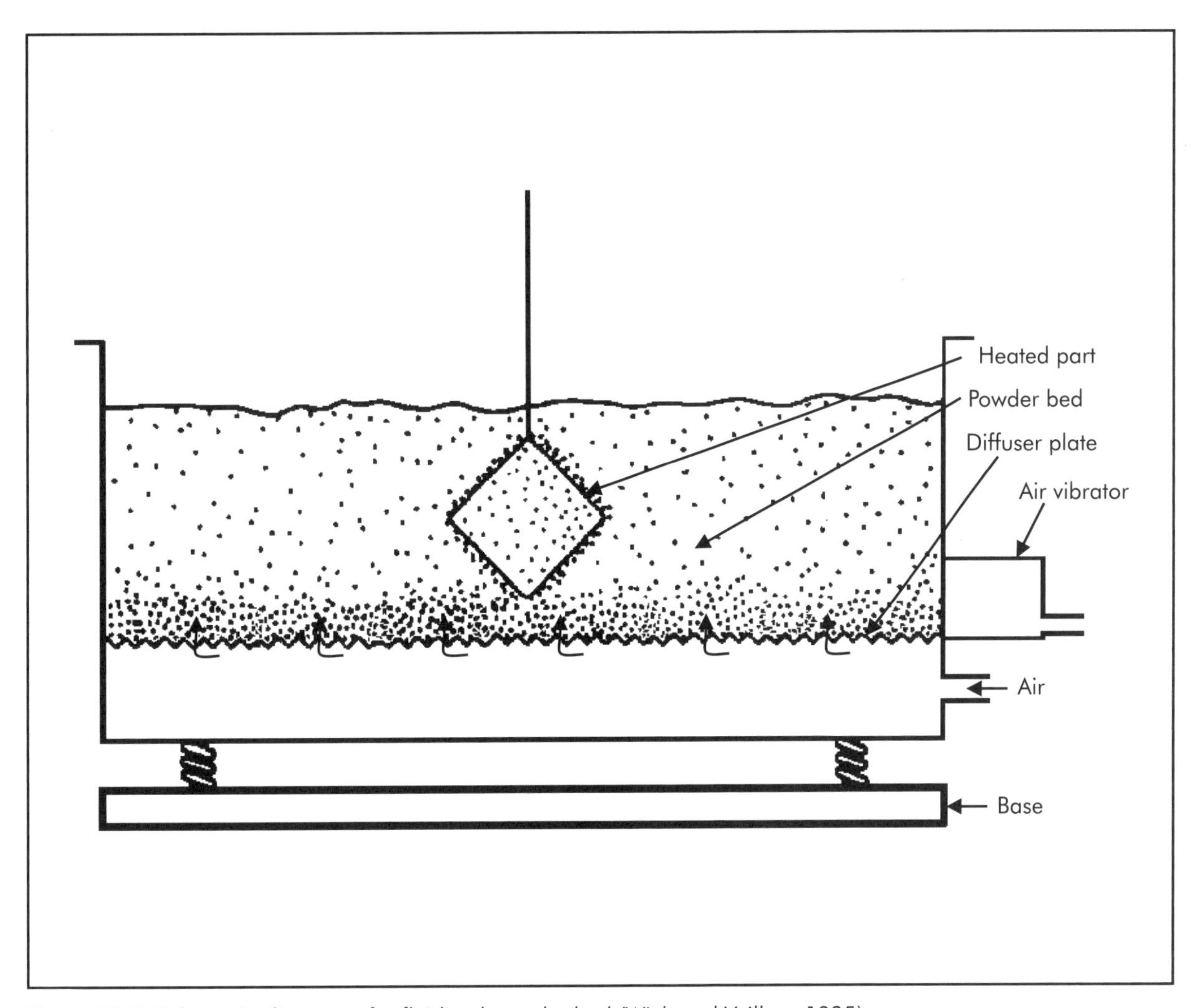

Figure 52-8. Schematic diagram of a fluidized powder bed (Wick and Veilleux 1985).

REVIEW QUESTIONS

52.1) What deburring method utilizes a self-forming abrasive tool?

52.2) Given a choice between honing and lapping, which process has the heaviest stock removal rate?

52.3) What is the difference between hard anodizing and regular anodizing?

52.4) What are the components of an organic coating?

52.5) What does VOC stand for?

52.6) What are the methods for applying powder coatings?

REFERENCES

Schrader, George F., and Elshennawy, Ahmad K. 2000. *Manufacturing Processes and Materials*, Fourth Edition. Dearborn, MI: Society of Manufacturing Engineers.

Wick, Charles and Veilleux, Raymond, eds. 1985. *Tool and Manufacturing Engineers Handbook*, Fourth Edition. Volume 3: *Materials, Finishing and Coating*. Dearborn, MI: Society of Manufacturing Engineers.

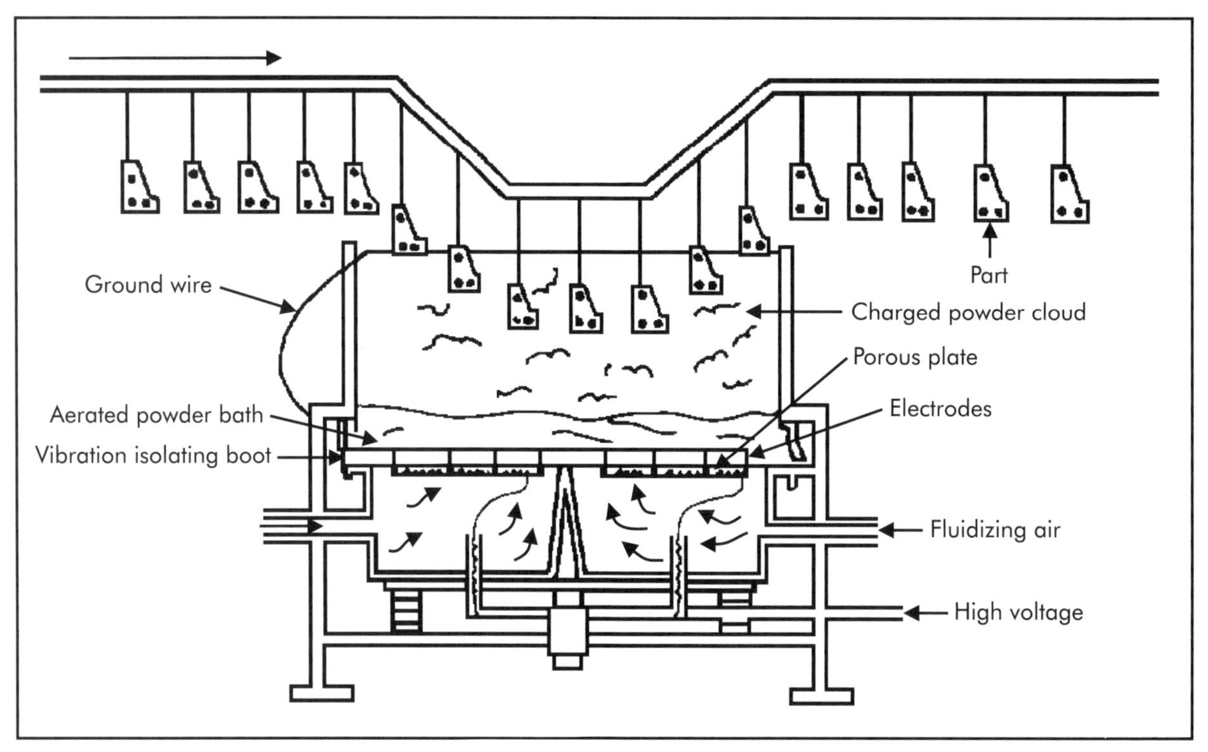

Figure 52-9. Schematic of a typical electrostatic fluidized bed (Wick and Veilleux 1985).

Chapter 53

Fixture and Jig Design

This chapter has been adapted from SME's *Fundamentals of Tool Design*, Fifth Edition (2003).

53.1 WORKHOLDERS

The term *workholder* includes all devices that hold, grip, or chuck a workpiece to perform a machining operation. The holding force may be applied mechanically, electrically, hydraulically, or pneumatically. A workholder must position or locate a workpiece in definite relation to the cutting tool, and it must withstand holding and cutting forces while maintaining a precise location. Workholders also may be used for assembly, welding, and inspection operations.

Workholders are commonly classified as fixures or jigs. *Fixtures* are workholders designed to hold, locate, and support the workpiece during a machining cycle, joining, or assembly operation. Fixtures do not guide the cutting tool, but rather provide a means to reference and align the cutting tool to the workpiece. *Jigs*, like fixtures, hold, locate, and support the workpiece, but also guide the cutting tool throughout its cutting cycle. Drill jigs are the most common type of jig. They are used for drilling, tapping, reaming, countersinking, counterboring, chamfering, and spot-facing.

The design or selection of a workholder is governed by many factors, the first being the physical characteristics of the workpiece. The workholder must be strong enough to support the workpiece without deflection. Cutting forces and vibration imposed by machining operations vary in magnitude and direction, and may impart torque as well as straight-line forces on the workpiece. The workholder must support the workpiece in opposition to the cutting forces. A good workholder maximizes the number of operations performed without removing the workpiece. The degree of precision required in the workholder will usually exceed that of the workpiece, typically by a factor of two to five, but sometimes by a factor of ten.

A workholder should be designed to receive the workpiece in only one position. If a workpiece with some symmetric features can be clamped in more than one position, it is probable that a percentage of workpieces will be incorrectly clamped and machined. Workholders should be designed to prevent incorrect placement and clamping by using poka-yoke (error proofing) techniques.

It is advisable to use standard workholders and commercially available components whenever possible. These items can be purchased for less than the cost of making them. They speed up implementation, generally have adequate strength and accuracy, and can be reused.

53.2 LOCATING PRINCIPLES

To ensure successful operation of a workholding device, the workpiece must be accurately located to establish a definite relationship between the cutting tool and some critical points or surfaces of the workpiece.

This relationship is established by locators in the workholding device, which position and restrict the workpiece to prevent its moving from its predetermined location. The workholding device presents the workpiece to the cutting tool in the required orientation. The locating device should be designed so that each successive workpiece, when loaded and clamped, will occupy the same position in the workholder. The locating design selected depends on the nature of the workpiece, the requirements of the metal-removing operation to be performed, and other restrictions on the workholding device.

TYPES OF LOCATION

Basic workpiece location can be divided into three fundamental categories: plane, concentric, and radial. In many cases, more than one style of location may be used to locate a particular workpiece. Most workholders use a combination of location methods to completely locate a workpiece. However, for the purpose of identification and explanation, each will be discussed individually.

Plane location is normally considered the act, or process, of locating a flat surface. However, irregular surfaces also may be located in this manner. Plane location is simply locating a workpiece with reference to a particular surface or plane, as shown in Figure 53-1.

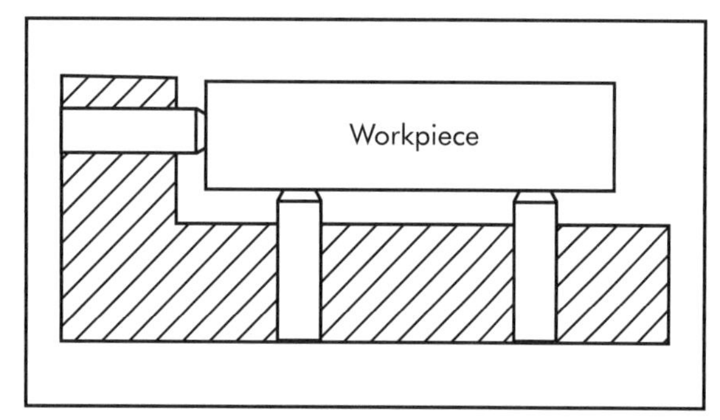

Figure 53-1. Plane location (SME 2003).

Concentric location is the process of locating a workpiece from an internal or external diameter as illustrated in Figure 53-2.

Radial location is normally a supplement to concentric location. With radial location, shown in Figure 53-3, the workpiece is first located concentrically and then a specific point on the workpiece is located to provide a specific fixed relationship to the concentric locator.

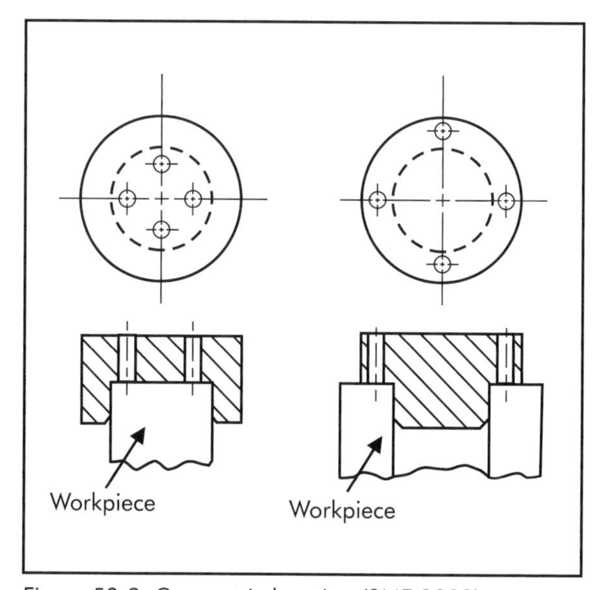

Figure 53-2. Concentric location (SME 2003).

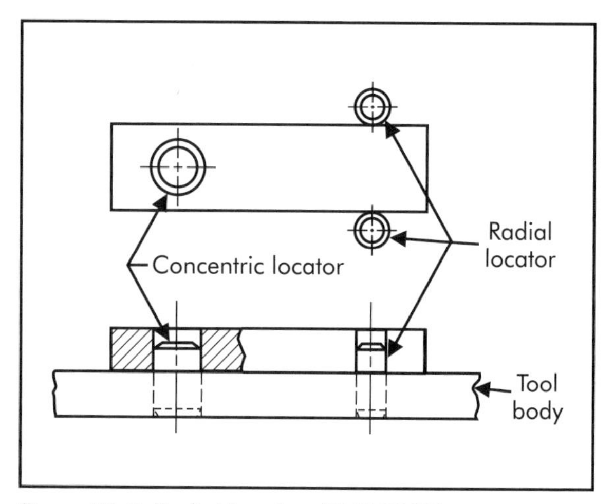

Figure 53-3. Radial location (SME 2003).

DEGREES OF FREEDOM

A workpiece in space, free to move in any direction, is designed around three mutually perpendicular planes and is said to have six degrees of freedom. The six degrees of freedom as they apply to a rectangular prism are shown in Figure 53-4. The prism is shown in orthographic projection, with all twelve positive and negative directions indicated in their respective positions. To accurately locate a workpiece, it must be confined to restrict it against movement in all of the six degrees of freedom (12 directions), except those called for by the operation. When this condition is satisfied, the workpiece is accurately and positively confined in the workholding device.

3-2-1 METHOD OF LOCATING

A workpiece may be positively located by means of six pins positioned so that collectively they restrict the workpiece in nine directions. This is known as the *3-2-1 method of location*. Figure 53-5 shows the prism resting on six pins, *A-F*. Thus by means of six locating points, three in a base plane, two in a vertical plane, and one in a plane perpendicular to the first two, nine directional movements are restricted.

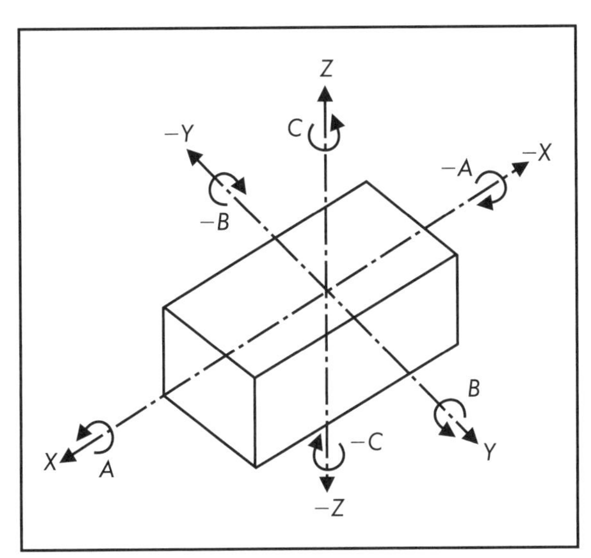

Figure 53-4. Six degrees of freedom and 12 directions (SME 2003).

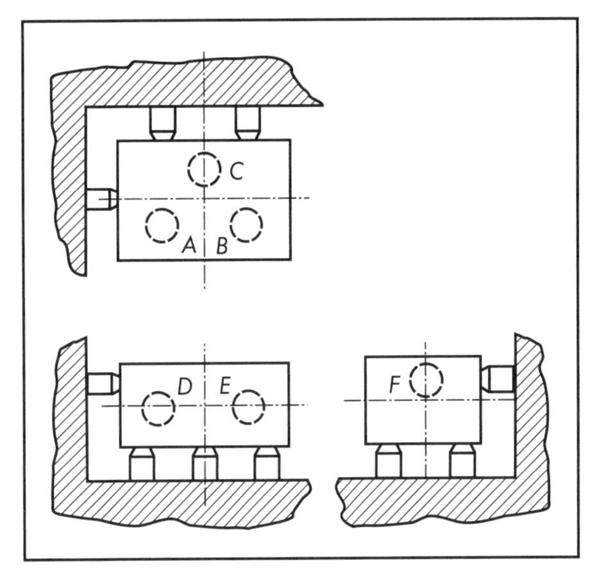

Figure 53-5. The 3-2-1 location principle (SME 2003).

Three directions remain unrestricted due to loading the workpiece into the workholding device. The remaining three directions may be restricted with clamps. Any combination of three clamping devices and locating pins may be used if this is more suitable to the design of a particular workholding device.

BASIC LOCATING RULES

To function properly, locators must be positioned correctly, properly designed, and accurately sized. To do all of this, in addition to permitting easy loading and unloading of the workpiece and clearance for access by the tool, requires forethought when planning the locational elements of the workholder. The following sections explain a few basic principles every designer should keep in mind when planning a part's location.

Position and Number of Locators

Locators and part supports should always contact a workpiece on a solid, stable point.

When possible, the workpiece surface should be machined to ensure accurate location. Locating points should be chosen as far apart as possible on any one workpiece surface.

The number of locators used to reference a part normally depends on the part. But no more points than necessary should be used to secure location in any one plane. The 3-2-1 principle determines the minimum number required. More can be used, but only if they serve a useful purpose; care must be taken that the additional points do not impair the locating function.

Always avoid duplicate or *redundant locators* on any part. Redundant location occurs when more than one locator is used to locate a particular surface or plane of a workpiece. The principle objection to using more than one locator or series of locators to reference the location is variance in part size. Any variation in the part size, even within the tolerance, can cause the part to be improperly located or bind between the duplicate locators. Besides these obvious problems, it is not cost effective to use more locators than necessary.

Locational Tolerances

Locational tolerance is one point that must always be considered when specifying locators for any workpiece. As a general rule, the locational tolerance should be approximately 20% (typical) to 50% (roughing operations) of the part tolerance, but may be as small as 10%.

An attempt to achieve excessively tight tolerances only increases costs. Likewise, overly large tolerances can shorten the life of a workholder and risks unacceptable process capability. The designer must balance the cost against the expected life of the tool and the required accuracy of the parts to determine a locational tolerance that will provide the required number of parts without excessive tooling costs.

BASIC TYPES OF LOCATORS

Locators are made in a wide variety of shapes and sizes to accommodate the large range of workpiece configurations. In addition, commercial locators are available in many styles to suit their ever-increasing use. To properly design and specify an appropriate locator, the designer first must be familiar with the different types of locators commonly used in jig and fixture applications. Fundamentally, there are two types of locators: external and internal. *External locators* are those devices used to locate a part by its external surfaces, whereas *internal locators* are used to locate a part by its internal surfaces, such as holes or bored diameters.

External Locators

External locators are normally classified as either locators or supports. *Locators* are those elements that prevent movement in a horizontal plane. *Supports* are locating devices that are positioned beneath the workpiece to prevent downward movement of the part as well as rotation around the horizontal axis.

The two basic forms of external locators or supports are fixed and adjustable. *Fixed locators* are solid locators that establish a fixed position for the workpiece. Typical examples of fixed locators include integral locators, assembled locators, locating pins, V-locators, and locating nests.

Adjustable locators are movable locators frequently used for rough-cast parts or similar parts with surface irregularities. Examples of adjustable locators are threaded locators, spring-pressure locators, and equalizing locators. Adjustable locators are used in conjunction with fixed locators to permit variations in part sizes while maintaining the fixed relative position of the part against the fixed locators.

Integral locators. *Integral locators* are machined into the body of the workholder as illustrated in Figure 53-6. In most instances, this type of locating, or supporting device, is the least preferred. The principal objections to using integral locators are the added time required to machine the locator and the problem of replacing the locator if it becomes worn or damaged. Another drawback to using integral locators is the additional material required to allow for machining the locator.

Assembled locators. *Assembled locators* are similar to integral locators in that they too must be machined. However, these locators have the advantage of being replaceable as shown in Figure 53-7. Assembled locators may be used as locators or supports. Since they are not part of the major tool body, using assembled locators does not require additional material for the tool body. Assembled locators are frequently made of tool steel and hardened to reduce wear.

Locating pins. *Locating pins* are the simplest and most basic form of locator. They may be made in-house from steel drill rod or purchased commercially. Commercial locating pins are available in several styles and types as shown in Figure 53-8.

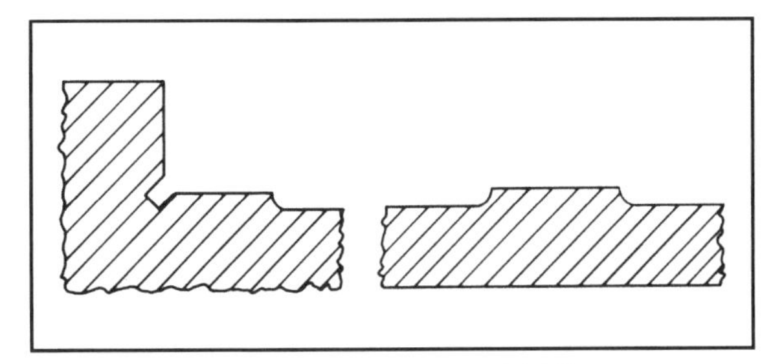

Figure 53-6. Integral locators (SME 2003).

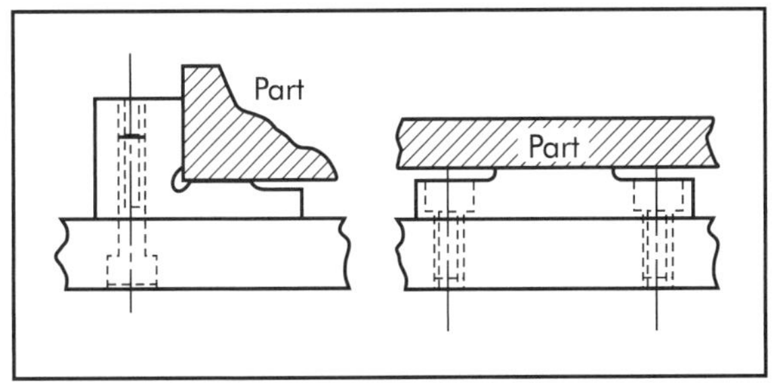

Figure 53-7. Assembled locators (SME 2003).

Available commercially, standard hardened *dowel pins* are frequently used as locating devices due to their simplicity, easy application, and ability to be replaced. Round pins are the most commonly used form of locating device. The location and number of locating pins is generally determined by the size, shape, and configuration of the part. However, in most cases the 3-2-1 principle is applied.

V-locators. A cylinder, like the prism, also has six degrees of freedom. To accurately locate a cylindrical workpiece, it must be confined to restrict its motion in each direction. Figure 53-9 illustrates a V-type locator. Clamping restricts the other directions of movement the locator does not restrict.

Locating nests. The nesting method of locating features a cavity in the workholding device into which the workpiece is placed and located. If the cavity is the same size and shape as the workpiece, this is an effective means of locating. Figure 53-10 illustrates a nest that encloses a workpiece on its bottom surface and around its entire periphery. The only degree of freedom remaining is in an upward direction. A similar nest can be used to locate cylindrical workpieces. Cavity nests are used to locate a wide variety of workpieces regardless of the complexity of their shape. All that is necessary is to provide a cavity of the required size and shape. Supplementary locating devices, such as pins, are not normally required. However, they increase locating speed and are common for welding and locating weak parts.

The cavity nest has some disadvantages. Since the workpiece is completely surrounded, it is often difficult to lift it out of the nest. When the workpiece tends to stick, an ejecting device can be incorporated in the workholder. Another disadvantage is that the operation performed may produce burrs on the workpiece, which tend to lock it into the

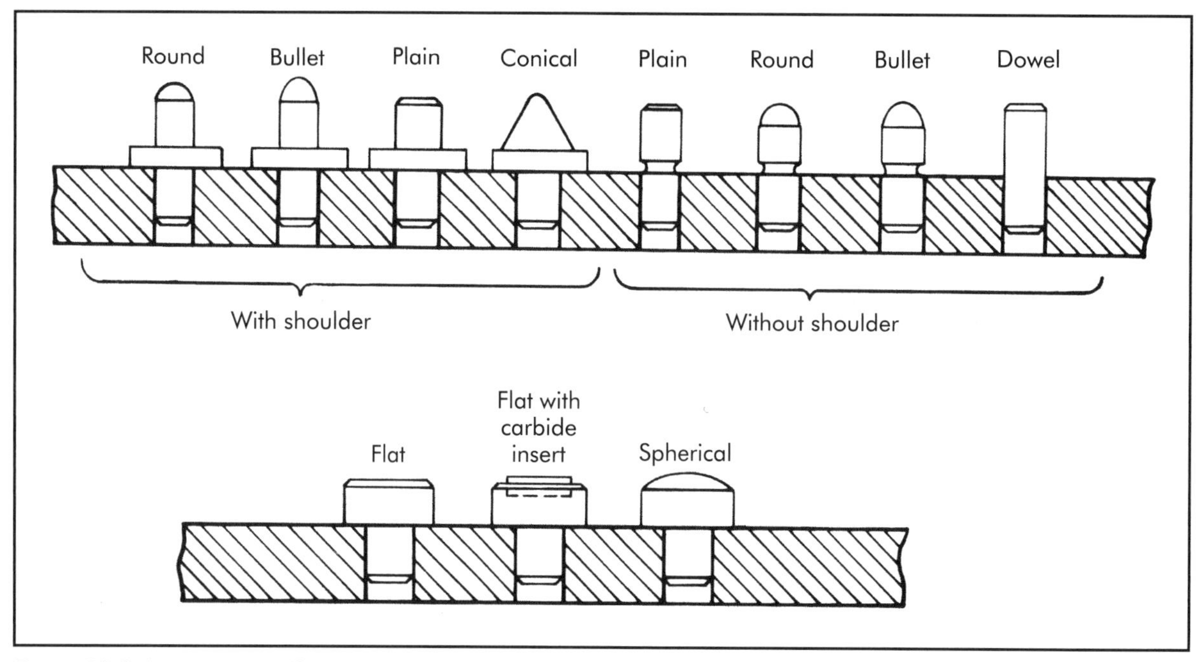

Figure 53-8. Locating pins (SME 2003).

nest. Chips from the cutting operation may lodge in the nest and must be removed before loading the next workpiece. Any chips remaining may interfere with proper positioning of the next workpiece. The last disadvantage is that the nest must be sized for the maximum allowable workpiece size. Smaller-sized workpieces will have room to move.

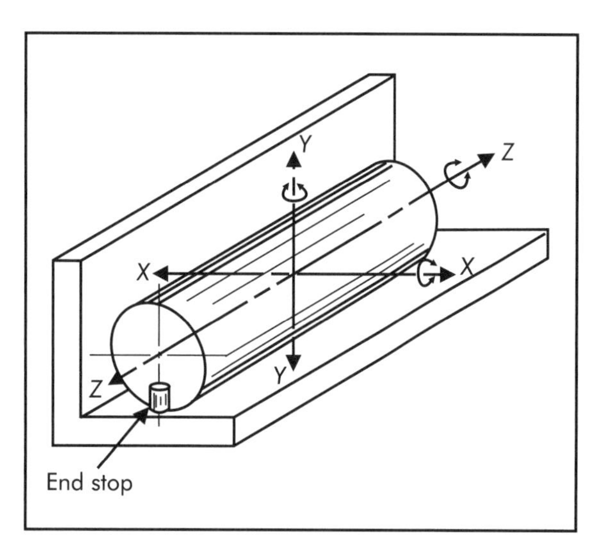

Figure 53-9. V-locator (SME 2003).

To avoid the disadvantages of the cavity type of nest, partial nests are often used for locating. Flat members, shaped to fit portions of the workpiece, are fastened to the workholding device to confine the workpiece between them. Figure 53-11 shows two partial nests, each confining one end of the bow-shaped workpiece. Each nest is fastened to the flat supporting surface of the workholder by means of two screws. Accurate positioning of the nests is ensured by dowels that prevent each nest from shifting from its required position. If workpieces are produced smaller than their maximum allowable size, adjustable locators can be used in combination with the nest.

Adjustable locators. Adjustable locators are widely used in applications where the workpiece surface is irregular or where large variations between parts make solid locators impractical. In some cases, this type of loca-

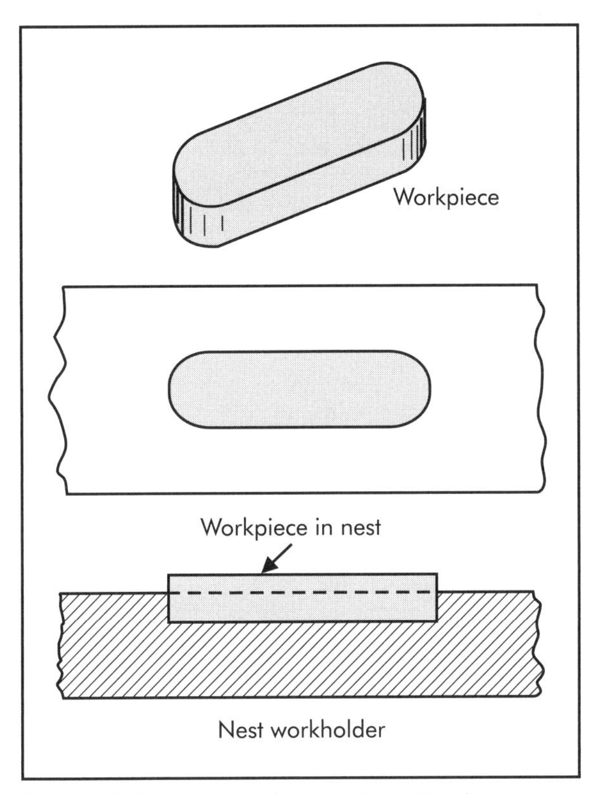

Figure 53-10. Nest-type locator (SME 2003).

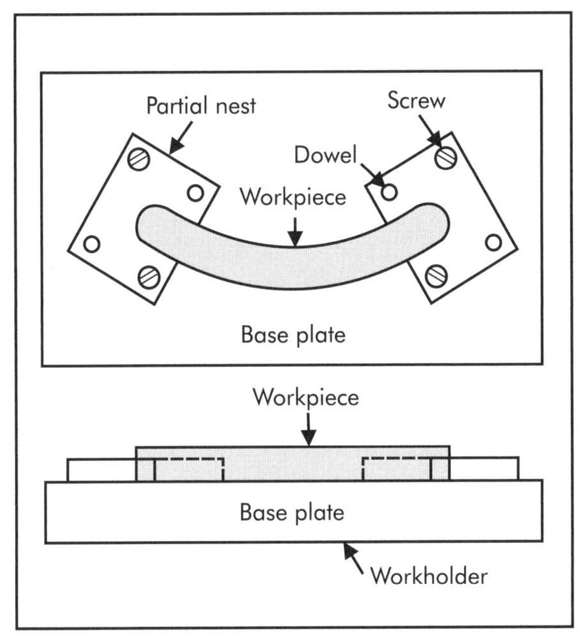

Figure 53-11. Workholder with partial nests (SME 2003).

tor can be used as a clamping device rather than a locator.

When adjustable locators are specified for a workholder, the position of the locator is not as critical as with solid locators, so the relative cost is greatly reduced. Frequently, adjustable locators are actually used as solid locators by simply adding a locknut or screw to secure the adjusting screw.

Adjustable supports. *Adjustable supports* are simply adjustable locators that are positioned beneath the workpiece. The primary variations of adjustable supports include threaded, spring, and equalizing.

Threaded supports. Threaded supports are used along with solid supports to permit easy leveling of irregular parts in the workholder.

Spring supports. *Spring supports* are also used with solid supports to level the workpiece. However, rather than using threads to elevate the locator, a secondary threaded element, such as a thumb screw, is used to lock the position of the spring support.

Equalizing supports. *Equalizing supports* are normally self-adjusting. That is, as one is depressed, the other rises. They are often used for castings.

Internal Locators

Internal locators use internal part features, such as holes or bored diameters, to locate the workpiece. The two basic forms of internal locators are fixed size and compensating. *Fixed-size locators* are made to a specific size to suit a certain hole diameter. Typical examples of this type of internal locator include machined locators, commercial pin locators, and relieved locators. *Compensating locators* are generally used to centralize the location of a part or to allow for larger variations in hole sizes. The two typical forms of compensating locators are conical and self-adjusting.

Machined internal locators. Machined internal locators are made to suit special-size hole diameters. In most cases, they are made for larger hole diameters. The exact form and shape of these locators are normally determined by the part to be located.

When round plugs are used in holes for locating, there is a tendency to stick when a close-fitting workpiece is applied. A method of reducing the tendency for workpieces to stick on a locating plug extending from a face plate is to relieve the plug by cutting away three equal segments as illustrated in Figure 53-12. The disadvantage is that a workpiece can be displaced in three directions on a relieved plug, with about a 20% loss in locating accuracy.

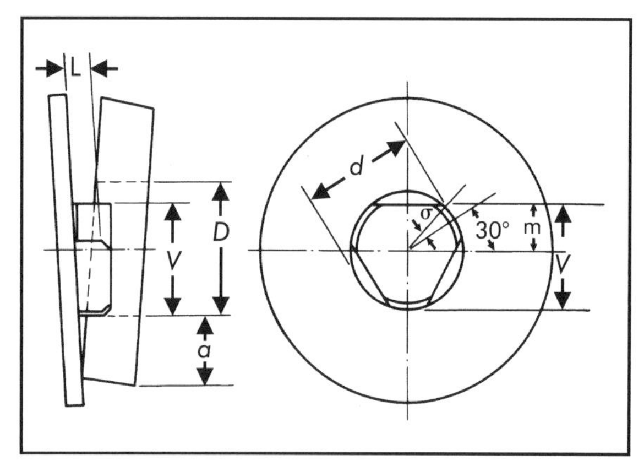

Figure 53-12. Nonsticking locator design (SME 2003).

Commercial pin locators. Commercial pin locators are made in two general styles, plain and shouldered. The ends of these locators are made in either round, flat, or bullet shapes to facilitate easy loading and unloading of parts. These locating pins are normally made undersize to prevent jamming and binding in the locating hole. The installed end of these pins is generally smaller than the location end to prevent improper installation.

Relieved locators. *Relieved locators*, as their name implies, are designed to minimize the area of contact between the workpiece and the locating pin. This reduces the chances of the locator sticking or jamming in the part. Figure 53-13 shows several examples of relieved locators. The most commonly used form of relieved locator is the diamond pin as shown in Figure 53-14.

Floating pin locators. One other style of locating pin that will correct slight differences between locating holes is the *floating pin locator*. The pin provides precise location in one axis while allowing up to 0.125 in. (3.18 mm) movement in the perpendicular axis.

The floating pin locator generally works like a diamond pin. Due to the increased

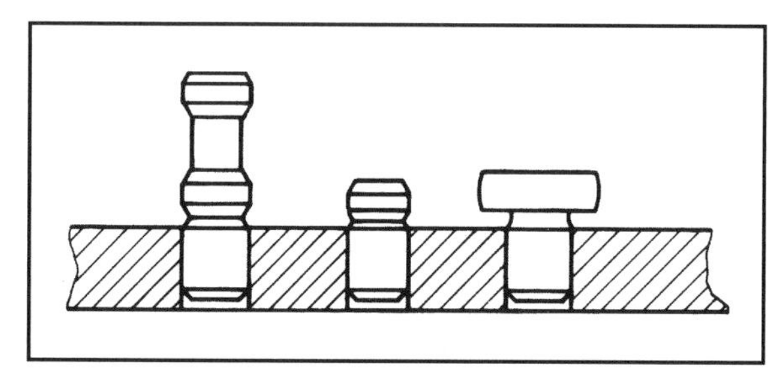

Figure 53-13. Relieved locators (SME 2003).

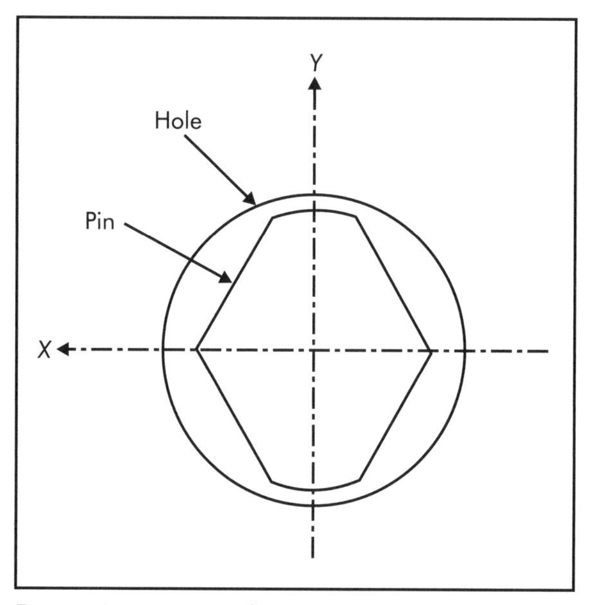

Figure 53-14. Diamond pin locator (SME 2003).

movement, however, this pin should be used for parts with somewhat looser locational tolerances of the mounting holes. The floating pin locator permits greater variation than that typically allowed by a diamond pin.

Conical locators. *Conical locators* are centralizing locators that compensate for variations in part sizes as well as centering a part in the workholder. The most efficient types of conical locators are spring loaded or threaded. Conical locators, while normally used for internal location, also may be used for external location with a conical cup.

53.3 CLAMPING PRINCIPLES

For a specific operation, the selection of general clamping—simple hand-operated clamps, quick-acting hand-operated clamps, power-operated clamps, etc.—should primarily be a function of operation analysis. The selection is based on an effort to balance the cost of the clamp against the cost of the operation to obtain the lowest possible total cost of both fixture and operation. Sound judgment of the tool designer in the application of specific clamping principles to the job at hand is essential. In general, clamping arrangements should be as simple as possible. Complicated arrangements tend to lose their effectiveness as the parts become worn, necessitating excessive maintenance, which might readily offset the savings of a faster operation.

The purpose of a clamp is to exert a force to press a workpiece against the locating surfaces and hold it securely. Primary cutting forces should be directed toward the locators; however, clamps often resist some processing forces. Clamping forces should be directed within the locating area, preferably through heavy sections of the workpiece directly upon locating spots or supports. Cutting forces should be borne by the fixed locators in a jig or fixture as much as possible; but generally some components of or moments set up by the cutting forces must be counteracted by clamping forces. To be effective, a clamp should be designed to exert a minimum force equal to the largest force imposed upon it in the operation.

The following clamping design and operational factors should be considered.

- Simple clamps are preferred because complicated ones lose effectiveness as they wear.
- Some clamps are more suitable for large and heavy work, others for small workpieces.
- Rough workpieces require longer travel of the clamp in the clamping range.
- The type of clamp required is determined by the type of operation to which it is applied.
- Clamps should not make loading and unloading of the work difficult, nor should they interfere with the use of hoists and lifting devices for heavy work.
- The number of clamps and supports increases as the rigidity of the workpiece decreases. The clamping system must not deform the workpiece.
- Clamps that are apt to move on tightening, such as plain straps, should be avoided for production work.
- The anticipated frequency of setups may influence the clamping means.
- Whenever possible, clamps should be located directly over a supported region of the workpiece.

TOOL FORCES

A clear understanding of the direction and magnitude of cutting forces may eliminate the need to restrain all 12 directional movements of a workpiece. Figure 53-15 shows how two pins and a table absorb the torque and thrust of a drilling operation. Theoretically, there is no need to hold the workpiece down as this is accomplished by the thrust of the drill. However, when the drill breaks through the thickness of the workpiece, an upward force may be created by interaction

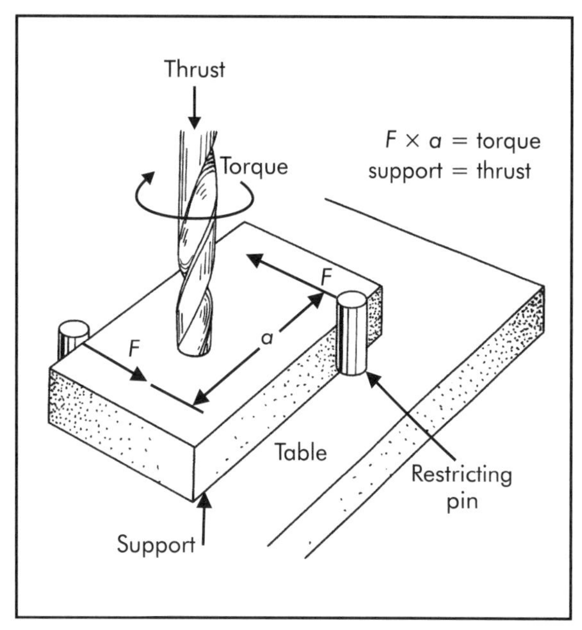

Figure 53-15. Pin-type drill fixture resisting torque and thrust (SME 2003).

between the drill flutes and material remaining around the periphery of the hole. If there is no restraint in this upward direction, the workpiece may be lifted above the pins, creating a very dangerous condition. An upward force also may be produced when a drill or reamer gets lodged in a workpiece and the tool is to be withdrawn.

Once the designer of a workholder has identified the possible directions and magnitude of the forces, he or she has two ways to restrain the workpiece to counteract these forces. One utilizes the strength and rigidity of some part of the workholder against which the workpiece rests or is forced by a clamp, screw, or wedge. The other utilizes friction between the workpiece's surface contacting under pressure a surface of the workholder.

Figure 53-16 shows a workpiece held in a vise. The horizontal component of the cutting force is absorbed by the solid jaw of the vise. The vertical component is resisted by friction between the workpiece and the jaws.

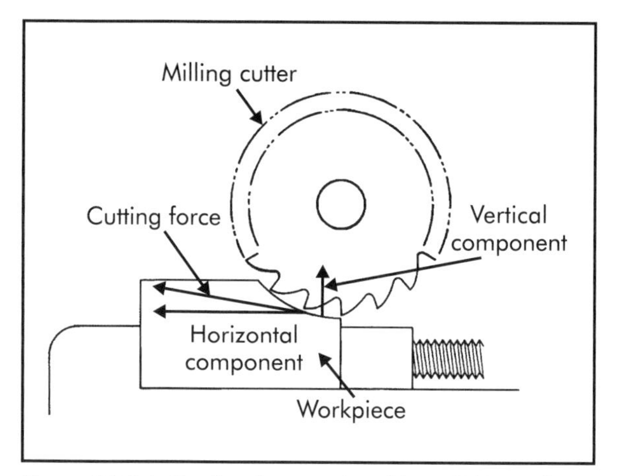

Figure 53-16. Cutting force resisted by solid jaw of vise (SME 2003).

CLAMPING FORCES

Complete analysis of the tool forces in a proposed operation will disclose which of the 12 directional movements must be restrained and to what extent.

Quite often, tool forces are of such magnitude and direction that a workpiece may be dislodged or moved from its required location. If the locating elements of a fixture cannot assure adequate restraint, it may be necessary to clamp the workpiece against them.

Clamps hold a workpiece against a locator. Perhaps the most common application is the bench vise, where a movable jaw exerts pressure on a workpiece, thereby holding it in a precise location determined by a fixed jaw. The bench vise uses a screw to convert actuating force into holding force. A number of commonly used mechanical methods for transmitting a multiplying force, such as the cam, lever, and wedge, are discussed later.

The clamping forces applied against the workpiece must be sufficient to hold the workpiece securely against the locators. Having accomplished this, further force is unnecessary and may be detrimental. The physical characteristics of the workpiece greatly influence clamping pressure. Hard vise jaws can crush a soft, fragile workpiece. The clamping pressure must hold, but not damage, deform, or impose too great a load on the workpiece.

The direction and magnitude of clamping pressure must be consistent with the purpose of the operation. An example is the boring of a precise round hole with the workpiece clamped in a heavy vise. Excessive clamping pressure can compress the workpiece. The bored hole may be perfect in size and roundness while the workpiece is compressed. The release of clamping pressure might permit the workpiece to return to its normal rather than compressed condition, and the hole might then be off-size and elongated.

Clamping pressure should not be directed toward a cutting operation, but should, wherever possible, be parallel to it. It should never be great enough to change any dimension of the workpiece.

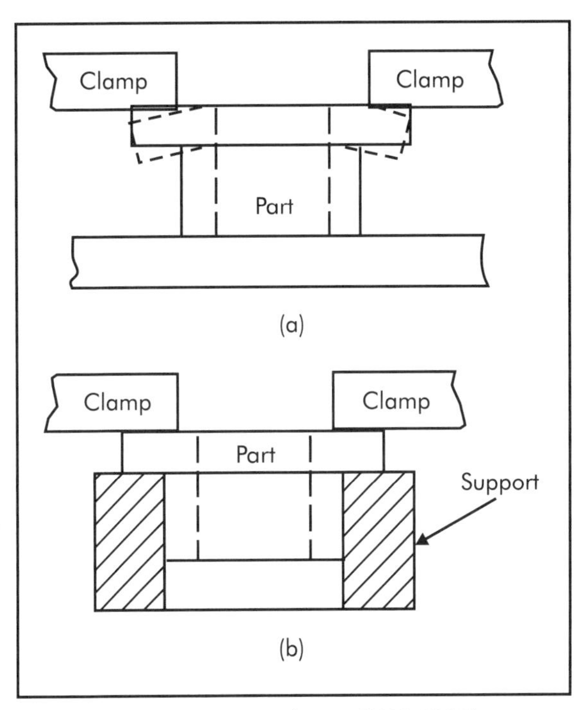

Figure 53-17. Positioning clamps (SME 2003).

Positioning Clamps

Clamps must be positioned to contact a workpiece at its most rigid point. When possible, they should be located over a locator or support. In cases where a part cannot be clamped over a locator, a secondary support must be installed to counteract the clamping forces and prevent damage to the part. As shown in Figure 53-17, the flanged part is located by its center. If the part was clamped as shown in (a), the part would distort. Therefore, a secondary support must be added (b). This additional support provides the required backing to prevent distorting the part. Remember, when adding a secondary support, allow enough space between the support and the part to prevent redundant locating.

BASIC TYPES OF CLAMPS

The specific type of clamp selected for a particular application normally depends on the type of tool, part shape and size, and the operation to be performed. Other considerations such as speed of the operation and permanence also must be considered for long or high-speed, high-volume production runs. There are several basic types and styles of clamps and clamping devices commonly used for jigs and fixtures.

- *Strap clamps* are the simplest and least expensive type of clamping device. As shown in Figure 53-18, the basic strap clamp consists of a bar, a heel pin (or block), and either a threaded rod or cam lever to apply the holding force.
- *Screw clamps* are a type of mechanical clamp that uses a screw thread to apply the holding force.
- *Cam-action clamps* are frequently used for fast-operating clamping devices. Figure 53-19 shows a cam-action clamp.
- *Toggle-action clamps* offer fast clamping and release actions. They move completely clear of the workpiece and

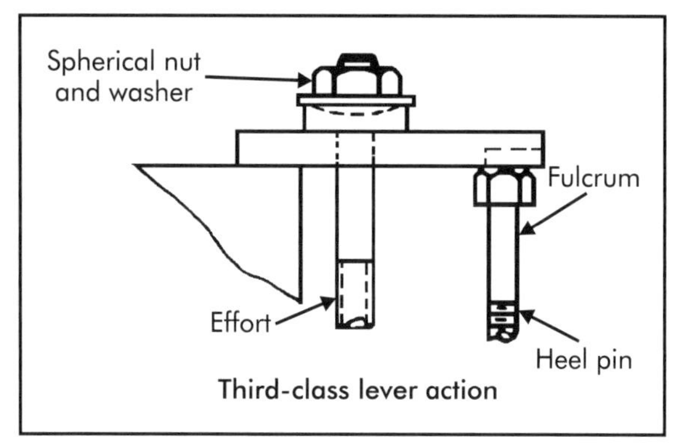

Figure 53-18. Strap clamp (SME 2003).

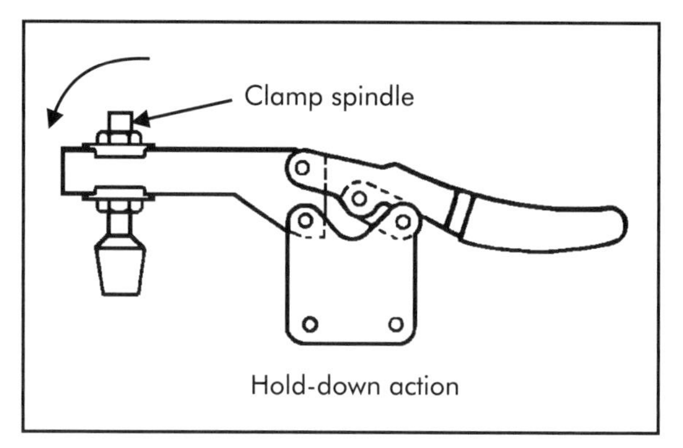

Figure 53-20. Toggle clamp (SME 2003).

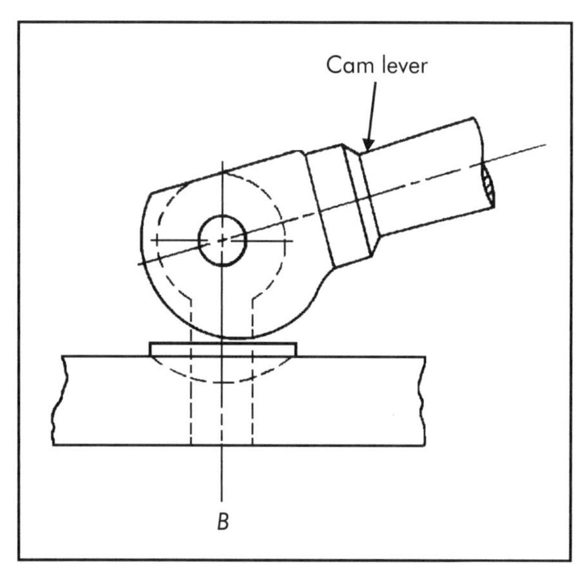

Figure 53-19. Cam-action clamp (SME 2003).

have a high ratio of holding force to actuation force. Several variations of toggle clamps are available to suit almost every workholding application. Figure 53-20 shows an example of a toggle clamp. For all their advantages, standard toggle clamps have always caused problems because of their limited range of movement and inability to compensate for different thicknesses. Once set to a clamping height, the standard toggle clamp can only suit very slight changes in workpiece thickness. Larger variations usually require adjustment of the clamp spindle.

- *Wedge-action clamps* use the basic principle of the inclined plane to securely hold and clamp a workpiece.

53.4 FIXTURE DESIGN

Fixtures are workholders designed to hold, locate, and support the workpiece during processing. Unlike jigs, fixtures do not guide the cutting tool, but rather provide a means to reference and align the cutting tool to the workpiece. Fixtures are normally classified by the machine with which they are designed to be used. A sub-classification is sometimes added to further specify the fixture classification. This sub-classification identifies the specific type of machining operation the fixture is intended to perform. For example, a fixture used with a milling machine is called a milling fixture; however, if the operation it is to perform is gang milling, it also may be called a gang-milling fixture.

The similarity between jigs and fixtures normally ends with the design of the tool body. For the most part, fixtures are designed to withstand much greater stresses and tool forces than jigs, and are always securely clamped to the machine. For these reasons, the designer must always be aware of proper locating, supporting, and clamping methods when fixturing a part.

In designing any fixture, there are several considerations in addition to the part that must be addressed to complete a successful design. Cost, production capabilities, production processing, and tool longevity are some of the points that must share attention with the workpiece when a fixture is designed.

TYPES OF FIXTURES

Fundamentally, fixtures are classified as either for general or special purposes. An advantage of general purpose fixtures is that they are usually relatively inexpensive, and can be used to hold a wide variety and range of sizes of workpieces on a wide variety and range of sizes of machines. However, they may not provide the accuracy of locating and speed of loading and unloading that special purpose fixtures provide.

General Purpose Fixtures

Vises, chucks, and split collets are some of the most popular general purpose fixtures. These commercially made workholders allow a single fixture to service an infinite number of parts by simply changing or modifying the holding and locating surfaces of the tool.

Vises are perhaps the most widely used and best-known fixture. All vises have one fixed and one movable jaw that hold a workpiece between them. The ability to configure the jaws to grip particular workpieces, the means of actuation and mounting, the ability to differ positioning of the workpiece, an array of available sizes, etc., present an endless variety of commercially available vises. By reworking the jaws or making special jaws, and adding such details as locating pins, bushings, and plates, vises can be easily converted into efficient, specialized workholders.

A *chuck* is a workholder generally used for gripping the outside or end of a workpiece or tool. It is usually attached to a machine-tool spindle, such as a lathe. A lathe chuck consists of a body with inserted workholding jaws, which slide radially in slots and are actuated by various mechanisms, such as screws, scrolls, levers, and cams, alone and in a variety of combinations. Chucks in which all the jaws move together, such as a three-jaw chuck, are self-centering and used primarily for round work.

An independent jaw chuck, such as a four-jaw chuck, permits each jaw to move independently for chucking irregular-shaped workpieces or to center a round workpiece. Independently movable jaws are typically more accurate than self-centering jaws.

The jaws of most lathe chucks can be reversed to switch from external to internal chucking. Jaws may be adapted to fit workpiece shapes that are not round. Standardized ways of attaching a lathe chuck to different machine tools allow chucks made by different manufacturers to be easily interchanged.

In addition to their standard jaws, lathe chucks may be fitted with a variety of special purpose jaws to accommodate different types of workpiece surfaces and configurations. The principal types of chuck jaws used for these purposes are called *soft jaws* and are generally made of aluminum.

The accuracy of a chuck deteriorates with usage due to wear, dirt, and deformation caused by excessive tightening.

A *split collet* is a type of spindle-mounted chucking device that has a slotted shell. The shell is split and acts as a spring. Collets may serve as workholders for a wide variety of parts. Collets are typically more accurate than three-jaw chucks.

Special Purpose Fixtures

Special purpose fixtures are designed and built to hold a particular workpiece for a specific operation on a specific machine tool. They are classified either by the machine they are used on, or by the process they perform on a particular machine tool, such as milling, turning, grinding, boring, broaching, and sawing. However, they also may be identified by their basic construction features. For example, a lathe fixture made to turn radii is classified as a lathe radius turning fixture. But if this same fixture is a simple plate with a variety of locators and clamps mounted on a face plate, it is also classified as a plate fixture. While many fixtures use a combination of different features, almost all can be divided among five distinct types: plate fixtures, angle-plate fixtures, vise and chuck-jaw fixtures, indexing fixtures, and multi-part or multi-station fixtures.

Plate fixtures. *Plate fixtures*, as their name implies, are constructed from a plate with a variety of locators, supports, and clamps. They are the most common type of fixture. The versatility of plate fixtures makes them adaptable for a wide range of machine tools.

Angle-plate fixtures. The *angle-plate fixture* is a modified form of a plate fixture. Here, rather than having a reference surface parallel to the mounting surface, the angle-plate fixture has a reference surface perpendicular to the mounting surface or at some other angle relative to it.

Modifying a general purpose vise or chuck. A general purpose vise or chuck can be modified for use as a special purpose fixture by using specially machined vise or chuck jaws in place of the standard, hardened jaws normally furnished with a vise. Vise and chuck-jaw fixtures are one of the least expensive types of fixtures to produce and usually the simplest to modify.

Indexing fixtures. *Indexing fixtures* are used to reference workpieces that must have machine details located at prescribed spacings. The typical indexing fixture will normally divide a part into any number of equal spacings, such as those used for geometric shapes or gears. However, some indexing fixtures may be used to locate and reference a workpiece with unequal spacings. Regardless of the configuration of the workpiece, indexing fixtures must have a positive means to accurately locate and maintain the indexed position.

Multi-part or multi-station fixtures. *Multi-part* or *multi-station fixtures* are normally used for one of two purposes: either to machine multiple parts in a single setting, or to machine individual parts in sequence, performing different operations at each station.

Modular Fixturing

Modular tooling fixtures use standard commercial components to quick-build special fixture designs. Modular tooling systems enable a company to have a large number of individual fixture designs for many different workpieces, without having to design, build, and store dedicated special fixtures for each one of those workpieces. Design, build, and setup time are all greatly reduced, along with overall fixturing cost.

Modular tooling systems are kits of tooling components that can be used together in various combinations to locate and clamp workpieces for machining, assembly, and inspection operations. A kit consists of mounting plates, angle plates, locators, clamps, and mounting accessories. Adapters are also available to permit the use of many standard and power workholding devices.

The two primary forms of modular tooling are those with grid-pattern holes and those with T-slots. T-slots are generally spaced farther apart than grid holes. Therefore, they offer fewer mounting positions even though they permit movement along the slot. The T-slot system relies on friction to hold parts in place, thereby permitting workpiece movement given sudden or excessive cutting forces.

Grid-pattern modular fixturing systems are available in two styles: alternating dowel and tapped holes, and multi-purpose holes.

The multi-purpose holes have an alignment bushing and threaded insert in each hole, thereby allowing each hole to be used as an alignment hole, mounting hole, or both.

Fixtures made from modular tooling kits can be used on standard machines, CNC machines, and CNC machining centers. Modular tooling systems are invaluable when confronted with short lead time or small production quantities that do not warrant the design and construction of a special fixture. They are also useful for infrequent production runs, prototype parts, and trial fixturing. Additionally, modular fixtures are reusable.

53.5 JIG DESIGN

Jigs, like fixtures, are designed to hold, locate, and support a workpiece. However, they also guide the cutting tool into the workpiece throughout the cutting cycle, which is important, for example, in small diameter and/or long drilling situations. Jigs can be divided into two general classifications: drill jigs and boring jigs. Of these, drill jigs are, by far, the most common. *Drill jigs* are generally used for drilling, tapping, and reaming, but also may be used for countersinking, counterboring, chamfering, and spot-facing. *Boring jigs*, on the other hand, are normally used exclusively for boring holes to a precise, predetermined size. The basic design of both classes of jigs is essentially the same. The only major difference is that boring jigs are normally fitted with a pilot bushing or bearing to support the outer end of the boring bar during the machining operation.

Since all jigs have a similar construction, the points covered for one type of jig normally apply to the other types as well. This section focuses on drill jig design.

Jig selection and design begins with an analysis of the workpiece and manufacturing operation to be performed. The workpiece, production rates, and machine availability normally determine the size, shape, and construction details of a jig. However, all jigs must conform to certain design principles. For efficient and productive manufacture of quality workpieces, a jig must provide a method to:

- Correctly locate the workpiece with respect to the tool.
- Securely clamp and rigidly support the workpiece during the operation.
- Guide the tool.
- Position and/or fasten it on a machine.

Jigs are often divided into two broad categories, open and closed. *Open jigs* are generally used when machining a single surface of a workpiece, whereas *closed jigs* are used when machining multiple surfaces. More often, jig types are identified by the method used to construct the jig (for example: template, plate, leaf, channel, etc.). The main types of jigs are discussed in the following sections.

TEMPLATE DRILL JIGS

Template drill jigs are not actually true jigs because they do not incorporate a clamping device. However, they can be used on a wide variety of parts and are among the simplest and least expensive drill jigs to build. A template drill jig is simply a plate containing holes or bushings to guide a drill. It is usually placed directly on a feature of a part to permit the drilling of holes at the desired location. When this is impractical, it is located on the part by measurement or sight lines scribed on the template.

A template drill jig is often used to drill holes in one portion of a large workpiece where a conventional jig large enough to hold the entire part would be impractical and costly. Template jigs usually cost much less than conventional jigs. Often making use of two or three template drill jigs is more economical than using one large conventional jig.

Template jigs, however, are not as foolproof as most other types, which may result in inaccurate machining by a careless operator. Orientation of the hole-pattern-to-workpiece datums may not be as accurate as with other types of jigs. However, the accuracy of the hole pattern within the template jig itself is comparable to that of any conventional jig.

PLATE JIGS

Plate jigs are basically template jigs equipped with a workpiece clamping system. Initial construction costs are greater for plate jigs than for template jigs, but plate jigs are generally more accurate and last longer.

A plate jig incorporates a plate, which is generally the main structural member that carries the drill or liner bushings. Slip bushings of various sizes can be used with liner bushings, allowing a series of drilling and related operations without the need to relocate or reclamp the workpiece. The plate jig's open construction makes it easy to load and unload a workpiece and dispose of chips. An example of an open-plate jig is shown in Figure 53-21.

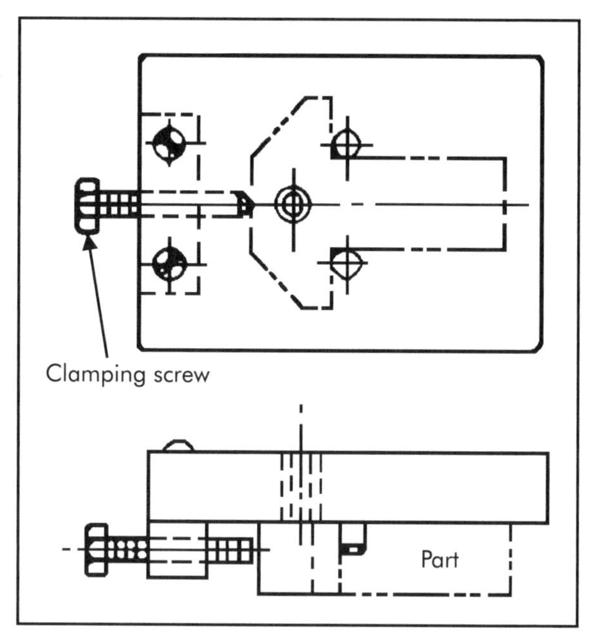

Figure 53-21. Open-plate jig (SME 2003).

The *angle-plate jigs* shown in Figure 53-22 are primarily used to drill workpieces at an angle to the part locators. The plain angle-plate jig, (a), is designed to drill holes perpendicular to the locating surface, while the modified angle-plate jig, (b), is designed to drill holes at angles other than 90° to the locating surface.

Plate-type jigs are usually moved around the table by hand. Therefore, special safety precautions should be taken to prevent the jig from spinning around the spindle whenever a cutting tool jams. The best way to prevent this is to build the jig with an extension handle long enough for the machine operator to overcome the torque of the jammed tool. When a plate jig is used with a radial drill, provision can be made to clamp the drill jig or the workpiece to the machine table.

A *leaf jig* is generally small, and incorporates a hinged leaf that carries the bushings, and through which clamping pressure is applied. Although the leaf jig can be used for large and cumbersome workpieces, most designs are limited in size and weight for easy handling. A leaf jig can be box-like in shape, with four or more sides for drilling holes perpendicular to each side.

The leaf jig shown in Figure 53-23 was specifically designed and built to drill two holes in a small connecting rod. The hinged drill plate contains the drill bushings. It is precisely located at both ends by the slots in the body of the jig. The workpiece is located and clamped between two V-blocks, one fixed, the other movable. The V-blocks are tapered to force the workpiece down against the base of the jig body.

Channel and *tumble box jigs* permit drilling into more than one surface of a workpiece without relocating the workpiece in the jig. This results in greater accuracy with less handling than required when using several

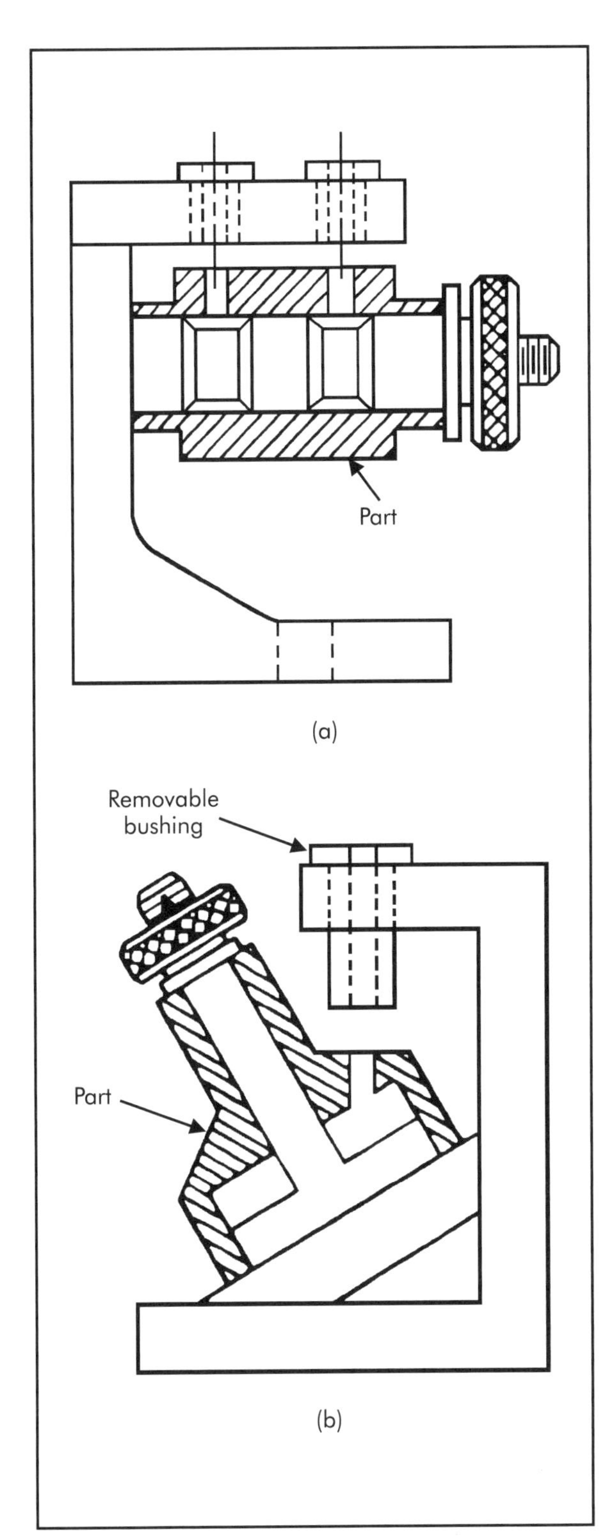

Figure 53-22. Angle-plate jigs (SME 2003).

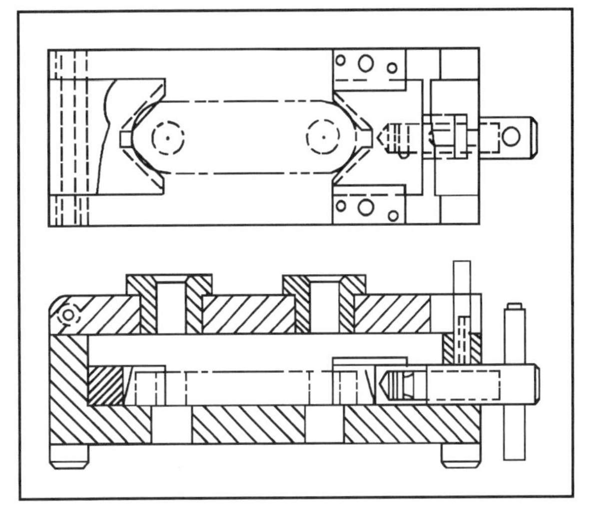

Figure 53-23. Leaf jig (SME 2003).

separate jigs. These jigs can be complicated and more expensive to build than several simpler types, but they can be very cost-effective if properly designed. Figure 53-24 illustrates a typical channel jig.

Indexing jigs are used to drill holes in a pattern, usually radial. Location for the holes is generally taken from the first hole drilled, a datum hole in the part, or from registry with an indexing device incorporated in the jig.

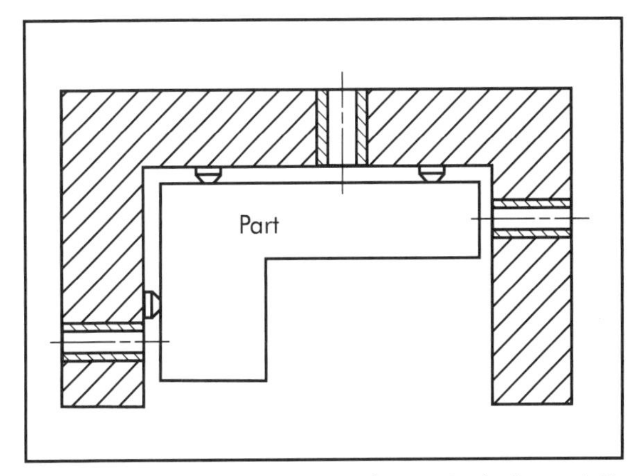

Figure 53-24. Cross-section of a typical channel jig (clamping details not shown) (SME 2003).

REVIEW QUESTIONS

53.1) What is the purpose of a workholder?

53.2) What is the advantage of using assembled locators over integral locators?

53.3) What part of the fixture must the tool forces be directed to?

53.4) Modular tooling systems are best suited for what types of production needs?

53.5) How is a jig different than a fixture?

REFERENCE

Society of Manufacturing Engineers. 2003. *Fundamentals of Tool Design*, Fifth Edition. Dearborn, MI: Society of Manufacturing Engineers.

Chapter 54

Advanced Quality Analysis

54.1 CONTINUOUS IMPROVEMENT

Prior to the introduction of total quality management (TQM), one of the most unequivocal differences between Japanese and U.S. firms was the emphasis placed on business improvements. While U.S. organizations were exhausting enormous amounts of time and money to maintain the status quo in products and processes, Japanese firms were investing in ways to continually improve the quality and reliability of their products and processes. The results catapulted Japan to a world-class position in many areas of manufactured goods and services. In the ensuing years, U.S. firms were forced to adopt a new philosophy, one that incorporated the principles of continuous improvement (CI) as promoted by Deming, Juran, Crosby, and Feigenbaum.

Today, there is no argument about the need to engage in continuous improvement activities if a company is to remain viable and competitive in the global marketplace. U.S. managers are charged with the promotion of CI, supplying the necessary resources, and rewarding proactive thinking. Manufacturing engineers, on the other hand, are expected to know and use the various techniques and tools available to engage in dramatic and incremental improvements. Some of the tools and techniques include:

- statistical process control (SPC),
- reliability,
- design of experiments (DOE),
- human factor principles,
- benchmarking,
- process capability analysis,
- human resource training,
- auditing,
- supplier certification,
- ISO 9000 and QS 9000 (replaced by TS16949) certification,
- Malcolm Baldrige award criteria, and
- problem solving.

Reaching the stage of world-class quality does not, however, give an organization the option to minimize improvement efforts. CI must be sustained to become better and better; thus, there is a need for advanced quality analysis. This chapter describes a number of tools used in CI efforts. These include:

- control charting,
- reliability,
- gage capability,
- design of experiments,
- Taguchi concepts,
- ISO 9000 and QS 9000 certification,
- six sigma fundamentals, and
- nondestructive testing.

It is important for manufacturing professionals to practice CI consistently. It is just as important to possess the right attitudes about CI as it is to have the knowledge and skills to affect it. CI is more than a proven way to gain market share, increase throughput, and secure profits. It is a philosophy, a way of thinking about the present as well as the future.

54.2 CONTROL CHART INTERPRETATION

Chapter 43 of *Fundamentals of Manufacturing*, Second Edition discussed x-bar and R-chart fundamentals. When just one subgroup average (or range) is beyond a control limit, the process is considered out of control. Further, because the distribution of the x-bar tends to be normal, serious departures from normality can signal the presence of special causes even if all points are within the control limits. Too many points near the limits or near the centerline may signal problems with the process, such as over-control or improper methods of sampling. Variation within the control limits is expected and natural for any process.

The appearance of a trend or recurring cycles in the data pattern can indicate that the process is experiencing a drift or cyclical change with respect to its mean or range. Runs of points above or below the centerline may indicate small shifts in the mean or level of variability. Whenever an out-of-control condition is indicated, it is important to determine the basic process fault producing it. Some useful generic conditions to look for include:

- trends/cycles, which indicate systematic changes in the process environment, worker fatigue, maintenance schedules, wear conditions, accumulation of waste material, and contamination;
- a high proportion of points near or beyond the control limits, which may indicate over-control of the process, large differences in incoming raw material, or the charting of more than one process on a single chart; and
- sudden shifts in the level, which may indicate a new machine, die, or tooling, a new worker, a new batch of raw material, a change in the measurement system, or a change in the production method.

Often, special causes of variation produce patterns that are not obvious. Several useful tests for the presence of unnatural patterns (special causes) can be performed by dividing the distance between the upper and lower control limits into six zones; each zone is one standard deviation wide. The zones for the upper half of the chart are referred to as A (outer third), B (middle third), and C (inner third). The lower half is considered a mirror image as seen in Figure 54-1. The probabilistic basis for the tests discussed here using the zones is derived from a normal distribution. Therefore, the tests are applicable to an x-bar chart.

The various tests are illustrated in Figure 54-2. Although the tests can be considered basic, they are not totally comprehensive. Analysts should be alert to any patterns of points that might indicate the influences of other special causes in a particular process. When the existence of a special cause is signaled by a test, the last point should be circled. Points can contribute to more than one test. In this case, however, a point should be circled each time a test is violated.

Runs test violations signify trends inside the control limits, which are indicative of special-cause variation. A violation implies a short-term shift in either the mean or dispersion. Violations are not always unfavorable. For example, if Test 7 were violated, it would indicate that at least 15 subgroups in

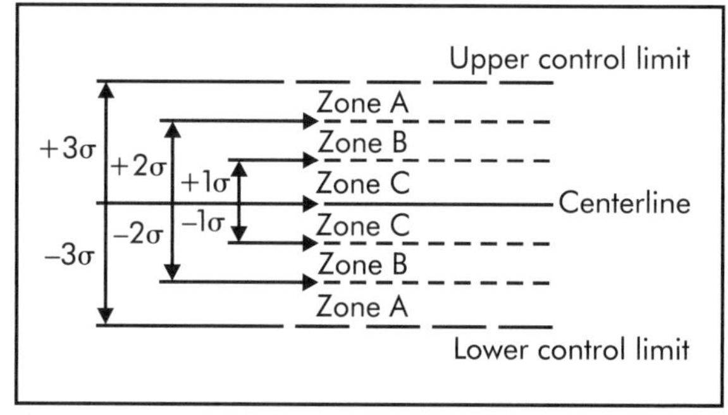

Figure 54-1. Dividing control regions into zones (Wick and Veilleux 1987).

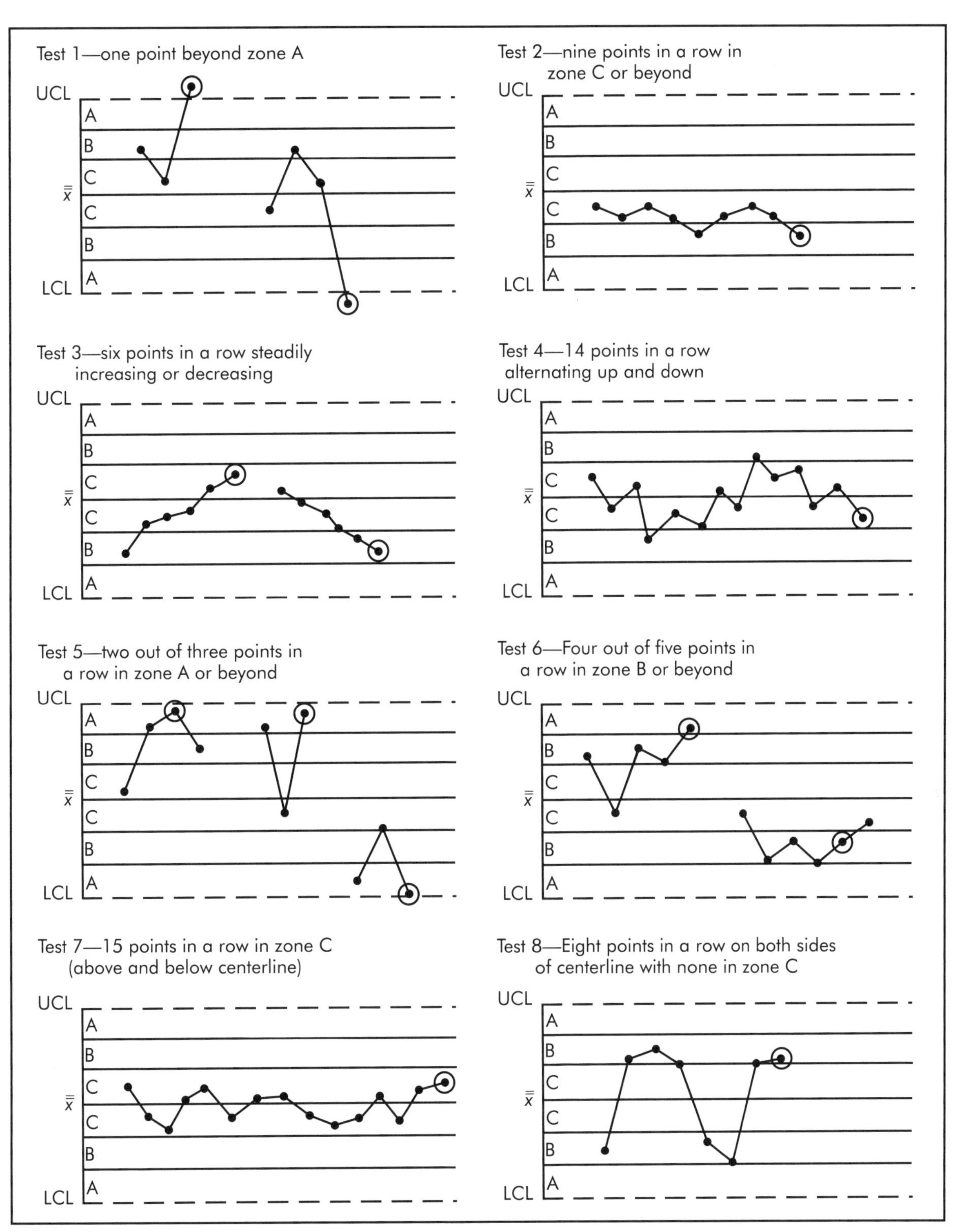

Figure 54-2. Pattern analysis of x-bar charts (Wick and Veilleux 1987).

a row displayed dispersion values less than or equal to 1σ. It would be beneficial to know what conditions led to these desirable results so they can be repeated in the future.

54.3 ATTRIBUTE CONTROL CHARTS

Many quality attributes of manufactured goods are measured. Examples include surface flaws on a sheet-metal panel, cracks in drawn wire, color inconsistencies on a painted surface, voids, flash or spray on an injection-molded part, wrinkles on a sheet of vinyl, or go/no-go gages. These defects or nonconformities are often simply observed visually or by some sensory device, which causes a part to be classified as defective. Products that fail an attribute test are labeled "defective" or "nonconforming." Such data may be initially studied according to attributes to determine the presence of certain key factors, which can be further analyzed using x-bar and R charts.

There are four basic attribute control charts commonly employed: *p charts* for determining the fraction defective using a constant or varying subgroup (sample) size; *np charts* for determining the number defective using a fixed subgroup (sample) size; *c charts* for the count of defects with subjects of the same size; and *u charts* for the number of defects per part for a varying subgroup (sample) size.

It is important to note that a subgroup is also referred to as a "sample." Subgroup size and sample size refer to the same quantity. For example, if a subgroup contains five observations, the subgroup or sample size would be five.

FRACTION DEFECTIVE (P CHART)

A *p chart* is used to determine the fraction of nonconforming parts in a subgroup (sample). It is governed by the normal distribution approximation to the binomial distribution. The average fraction defective is:

$$\bar{p} = \frac{\sum np}{\sum n} \tag{54-1}$$

where:

$\bar{p}$ = average fraction defective
np = number defective
n = subgroup size

The control limits for the p chart are:

$$\bar{p} \pm 3\sqrt{\frac{\bar{p}(1-\bar{p})}{n}} \tag{54-2}$$

As the subgroup size, n, changes, the control limits must be recalculated. As a consequence, stair-cased control limits are expected if the actual value of n is used for each sample. This, in turn, might be construed as abnormal behavior. To avoid such problems, one common practice—particularly when p charts are made by hand—is to use the average value of n (Eq. 54-3), except when data is near the control limits. For these points, the control limits are recomputed using the actual value of n. Although this creates stair-cased control limits, it is the only 100% accurate method and only one set of computations is required. The use of statistical process control (SPC) software eliminates such problems.

$$\bar{n} = \frac{\sum n}{g} \tag{54-3}$$

where:

$\bar{n}$ = average subgroup size
n = number inspected
g = number of subgroups

Example 54.3.1. Table 54-1 shows the results of 14 runs (subgroups) involving the number of defects in a manufacturing process with variable subgroup size, n. Calculate the control limits using the average subgroup size method.

Solution. The average number of defects, $\bar{p}$, is given by:

Table 54-1. Data for p chart

Subgroup	Subgroup Size, n	Number Defective, np	Fraction Defective, p
1	343	5	0.015
2	392	7	0.018
3	453	7	0.015
4	286	8	0.028
5	512	13	0.025
6	398	5	0.013
7	559	12	0.021
8	265	6	0.023
9	238	10	0.042
10	321	5	0.016
11	436	7	0.016
12	497	9	0.018
13	528	8	0.015
14	446	6	0.013

$$\bar{p} = \frac{\sum np}{\sum n} = \frac{5+7+....+6}{343+392+....+446} = \frac{108 \text{ defective}}{5{,}674 \text{ parts}} = 0.019 \quad \text{(Eq. 54-1)}$$

The average subgroup size is given by:

$$\bar{n} = \frac{\sum n}{g} = \frac{343+392+....+446}{14} = \frac{5{,}674 \text{ observations}}{14 \text{ subgroups}} = 405.3 \approx 405 \quad \text{(Eq. 54-3)}$$

The control limits using the average subgroup size are given by:

$$UCL_p = \bar{p} + 3\sqrt{\frac{\bar{p}(1-\bar{p})}{\bar{n}}} = 0.019 + 3\sqrt{\frac{0.019(1-0.019)}{405}} = 0.039 \approx 0.04 \quad \text{(Eq. 54-2)}$$

$$LCL_p = \bar{p} - 3\sqrt{\frac{\bar{p}(1-\bar{p})}{\bar{n}}} = 0.019 - 3\sqrt{\frac{0.019(1-0.019)}{405}} = -0.001 \approx 0 \quad \text{(Eq. 54-2)}$$

where:

UCL = upper control limit
LCL = lower control limit

A computer-generated p chart using individual subgroup sizes for this process is shown in Figure 54-3. Note the stair-cased control limits (shown as dotted lines). No points are out of control when the actual sample size is used. When the average sample size is used, Subgroup 9 is out of control.

NUMBER OF UNITS DEFECTIVE WITH FIXED SUBGROUP SIZE (*NP* CHART)

The np chart counts the number of units defective in a fixed subgroup size. It plots the number defective per subgroup in contrast to the p chart, which plots the fraction defective. As with the p chart, such situations are governed by the normal approximation to the binomial distribution. The average fraction defective, $\bar{p}$, is:

$$\bar{p} = \frac{\sum \text{defective parts}}{\sum n} \quad (54\text{-}4)$$

where:

$\bar{p}$ = average fraction defective
n = subgroup size

The control limits for the np chart are given as:

$$n\bar{p} \pm 3\sqrt{n\bar{p}(1-\bar{p})} \quad (54\text{-}5)$$

COUNT OF DEFECTS WITH FIXED SUBGROUP SIZE (C CHART)

Previous cases were concerned with whether a part was defective. No attention

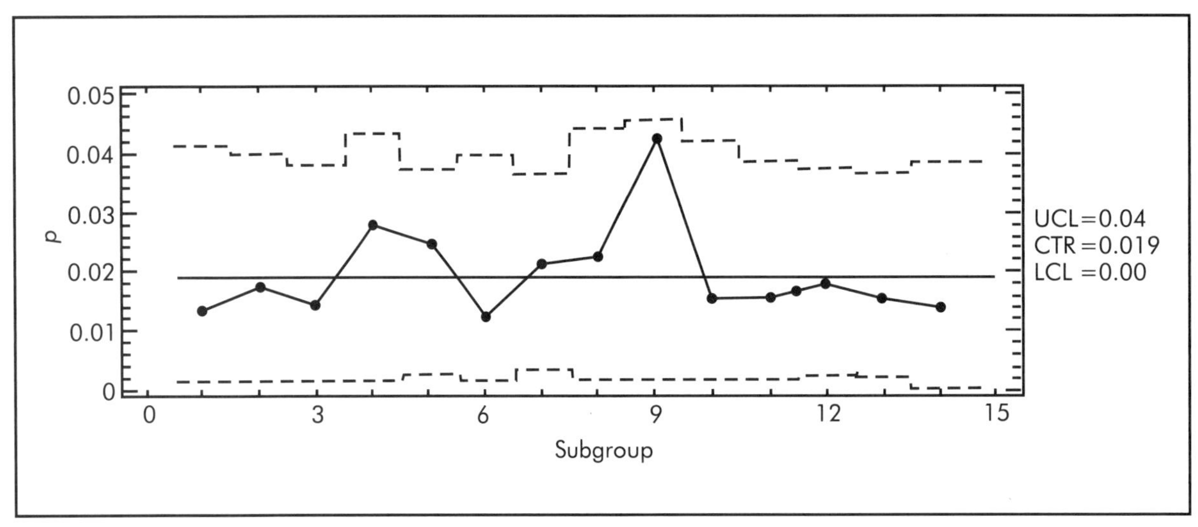

Figure 54-3. A p chart using actual sample size (Example 54.3.1).

was given as to how many, or what, defects were present. Now, consider situations where the number of defects or nonconformities per part are counted. This makes a difference in the control limits, namely, the number of defects per part is tabulated instead of the number of defective parts. For example, a car part painting operation is concerned with the following defects: fisheyes, runs, and orange peel. The sum of these defects per part is calculated when making a control chart. In general, no consideration is given as to where on the part these defects occurred.

The c chart is employed when a fixed subgroup or sample size is used, such as when considering the sum of missing rivets on an aircraft wing or voids in injection-molded parts. In the case of the c chart, each wing or plastic part inspected is of the same size.

The average number of defects, $\bar{c}$, is given as:

$$\bar{c} = \frac{\sum c}{g} \tag{54-6}$$

where:

$\bar{c}$ = average number of defects
c = number of defects per subgroup
g = number of subgroups

The control limits for the c chart are given as:

$$\bar{c} \pm 3\sqrt{\bar{c}} \tag{54-7}$$

DEFECTS PER UNIT WITH VARIABLE SUBGROUP SIZE (*U* CHART)

As with p versus np charts, compensation must be made if the part size changes. For example, if the car part operation paints different size parts, for example, hoods, doors, and rear quarter panels, the area of each part is measured and the number of defects per unit area is reported. In this case, a u chart is employed.

$$u = \frac{c}{n} \tag{54-8}$$

where:

u = number of defects per unit
c = number of defects per subgroup
n = subgroup size

The average number of defects per unit, $\bar{u}$, is given by:

$$\bar{u} = \frac{\sum c}{\sum n} \tag{54-9}$$

The control limits for the u chart are:

$$\bar{u} \pm 3\sqrt{\frac{\bar{u}}{n}} \qquad (54\text{-}10)$$

A parallel exists between the u chart and the p chart in that both consider cases where a variable sample size is involved. As a result, stair-cased control limits are expected for both. In summary, the difference between a c chart and a u chart is the use of a fixed versus variable subgroup size, respectively.

Example 54.3.2. Table 54-2 presents the number of defects from 20 samples in a painting operation using parts of a fixed size (surface area). Calculate the control limits for the appropriate control chart.

Table 54-2. Data for c chart

Part	Defects	Part	Defects
1	4	11	3
2	5	12	8
3	1	13	16
4	3	14	8
5	4	15	8
6	8	16	9
7	6	17	10
8	8	18	8
9	6	19	12
10	6	20	14

Solution. Because parts of constant size are evaluated, a c chart should be employed.

$$\bar{c} = \frac{\sum c}{g} = \frac{4+5+....+14}{20}$$

$$= \frac{147 \text{ defects}}{20 \text{ subgroups}} = 7.35 \qquad \text{(Eq. 54-6)}$$

The control limits are:

$$UCL_c = \bar{c} + 3\sqrt{\bar{c}}$$

$$= 7.35 + 3\sqrt{7.35} = 15.48 \qquad \text{(Eq. 54-7)}$$

$$LCL_c = \bar{c} - 3\sqrt{\bar{c}}$$

$$= 7.35 - 3\sqrt{7.35} = -0.78 \approx 0$$

As shown in Figure 54-4, one point (13) is out of control.

54.4 RELIABILITY

Reliability is concerned with product failures over time. It can be defined as "The probability that an item will perform a required function without failure under some stated conditions for a stated period of time."

Although it often costs a great amount to obtain reliability data, it has been found that the cost benefit is positive; that is, the long-

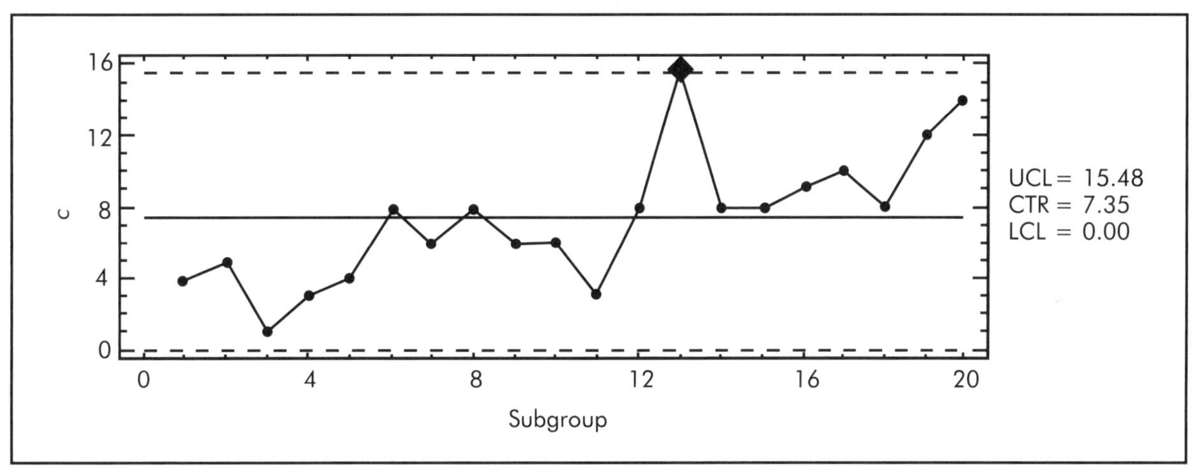

Figure 54-4. A c chart for Example 54.3.2 (point 13 is out of control).

term consequences of not studying product reliability usually outweigh the initial costs. As in other quality disciplines, the earlier reliability concepts are applied, the greater the potential benefits. Reliability thinking is compatible with Taguchi concepts and the idea of "build it right the first time." Life-cycle cost considerations applied to reliability studies have shown that it costs less in the long run to make products as reliable as possible.

Figure 54-5 illustrates the old view where quality and reliability were thought to cost "too much." Costs initially decrease and then increase over time. Figure 54-6 depicts the modern view where quality and reliability have been shown to be cost beneficial. Total costs decrease over time.

Product reliability studies are often used for warranty, liability, design, and shelf-life purposes.

PRODUCT WARRANTY

It is common today to offer 50,000 mile, 70,000 mile, or even greater mileage warranties for new automobiles or component parts. How can auto manufacturers offer such warranties and be assured they will not go broke due to excessive customer claims? Reliability comes into play because the percent of product failures versus mileage can be modeled statistically with a high degree of confidence. Thus, a manufacturer can predict its warranty exposure (percentage of failures and associated costs) before a product is launched, assuming that reliability testing has been performed. This allows for sufficient warranty reserve funds to be laid aside. Product pricing also takes predicted warranty costs into account. Extended product warranties are common with some products. Reliability studies are used to set extended warranty pricing, restrictions, and time limits.

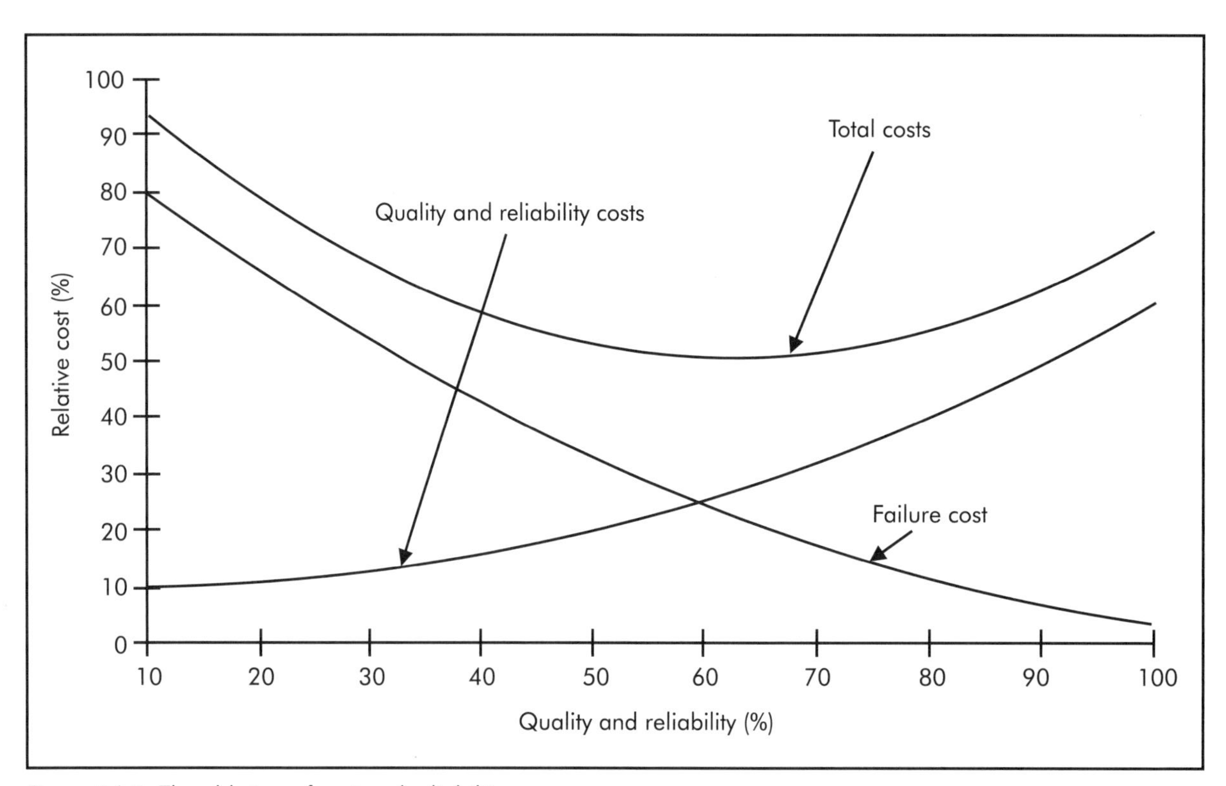

Figure 54-5. The old view of cost and reliability.

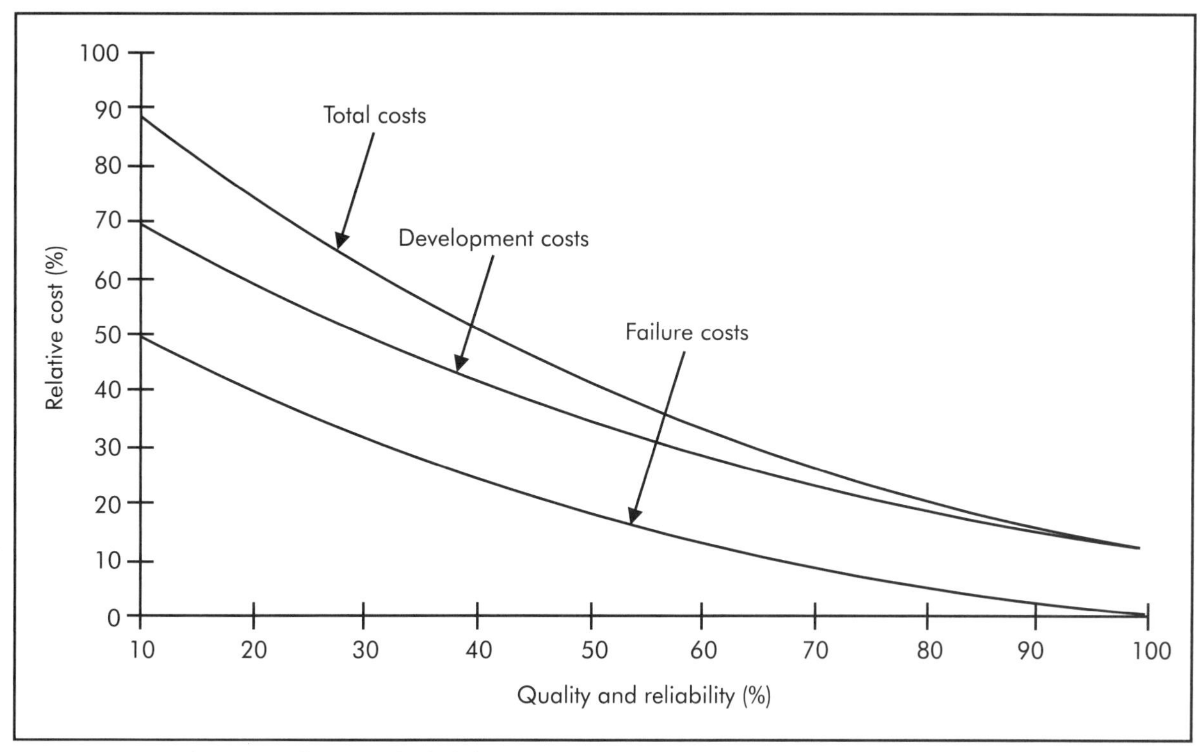

Figure 54-6. Modern view of cost and reliability.

The flip-side of the warranty/reliability relationship is to use reliability studies in product design to meet desired warranty specifications. For example, suppose that a maker of a consumer power tool wants to offer a 36-month warranty. Reliability studies show that the tool fails, as expected, at various times and due to various causes (different failure modes). Assume that all failure modes except one meet the 36-month requirement. To meet the desired warranty level, the manufacturer must increase the reliability of a single mode (component) either by redesign or upgrading the component.

PRODUCT LIABILITY

Today, it is not unusual for a manufacturer of a product, made at any time in the past, to suffer multi-million-dollar damage law suits for product failure where injury (perceived or actual) or death has occurred. Since product failures are modeled statistically, there always exists a finite probability of failure regardless of the manufacturer's best efforts. Thus, even one or two product failures leading to customer injury or death may place a company in financial jeopardy. Unless litigation reserves are set aside, and/or predicted litigation costs are built into pricing, bankruptcy could occur. In the litigation arena, reliability testing data is considered "defensive manufacturing." For example, a manufacturer with extensive reliability test data that support state-of-the-art practices, with few predicted failures, may incur less punitive financial damage in a product liability legal battle.

PRODUCT DESIGN

Many products often have two design phases: functional design and reliability design. The two phases are often performed separately using design engineers for the first and reliability engineers for the second. Certainly it is more beneficial to have both engineering disciplines work collaboratively. Both design functions must proceed together to meet functional and reliability specifications. Reliability contributes concepts such as redundant, parallel, and back-up components to meet product specifications at the lowest possible cost. Failure mode and effects analysis (FMEA), as discussed in Chapter 19 of *Fundamentals of Manufacturing*, Second Edition, is a popular method used to meet functional and reliability specifications.

PRODUCT SHELF LIFE

So far, the discussion of reliability has assumed complete product failure. However, some products do not fail in an all-or-none fashion. Rather, many ordinary products simply degrade in performance over time. Examples include foodstuffs, paints, drugstore sundries, photochemicals, pharmaceuticals, and household chemical products. In all such cases, *useful product life* is the issue of concern (that is, shelf life), not total product failure. When appropriate performance levels are known or required, reliability studies lead to predictions of a product's useful shelf life.

BATHTUB CURVE RELATIONSHIP OF NON-REPAIRABLE ITEMS

Over a product's complete lifetime, non-repairable items (for example, resistors or light bulbs) often show a failure (hazard) rate pattern called the *bathtub curve*. This represents the combined effects of an early decreasing failure rate stage representing the weakest items, a useful life with a somewhat constant failure rate, and an increasing failure rate stage due to wear-out. A typical bathtub curve is shown in Figure 54-7.

FAILURE OF NON-REPAIRABLE ITEMS

The failure pattern usually differs between repairable and non-repairable items. *Non-repairable items*—by definition—can only fail once; when this happens, their life is over. Examples include light bulbs, fuses, and transistors. While the device is still functioning, its hazard rate is of interest; this is the rate of failure at a particular point in time. The terms *mean life*, *mean time to failure* (MTTF), and *expected life* are usually applied to non-repairable items. Though non-repairable items may contain one or many components, the concern is the time to the first component's failure. For example: a light bulb may fail due to filament breakage or loss of vacuum in the globe; a transistor may fail due to over-voltage; or a non-repairable electronic part may fail due to capacitor leakage, bad connectors, etc.

Non-repairable items can fail in three basic ways. The failure (hazard) rate with time can be constant, increasing, or decreasing. A *constant hazard rate* is usually caused by exceeding load ratings. Moreover, it normally occurs in random fashion. Examples are over-voltage on electronic items and tire overloading. An *increasing failure rate* is exemplified by material fatigue due to cyclic loading. In this failure mode, there is usually an initial period without failures, that is, a latent period after which the failure rate increases. *Decreasing failure rates* are often exhibited by electronic items that have high failure rates in the beginning (for example, "infant mortality," a burn-in period) with decreasing failure rates thereafter.

FAILURE OF REPAIRABLE ITEMS

The conventional definition of the reliability for repairable items is "The probability that product/system failure will not occur in a given time period when more than one item can fail." A common way to express the failure rate for a repairable item is its *mean time*

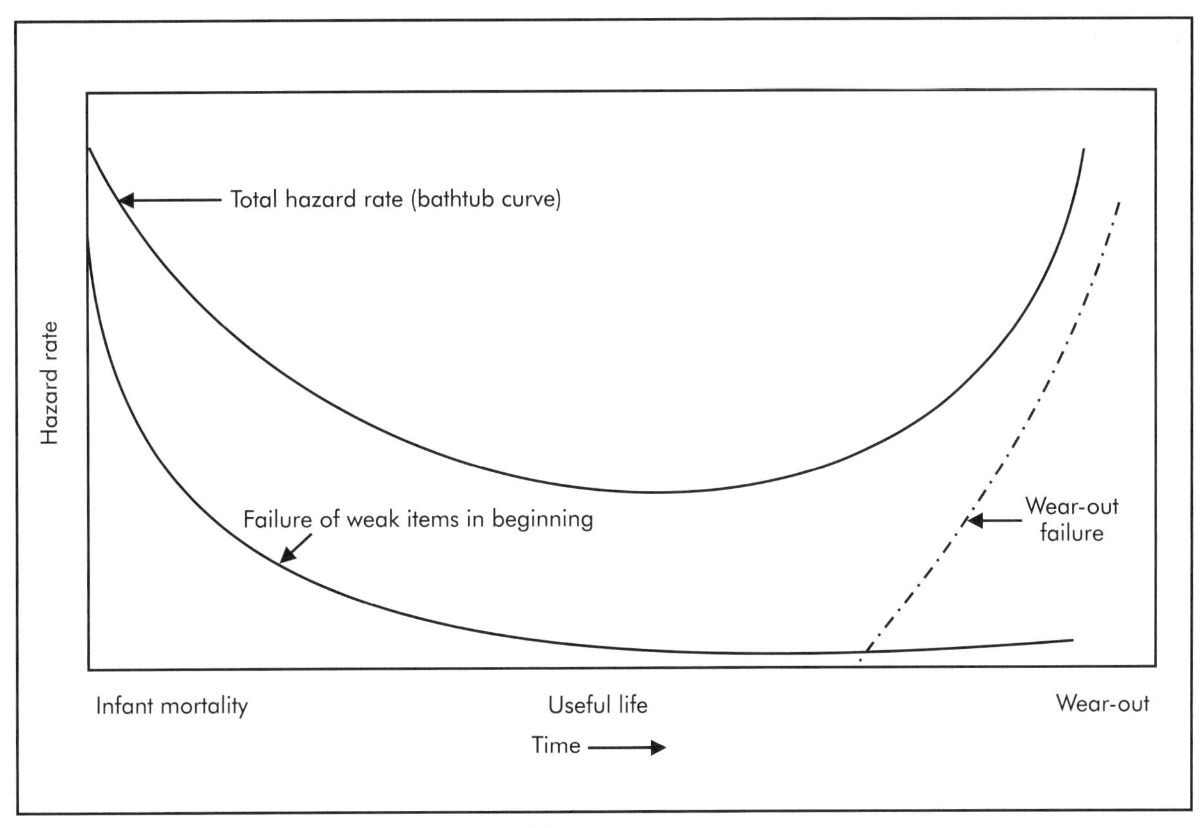

Figure 54-7. Typical bathtub curve for non-repairable items.

between failure (MTBF). A repairable item's life is ultimately dependent on maintenance schedules. As with non-repairable items, repairable items can display several failure rate trends. These are:

- a constant failure rate, which usually occurs due to external causes, for example, rust, evaporation of lubricating oils, etc.;
- an increasing failure rate, which is most often due to simple wear-out of components in the latter life stages, for example, those that are not part of scheduled repair/replacement maintenance; or
- a decreasing failure rate, which can occur if the reliability of repair parts increases over time.

PROBABILITY DISTRIBUTIONS

The overall goal in reliability studies is to mathematically model the reliability of products. Modeling is an empirical art in that various models (statistical distributions) are tested to determine the one that best fits the problem at hand. The exception to this rule would be when the best model for a particular product is already known. Although no ideal model is guaranteed for each situation, certain models seem to work best for certain products. The best model for each problem is usually determined by a combination of plotting, linear regression, and maximum likelihood fitting of lifetime reliability data.

Once the best-fit model has been obtained, predictions can be made about how long a product will last under various conditions.

The average life, standard deviation, and other conventional life statistics also can be determined. Reliability distribution fitting needs to be as accurate as possible because manufacturers often predict (extrapolate) product life behavior beyond the range where data was collected.

MODELING RELIABILITY DATA

Simple probability relationships are employed to model reliability data. Probability concepts were introduced in Chapter One of *Fundamentals of Manufacturing*, Second Edition.

If two events, A and B, are mutually exclusive, meaning two failure modes cannot occur at the same time, the probability of either event occurring obeys the addition or union rule:

$$P(A \text{ or } B) = P(A \cup B) = P(A) + P(B) \quad (54\text{-}11)$$

If two events, A and B, are *not* mutually exclusive (that is, two events or failure modes can occur simultaneously), the general addition or union rule is employed:

$$\begin{aligned} P(A \text{ or } B) &= P(A \cup B) \\ &= P(A) + P(B) - P(A \text{ and } B) \end{aligned} \quad (54\text{-}12)$$

As an aid to understanding, consider the Venn diagram in Figure 54-8. The area of the rectangle represents the total system probability, P. The probability of A or B occurring would seem to be $A + B$, however, the area AB would be counted twice; therefore, it is subtracted once.

If two events, A and B, are independent, meaning the occurrence of one event is not influenced by occurrence of the other, the probability of both occurring obeys the multiplication rule:

$$\begin{aligned} P(A \text{ and } B) &= P(A \cap B) \\ &= P(AB) \\ &= P(A) \times P(B) \end{aligned} \quad (54\text{-}13)$$

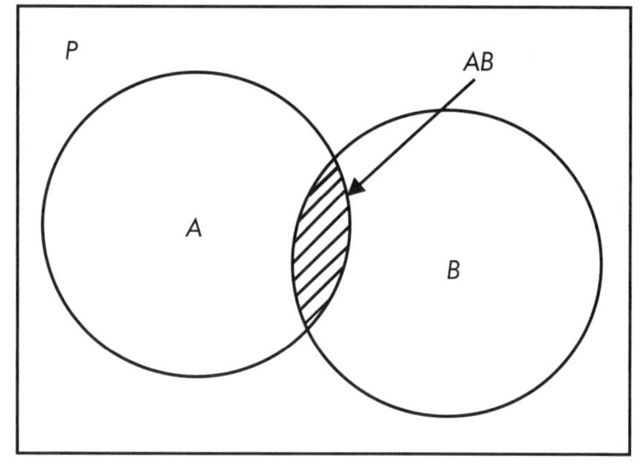

Figure 54-8. Venn diagram of P(A or B)—the general addition or union rule.

Example 54.4.1. Consider the system shown in Figure 54-9 as a block diagram. It is composed of two sections, I and II. Section I has a single component, A, with reliability = 0.99. Section II, on the other hand, has three components, B, C, and D, with reliabilities of 0.98, 0.97, and 0.95, respectively. Compute the system's reliability: $P(\text{system works}) = P(\text{I and II})$. Note that both sections must operate if the system is to work. The arrows illustrate the flow path for the system.

Solution. Obviously, A must be operable for a workable system. However, section II is operable if either components B and C are operable, or D alone. The probability relationships for the system are:

$$\begin{aligned} P(\text{system works}) &= P(\text{I and II}) \\ P(\text{I}) &= 0.99 \\ P(\text{II}) &= P(B \text{ and } C) \text{ or } P(D) \\ P(B \text{ and } C) &= 0.98 \times 0.97 \\ &= 0.9506 \end{aligned} \quad (\text{Eq. } 54\text{-}13)$$

$$\begin{aligned} P(\text{II}) &= P(B \text{ and } C) + P(D) \\ &= P(B \text{ and } C) + P(D) - P(B \text{ and } C \text{ and } D) \\ P(\text{II}) &= 0.9506 + 0.95 - (0.98 \times 0.97 \times 0.95) \\ &= 0.99753 \end{aligned} \quad (\text{Eq. } 54\text{-}12)$$

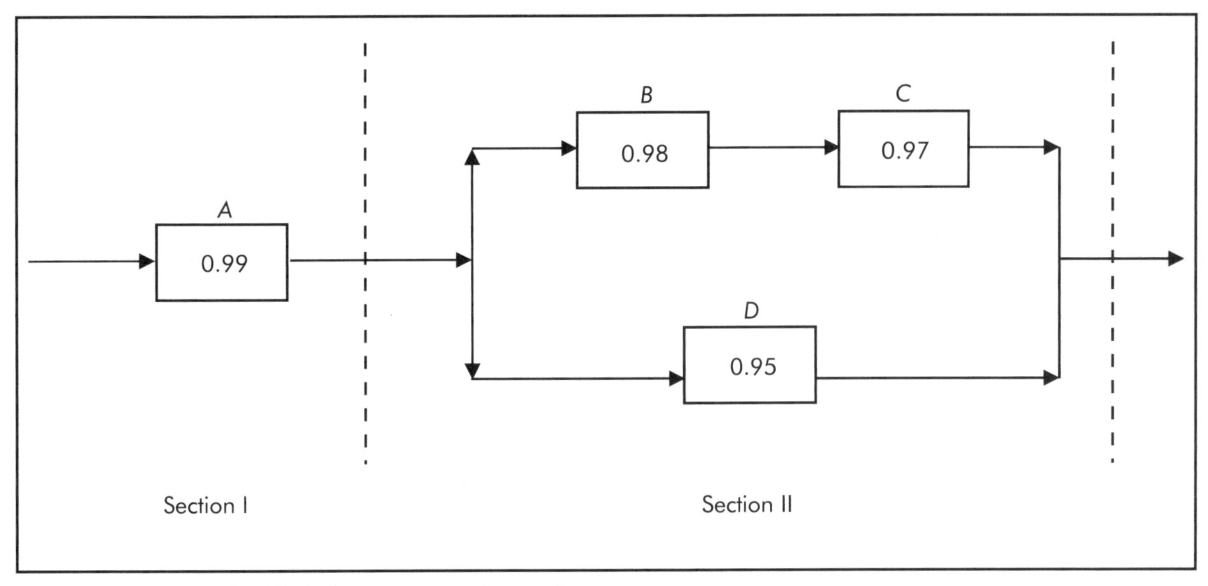

Figure 54-9. System for block diagram reliability analysis.

$$P(\text{system works}) = P(\text{I and II}) = 0.99 \times 0.99753 = 0.9876$$

$$P(\text{system works}) = 98.76\% \quad \text{(Eq. 54-13)}$$

54.5 GAGE CAPABILITY

The variable data used in statistical process control (SPC) studies must be measured. A potential problem arises because some of the product's variability may come from the manufacturing process, some from the data measuring instrument, and still more from the differences between operators. Confining the discussion to product and measurement (gage) errors, and recalling that variance, σ, is additive, the total product variance can be expressed as:

$$\sigma^2_{total} = \sigma^2_{product} + \sigma^2_{gage} \quad (54\text{-}14)$$

Two common ratios are used in gage studies. These are:

- the *P*/*T* ratio,

$$\frac{P}{T} = \frac{6\hat{\sigma}_{gage}}{USL - LSL} \quad (54\text{-}15)$$

where:

P = precision
T = tolerance
$\hat{\sigma}_{gage}$ = gage variation
USL = upper specified limit
LSL = lower specified limit

- and the *gage/product error* ratio:

$$\frac{gage}{product\ error} = \frac{\hat{\sigma}_{gage}}{\hat{\sigma}_{product}} \quad (54\text{-}16)$$

where:

$\hat{\sigma}_{gage}$ = gage variation
$\hat{\sigma}_{product}$ = product variation

REPEATABILITY AND REPRODUCIBILITY—GAGE R&R

Gage variance—or measurement error—is composed of two parts: one is intrinsic instrument error, the other differences between operators. The study of both is termed *gage R&R analysis*. In gage R&R, *repeatability* is the variance due to intrinsic instrument error; *reproducibility* is defined as the variance due to different operators. Because variance is additive, the following relationship is true:

$$\sigma^2_{gage} = \sigma^2_{repeatability} + \sigma^2_{reproducibility} \quad (54\text{-}17)$$

The reason gage R&R studies are conducted is because when any product is made—assuming that SPC methods are employed—there is always interest in the sources of product variation. As shown, the total variance is parsed into its separate components: a fraction due to operator differences, another due to instrument error, and the remainder due to undefined product error. In summary, if operator variance is the largest fraction, more operator training should be conducted; if instrument error is greatest, better instruments should be used; and finally, if product variance is too large, the manufacturing process should be changed to increase process capability.

54.6 DESIGN OF EXPERIMENTS

Experimental design is the intentional change of inputs (factors) in a process to observe the corresponding change in outputs (responses). A process creates a service or product and consists of some combination of machines, materials, methods, people, environment, and measurements. Design of experiments (DOE) is a scientific, statistically-based method that allows a researcher to better understand a process and determine how the inputs affect the outputs.

Traditionally, process variables are changed one at a time and then their respective effects are compared to a standard. Each variable is changed one at a time to see which has the greatest impact on the process. Changing one variable at a time does not identify the interaction of variables or the curvilinear effect of the variables. Many experiments or statistical trials (also called Monte Carlo simulations) are needed and random errors can be compounded or not considered. Experimental design allows multiple variables to be studied simultaneously, thereby shortening the experimental process and determining the interaction effects between variables.

SELECTING QUALITY CHARACTERISTICS (RESPONSES)

Guidelines for selecting quality characteristics are as follows.

- The characteristic(s) should be related as closely as possible to the basic engineering mechanism.
- Use continuous responses (variable data) when possible.
- Use stable quality characteristics that can be measured precisely and accurately.
- Make certain the quality characteristics encompass the important aspects of the product/process with respect to customer needs and expectations.
- Use all pertinent resources, for example, common sense, engineering knowledge, statistical methods, customer requirements, historical data, and product specifications.
- For complex systems, select responses at the subsystem levels and run experiments at these levels before trying to optimize the overall system.

CONDUCTING AN EXPERIMENT

Discipline is required when conducting a designed experiment. During brainstorming, when all possible experimental factors are considered, creativity—including "wild" ideas—is highly recommended. The results of brainstorming are often summarized in a cause-and-effect fishbone diagram such the one as illustrated in Figure 54-10. Conducting an experiment imposes a discipline that may be unfamiliar to many experimenters. Lack of discipline is often the cause of a failed experiment and/or ambiguous results. To those persons who are new to experimental

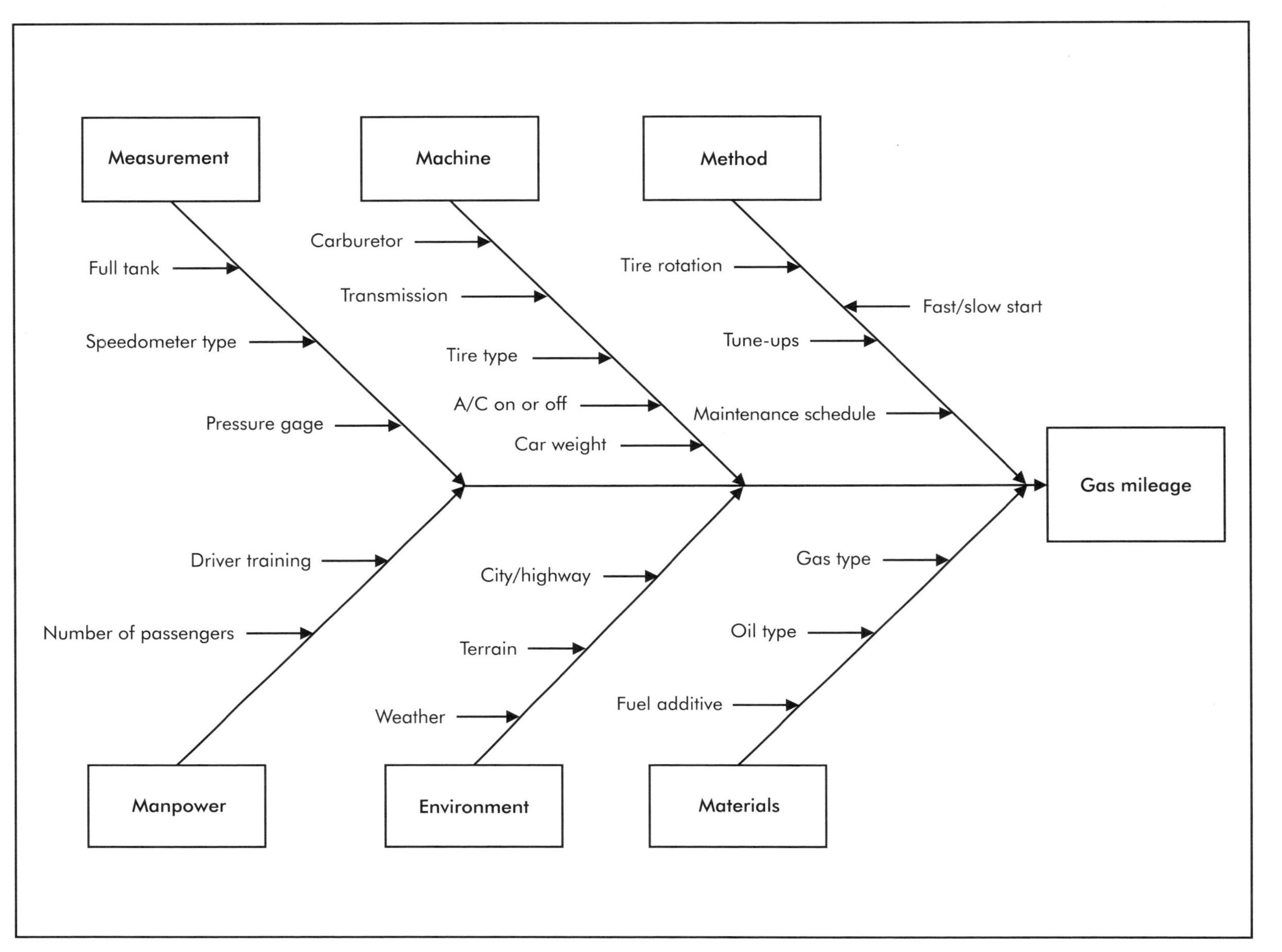

Figure 54-10. Cause-and-effect fishbone diagram of gas mileage.

design, it will sometimes be tempting to accept the results of a particular run as "good enough," suggesting, "why go further?" Examples of discipline include:

- The factor setting must be as close as possible to those specified in the design.
- The sample replicates must be prepared exactly the same each time.
- Data sheets must be prepared in advance.
- All persons involved should understand the experiment and allow no deviation from the plan.

EXPERIMENTAL DESIGNS

Conventional DOEs fall into two classes: two-level designs and three-level designs. In two-level designs, the experimental factors are studied at two levels—low and high. In three-level designs, the factors are studied at three-levels—low, middle, and high. The majority of DOEs are conducted at two levels.

Designed experiments are performed according to a predefined matrix or template that has several important properties. First, the factor space in a DOE is symmetrical. Second, the factors are orthogonal to each other. Orthogonality is a mathematical term that can be thought of as perpendicularity. Experiments comprised of orthogonal factors are called *factorial designs*. The use of orthogonal experiments allows simple pencil and paper analysis of data. Various symmetries are available depending on the goal of the experiment—screening design, testing for nonlinear (for example, quadratic) behavior, etc. The use of the proper symmetry gives the maximum amount of information for the least number of experiments. Figure 54-11 is an illustration of a symmetrical experiment for variables (factors) A and B. Each factor is used at two levels, 1 and 2. One run, for example, is conducted with $A = 1$ and $B = 1$; another run uses $A = 2$ and $B = 2$, etc.

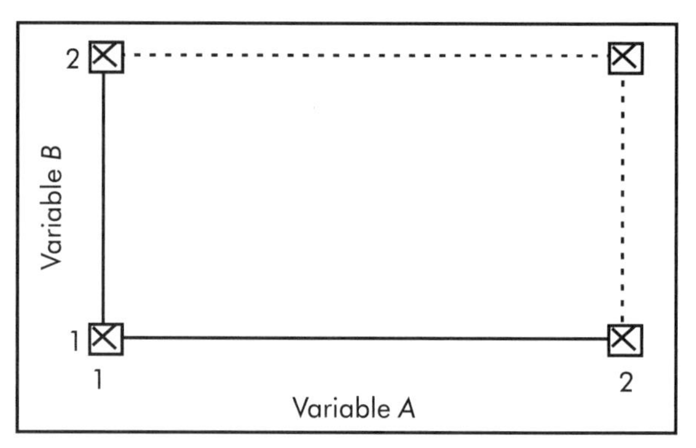

Figure 54-11. Example of a symmetrical experiment for two factors.

The number of runs in the common two-level designs can be expressed as 2^n, where n = the number of factors to be studied. For example, L-8 = 2^3 runs, L-16 = 2^4 runs, and so on.

When a DOE matrix of size 2^n is used to study n or fewer factors, it is possible to detect *factor interactions*, for example, $A \times B$, $B \times C$, etc. Indeed, factor interactions are the rule—rather than the exception—in many real-world experiments. Table 54-3 shows a standard order L-8 design matrix for three factors, A, B and C, as well as all possible interactions between these factors.

The most important aspect of the matrix is the ordered array of numbers (–1 and 1).

Table 54-3. Standard order L-8 design matrix for three factors, including interactions

	Factors						
Run	A	B	C	AB	AC	BC	ABC
1	–1	–1	–1	1	1	1	–1
2	1	–1	–1	–1	–1	1	1
3	–1	1	–1	–1	1	–1	1
4	1	1	–1	1	–1	–1	–1
5	–1	–1	1	1	–1	–1	1
6	1	–1	1	–1	1	–1	–1
7	–1	1	1	–1	–1	1	–1
8	1	1	1	1	1	1	1

These represent the low (–1) and high (1) factor levels used in the experiment. In run 1, for example, factors A, B, and C are used at their low levels. The remaining runs are interpreted similarly.

ANALYSIS OF DESIGNED EXPERIMENTS

Typically DOE analysis is performed with DOE software or other statistics-based software. Detailed descriptions of the analysis tools is beyond the scope of this book and the certification exam, however a brief description of each is presented.

DOEs are analyzed by several methods:

- contrast sums,
- t-test,
- F-test,
- analysis of variance (ANOVA), and
- linear regression.

Contrast sums are used to determine if there is a significant difference in the change of the quality characteristic when one factor is high versus low.

The *t-test* is used to compare two means to determine if they are significantly different at some confidence level, for example, the 95% level. "T-test" refers to the t-distribution, which is similar to the normal distribution but is slightly wider.

The *F-test* is used to compare two variances. It is named for Ronald Fisher who developed experimental design, hypothesis testing, and ANOVA in the 1930s. The F-test is also used in ANOVA.

ANOVA is a general statistical technique for detecting and quantifying the sources of variation in any test or experiment. The interest is in variation about the mean. In other words, the mean response is calculated first. Then it is determined if the leftover variance can be explained by one or more factors. After accounting for factor effects, the residual (error) remains. Problems analyzed by ANOVA may involve quantitative and/or qualitative variables, that is, attribute variables. Following are various types of ANOVA:

- no-way ANOVA—repeat tests yielding the simple average and standard deviation;
- one-way ANOVA—a series of tests involving one factor at multiple levels;
- two-way ANOVA—a series of tests involving two factors at multiple levels; and
- n-way ANOVA—the general technique for n factors at m levels.

Finally, linear regression is used to determine the specific settings for each variable to maximize the predetermined quality characteristic.

54.7 TAGUCHI CONCEPTS

Taguchi offers an array of quality-related concepts. Application of Taguchi concepts provides:

- a definition of quality;
- consistency to the evaluation of quality;
- quality improvement with a minimum cost increase; and
- unique use of experimental design.

The major goals of Taguchi methods are:

- higher quality with a minimum (or no) cost increase;
- more consistent product (less "noise");
- minimization of conditions that cause rework (rework is another result of product inconsistency—Taguchi's view: build it right the first time);
- shortening the product development cycle;
- identifying the critical variables in product design/production; and
- minimizing the effect of external noise (for example, humidity, or temperature).

Taguchi recommends these steps for product engineering development:

1. Start with the lowest-cost components. All components do not have to be toleranced to National Aeronautics and Space Administration (NASA) standards.
2. Determine the critical design parameters. All factors are not critical. Critical parameters should be based only on qualitative evidence, not on gut-feel instincts. Set the critical design parameters at nominal levels. The ideal nominal starting levels should be based on the results of designed experiments.
3. Determine the noise parameters. Noise parameters differ from one system to the next. However, they usually include factors such as the operator, shift, day, environmental conditions, etc.
4. Minimize the effects of noise parameters. DOE is used in specific ways to minimize the effects of noise parameters, that is, to make production robust.
5. Optimize with the lowest-cost components. The initial goal is to find out how good a product can be using DOE and the lowest-cost parts.
6. Tighten component tolerances only when required to meet specifications. Tolerances in aerospace and government work are often set arbitrarily high because those writing the specifications want to make errors on the side of "goodness." If DOE is not used, critical specifications may be neglected or ignored.
7. Make the product robust against customer use conditions. Note the two types of robustness: (a) consistent product regardless of noise effects, and (b) durability in terms of customer use.

LOSS FUNCTION (QUALITY DEFINITION)

Taguchi emphasizes the concept of loss in his definition of quality: *Quality* can be defined as the loss incurred by all of society through the use of a product, less the incremental cost to the manufacturer to prevent the loss. Losses are typically incurred through:

- time and money spent to repair/replace defective products,
- warranties,
- recalls,
- customer dissatisfaction,
- lost market share, and
- accidents and pollution.

Quality is often evaluated by the "goalpost" method shown in Figure 54-12. An example is the go/no-go gage. In this method, product quality is considered to be good if the measured characteristic is either greater than the lower control limit (LCL) or less than the upper control limit (UCL). Quality is considered to be poor only if the characteristic is less than the LCL or greater than the UCL.

Taguchi, on the other hand, uses the *loss function* to evaluate quality. When a nominal (target) value is required, Taguchi evaluates loss as:

$$L(y) = k(y - m)^2 \qquad (54\text{-}18)$$

where:

$L(y)$ = loss function (loss due to variation of the specified quality characteristic from the nominal value)
k = constant

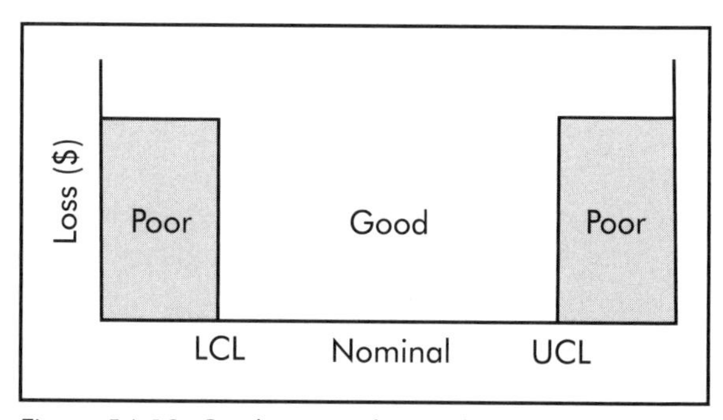

Figure 54-12. Goalpost quality evaluation method.

y = value of the specified quality characteristic
m = nominal (target) value of the specified quality characteristic

When $y = m$, the loss is 0. However, the more y deviates from m, the greater the loss. When quality is evaluated in this way, it means that quality is continuous rather than all or none as in the goalpost method. Figure 54-13 is an illustration of the loss function when a nominal value is desired.

TAGUCHI'S VIEWS ON PRODUCT/ PROCESS DEVELOPMENT

Taguchi suggests dividing product/process development into three stages:

1. System design—40% of the effort should be devoted to this phase.
   - Make prototype(s) based on engineering and scientific knowledge.
   - Use new ideas, innovations, etc.
   - Base nominal parameter settings on engineering judgment.
2. Parameter design—40% of the effort should be devoted to this phase.
   - Determine the optimal settings for the critical parameters.
   - Minimize the effects of noise (make product robust).
   - Exploit nonlinear relationships (that is, use L-9 or other three-level designs).
   - Quality will be improved without added cost; only factor settings are changed.
3. Tolerance design—the remaining 20% of the effort should be devoted to this phase.
   - Higher tolerances are required only if performance is unacceptable after parameter design.
   - Determine the contribution of each parameter to system variation.
   - Determine the allowable tolerances for the design parameters.
   - Some (usually, very few) tolerances will be tightened.
   - Quality will be improved, but with added cost.

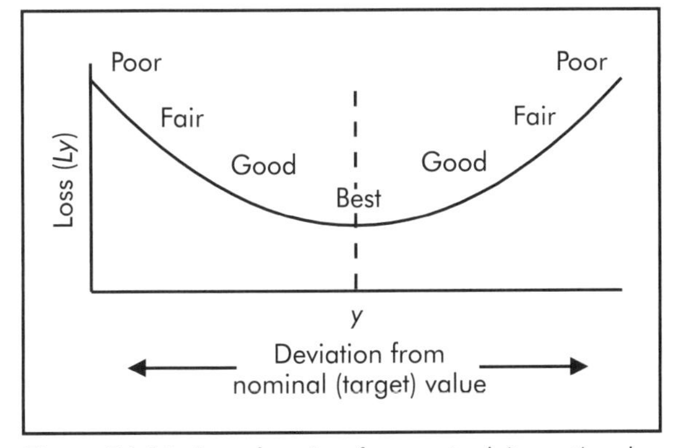

Figure 54-13. Loss function for nominal (target) value situations.

ROBUST PRODUCT DESIGN

Robustness means relative immunity to either manufacturing or use conditions by the customer. Immunity implies that product characteristics are unchanged, or minimally changed, by one or more outside conditions, which either can not be controlled or control is not warranted due to costs. In a manufacturing example, the effect of relative humidity in a production environment is an indicator of product robustness. The product is said to be robust against relative humidity if performance does not change when the humidity varies. In a customer example, an automobile is robust against temperature if it starts and runs well at most or all outside temperatures. Note that robustness can be detected only in three-level designs.

SIGNAL-TO-NOISE RATIO

Because problems involving the average response and response standard deviation are so common in manufacturing, Taguchi recommends forming a single response, termed the signal-to-noise (S/N) ratio. *Signal* refers to the desired level of performance. Factors that promote the desired level of performance are called *signal factors*. *Noise* refers to variability. Factors that cause product variability are called *noise factors*. The goal in all instances is to maximize S/N as greater S/N ratios indicate higher product robustness. Following is one example of an S/N equation.

$$\frac{S}{N} = \frac{\overline{Y}}{S_Y} \tag{54-19}$$

where:

S = signal
N = noise
$\overline{Y}$ = average response
S_Y = root mean square variation

54.8 ISO 9000 (2000)

The International Organization for Standardization (ISO) has produced technical standards since 1947. These standards, documented voluntarily, are regulatory specifications for such things as mechanical fasteners, springs, and drafting procedures. In 1987, "ISO 9000—Quality Management Systems" was published and distributed worldwide. This standard was different from previous standards in that it was designed to be more generally applied, whereas earlier standards were very specific in application. In particular, ISO 9000 gives any organization, regardless of type or size, guidelines for establishing and improving its management system.

ISO 9000 was first revised in 1994. However, continuous improvement is an evolutionary process and ISO 9000 is not exempt, not to mention that ISO directives require standards to be revised every five years. Consequently, the ISO 9000 (2000) standard was unveiled on December 15, 2000.

ISO 9000 (2000) is a family of standards containing three separate documents: "ISO 9000, Quality Management Systems—Fundamentals and Vocabulary"; "ISO 9001, Quality Management Systems—Requirements"; and "ISO 9004, Quality Management Systems—Guidelines for Performance Improvement." ISO 9000 defines the terms used in the standard; ISO 9001 describes the regulatory requirements; and ISO 9004 provides guidance on continually improving a company's management system. Three documents from the previous 1994 edition (ISO 9001, ISO 9002, and ISO 9003) were integrated into the ISO 9000 (2000) standard under ISO 9001 (2000). The U.S. adopted the ISO 9000 family of standards as the American National Standards Institute/American Society for Quality (ANSI/ASQ) QS-9000 series.

According to the International Forum for Management Systems, ISO 9000 is important for these major reasons:

- Product quality and business efficiency—the standards are designed to help an organization review its production processes and establish systems to improve and maintain the quality of its products and/or services. By documenting processes and production systems while implementing a quality management system, ISO 9000 provides a manufacturer with a methodical means of reviewing operations, identifying inefficiencies and gaps in systems, and implementing actions to eliminate the inefficiencies and gaps.
- Competition—in certain industries, ISO 9000 registration is widespread and is all but required to remain competitive against companies who are already registered. In fields where ISO 9000 is recognized as an effective management tool, registration offers a significant business advantage to a company over its competitors.

- Customer requirements—more companies, and certain industries, are requiring suppliers to become ISO 9000 registered to meet contract specifications for quality management. In certain countries, including members of the European Union (EU), ISO 9000 registration is a requirement for conducting trade with government entities and companies who must comply with industry and government regulations. Examples of industries required to register include medical devices and construction products.
- Productivity and profitability—by increasing organizational efficiency, retaining or enhancing competitive edge, and meeting customer and market requirements for doing business, an organization will increase productivity, maintain or enhance product quality, and maintain or increase sales, which will result in improved profitability.

Eight quality management principles form the basis on which ISO 9000 (2000) is founded. The principles are as follows.

1. Customer focus: the emphasis is on understanding and exceeding the expectations of the customer.
2. Leadership: the responsibility of an organization's leadership is to propose direction and objectives and then to create an atmosphere wherein employees are encouraged to participate in meeting those objectives.
3. Involvement of people: human resources are an invaluable asset, and it is important for an organization to utilize the abilities and contributions of individuals to the fullest extent.
4. Process approach: organizations that exhibit superior customer satisfaction are those that seek and maintain the quality and productivity of key processes.
5. Systems approach to management: management that views the organization as an aggregation of interrelated processes assumes a systems approach.
6. Continual improvement: it is critical for an organization to engage in unrelenting thought and activities to improve its process, products, and services.
7. Factual approach to decision making: to make informed decisions, organizations should collect and analyze appropriate data.
8. Mutually beneficial supplier relationships: supplier-customer relationship building is necessary to promoting value-added processes, products, and services.

ISO 9001 contains 20 elements, which are stated briefly as follows.

1. Management responsibility: the highest levels of management must determine the company's objectives and commitment to quality.
2. Quality system: each supplier must establish a quality system and manual to document how it will respond to QS-9000 requirements.
3. Contract review: the highest levels of management must determine customer requirements and translate these into measurable criteria that will assure customer satisfaction.
4. Design control: each supplier is responsible to create and maintain a system to control and verify product design.
5. Document and data control: suppliers must ensure that all documents and data comply with QS 9000 requirements.
6. Purchasing: organizations must ensure that purchased products conform to requirements.
7. Control of customer-supplied product: suppliers must verify storage and maintenance, including products that are lost, damaged, or noncompliant.

8. Product identification and traceability: suppliers are to document procedures to identify product from receipt through production, delivery, and installation.
9. Process control: suppliers must identify and plan production, installation, and servicing processes that directly impact quality.
10. Inspection and testing: organizations must establish procedures for inspecting and testing processes, products, and services to ensure conformance to requirements.
11. Control of inspection, measuring, and test equipment: organizations are required to establish a system to control, calibrate, and maintain inspection, measuring, and test equipment.
12. Inspection and test status: suppliers must identify the product's inspection and test status.
13. Control of nonconforming product: suppliers are required to ensure that product not conforming to specifications is not inadvertently shipped to customers.
14. Corrective and preventive action: organizations must describe how they will correct problems and prevent them from recurring.
15. Handling, storage, packaging, preservation, and delivery: suppliers must document procedures to ensure product integrity.
16. Control of quality records: organizations are required to have a documented system to identify, collect, access, store, and maintain records.
17. Internal quality audits: organizations must have a plan in place to schedule and conduct internal audits, which are performed by an internal auditor.
18. Training: organizations are required to train personnel to perform value-added, qualified functions.
19. Servicing: organizations must provide evidence that servicing, including repairs and maintenance, is carried out.
20. Statistical techniques: organizations are required to determine the need for statistical analysis to ensure sound decision-making.

An organization seeking certification should refer to the ISO 9000 family of standards for further information. However, "ISO 9001, Quality Management Systems—Requirements" is the document any organization would follow to evaluate its ability to meet customer and applicable statutory specifications. Furthermore, ISO 9001 is the only document that third-party auditors use to assess and register organizations as complying with the ISO 9000 series.

Third-party auditors conduct on-site visits to compare a company's operations relative to the requirements of ISO 9001. When selecting a third-party auditor, organizations should make certain the entity is duly accredited by the Registrar Accreditation Board (RAB). Established in 1989, the RAB is a nonprofit organization that provides assessment services to organizations seeking to conform to regulatory standards such as ISO 9001. It is financially self-supported and governed by a joint (ANSI-RAB) oversight board of directors that includes technical experts, business executives, industry representatives, and employees of registrar organizations. In 1996, the ANSI-RAB National Accreditation Program (NAP) was formed. The NAP provides accreditation of ISO 9001 registrars.

The registration process begins with the third-party auditor reviewing the company's quality management system (QMS), sometimes referred to as a quality manual. The auditor then undertakes a non-mandatory pre-assessment in an attempt to identify any noncompliance between the company's QMS and ISO 9001 requirements. This step is followed by an auditor team assessment of the QMS and its documentation.

The size of the audit team is based on several factors such as the size of the organization and the product produced. If the company passes this assessment, it is awarded an ISO 9001 certificate of registration. The last phase of the process involves periodic reviews to verify the company remains compliant to the standard. The original third-party auditor conducts a surveillance audit every six months by randomly selecting several elements to review the extent to which the company has remained in compliance. This process is repeated every six months thereafter until three years has expired. At the end of the three-year period following initial certification to the standard, the organization will undergo another full audit. This audit is as extensive as the original audit conducted when the company was first registered to ISO 9001.

54.9 QS-9000

QS-9000 is a set of quality management requirements for suppliers, whereas ISO 9000 is a generic quality management standard. QS-9000 was developed by DaimlerChrysler, General Motors, and Ford Motor Company specifically for the automotive sector. It was based on the 1994 edition of ISO 9001. It does not, however, pertain to all automotive suppliers, only to suppliers of production materials, production and service parts, heat treating, painting, and other finishing services.

The primary purpose of QS-9000 was to unify the supplier quality requirements and assessment standards of DaimlerChrysler, General Motors, and Ford Motor Company. Prior to this, all three companies had separate requirements. The multiplicity of standards became an overwhelming burden on suppliers. Thus, in 1994, QS-9000 was created to alleviate redundant requirements and procedures.

QS-9000 consists mainly of two sections. Section I includes the 20 elements of ISO 9001 (enumerated previously) and Section II is specific to the requirements of DaimlerChrysler, General Motors, and Ford. These company-specific requirements are not meant for all suppliers, only those with which DaimlerChrysler, General Motors, or Ford does business. If an organization supplies Ford only, for example, then it would be required to meet the specific requirements of Ford and not those of DaimlerChrysler or General Motors. If, however, an organization wants to supply all of the above automakers, then it must be certified on each of the respective company's separate requirements.

TS-16949 is the new internationally recognized quality systems specification for the automotive industry. It harmonizes the supplier quality system requirements of the U.S., Germany, Italy, and France. TS-16949 replaces QS-9000 with the planned phase-out of all QS-9000 certifications in 2006. This new specification takes an action-oriented approach to quality systems. TS-16949 focuses on process effectiveness versus documentation, increased top management involvement, customer satisfaction (good products delivered to the customer), employee motivation and empowerment, and supplier development.

54.10 SIX SIGMA FUNDAMENTALS

Motorola won the Malcolm Baldrige National Quality Award in 1988 by using a strategy they called six sigma. Since then, countless organizations have decided to incorporate six sigma. Generally speaking, *six sigma* is a highly structured methodology that incorporates techniques and tools for eliminating defects and variability from any process.

A statistical concept, sigma (σ) is a Greek letter used to represent standard deviation. The term "six sigma" is a measure of high quality, or, in other words, extremely low variability. It means the engineering tolerances must be at least plus or minus six standard deviations from the process mean. At six sigma, only 3.4 non-conformances per million opportunities will occur. As a comparison, many organizations use three sigma (66,810 non-conformances per million) or four sigma (6,210 non-conformances per million) as an acceptable level of quality.

The methodology of six sigma originated in 1985 with Bill Smith, quality engineer for Motorola, as part of a program entitled "Design for Manufacturability." It was this program that defined the steps to achieve six sigma performance. Over time this program came to be known as "Design for Six Sigma." Motorola also originated the functions and standards for Black Belts, trained and experienced six sigma practitioners who lead project improvement teams (Barney and McCarty 2003).

Different organizations and people use differing adaptations of Motorola's six sigma program. The reference section of this chapter lists several recent books, each taking a different approach to six sigma.

Even Motorola has produced a new edition of its six sigma book, *The New Six Sigma* (Barney and McCarty 2003). Realizing that six sigma is more than just defects and variability reduction, Motorola redesigned it. Now presenting more of a business improvement methodology, *The New Six Sigma* is organized around four leadership principles:

- Align—link customer requirements to core business processes and create appropriate goals and measures.
- Mobilize—train and empower teams so they can execute improvement projects.
- Accelerate—employ a learning-by-doing model to facilitate simultaneous education and execution.
- Govern—maintain a regimen of structured reviews and in-depth investigations of results.

In spite of Motorola's improvements, however, *Six Sigma: the Breakthrough Management Strategy Revolutionizing the World's Top Corporations* (Harry and Schroeder 2000) is, by far, the title with which most people identify today, as well as that described in the remainder of this section. While an employee of Motorola, Mikel Harry engaged in refining the six sigma strategy to focus on rapid dissemination of knowledge and putting improvement tools in the hands of all employees within an organization as opposed to a separate quality department (Maguire 1999).

Six Sigma: the Breakthrough Management Strategy Revolutionizing the World's Top Corporations is based on the define, measure, analyze, improve, and control (DMAIC) model. Every six sigma project to improve a process or service must go through these five phases in the following, specified sequence (*The Black Belt Memory Jogger* 2002):

1. Define—based on customer requirements, identify the process or service to be improved.
    - Select problem areas within that process or service.
    - Identify the key customer-preferred attributes of the process or service (define the specification).
    - Describe the performance criteria for the process or service.
2. Measure—measure the process or service to obtain a performance baseline.
    - Prescribe the measurable parameters for the process or service.

- Determine the extent to which the process or service is capable of meeting customer requirements (gage R&R and capability studies).
- Establish an improvement target.

3. Analyze—analyze the data to determine where errors are occurring.
   - Identify the sources of variation in the process or service (graphical analysis, fishbone diagram, and FMEA).
   - Test for corrective action on plausible causes (hypothesis testing).
4. Improve—improve the process.
   - Experiment to determine cause-and-effect relationships (DOE).
   - Institute acceptable variation from customer-defined targets (tolerances to the specifications).
   - Substantiate improvements to the process or service.
   - Recalculate process or service capabilities.
5. Control—control the process to ensure the errors do not return.
   - Establish techniques to monitor and control the process or service.
   - Record and archive all project data.

The discipline instituted by using this framework is unparalleled with other systems. Specifically, any six sigma improvement project requires the direct participation of upper management, the on-sight support of an experienced consultant (a master Black Belt), the seasoned guidance of in-house six sigma practitioners (Black Belts), and the participation of lesser trained and experienced practitioners (Green Belts) as they interact with employees involved in the process or service targeted for improvement.

The specific improvement tools and techniques of six sigma include the following:

- process mapping,
- process capability,
- design of experiments,
- reliability analysis,
- control charts,
- flow charts,
- cause-and-effect (fishbone) diagram,
- histograms,
- Pareto charts,
- scatter diagrams,
- run charts,
- affinity diagrams,
- gap analysis,
- gage R&R, and
- hypothesis testing.

Obviously, these tools are not new; some have been used since the days of Walter Shewhart in 1926. What is unique about *Six Sigma* is that these tools have been incorporated into an organized structure (DMAIC), which is applicable to any improvement project.

In a business context, the extensive use of statistical methods requires the direct involvement of management in their application. Management must also assess the impact of final results on cost, quality, productivity, and customer satisfaction.

With regard to company involvement, the CEO is responsible for creating a vision and mission for any improvement effort and providing the time and resources for project completion. A project champion is responsible for the logistical issues relative to making a project happen. Master Black Belts coach the Black Belts and work with the company's champion to facilitate, mentor, and lead the problem-solving efforts. Black Belts are responsible for leading project improvement teams. Green Belts work within their respective company areas to assist in project completion (Brue 2002). In addition to these particular individuals there are many other employees who serve in supportive and data-gathering functions.

The bottom line for the effectiveness of a six sigma improvement project is the extent to which it results in increasing productivity, eliminating waste, improving quality, and exceeding customer expectations. Consequently, the magnitude of these improvements is a required component of six sigma project justification.

Like other business improvement methodologies such as Total Quality Management (TQM), ISO 9000, and the Baldrige criteria for performance excellence, the success of six sigma requires an organization's total commitment of time, effort, and resources. This means extensive training, shared vision, designated responsibilities, and continuous active participation.

54.11 NONDESTRUCTIVE TESTING

Nondestructive testing (NDT) is a comprehensive term that refers to experiments performed on parts for the following purposes:

- to verify physical attributes such as size, shape, weight, color, or composition, and/or
- to identify defects.

As the term implies, NDT allows manufacturing inspectors to verify discrete parts without rendering them unusable. The primary function of nondestructive testing is to identify product nonconformities that violate customer expectations. The purpose of the nondestructive aspect of NDT is to increase throughput. Specifically, NDT permits tested parts to be salable. It is also used to identify those parts that are minimally compromised, yet acceptable in certain situations. Inspectors experienced with NDT are capable of testing and measuring products to determine the extent that flaws may jeopardize part reliability and maintainability.

NDT techniques are extremely versatile. For example, NDT is regularly used to check bonding, brazing, and welding integrity, in addition to castings, laminates, and fluids. Different NDT methods are capable of a variety of inspection tasks. Some of these include the identification or measurement of:

- interior cracks, porosity, holes, and ruptures;
- material thickness;
- conductivity;
- metallurgical properties;
- weld integrity; and
- surface flaws and texture.

There has been much interest and development in NDT because of increased requirements for product reliability and safety. In addition, intensified government regulations and global competition are driving research and continuous improvement in NDT. Manufacturing engineers and technicians are familiar with many of the more common NDT methods. These include visual inspection, direct measuring instruments, and coordinate measuring machines. Other nondestructive testing methods commonly used in industry include:

- liquid penetrant,
- magnetic particle,
- ultrasonic,
- radiographic,
- fluoroscopic,
- eddy current,
- leak testing, and
- acoustic emission.

LIQUID PENETRANT TESTING

In *liquid penetrant testing*, a test piece is coated with a solution (penetrant) that contains a visible or fluorescent dye. Excess penetrant is then removed from the surface of the object and a developer is applied to draw the penetrant out of any cracks or discontinuities. With fluorescent dyes, ultraviolet light is used to make the imperfections show up in a bright color, thus allowing any flaws to be readily seen as illustrated in Figure 54-14.

Liquid penetrant inspection can be used to detect all types of surface cracks, porosity, laminations, and bond joints in hard,

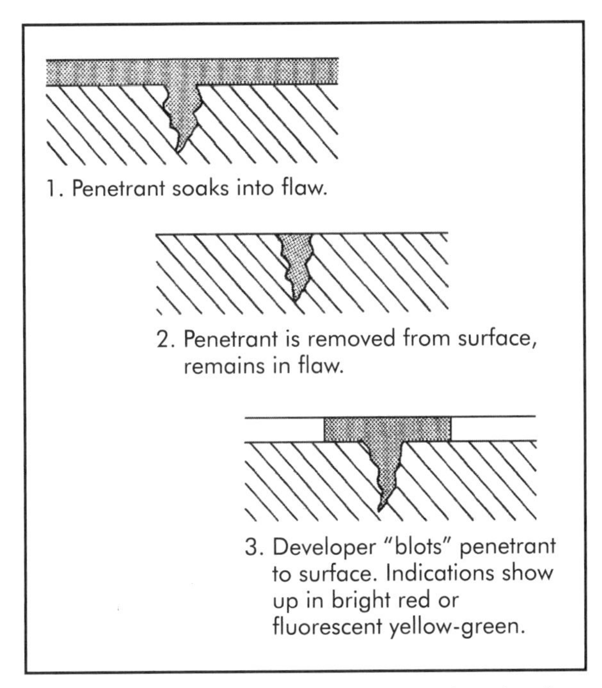

Figure 54-14. Liquid penetrant testing (Wick and Veilleux 1987).

nonporous materials. It is equally effective for testing nonmagnetic materials, magnetic materials, ceramics, and glass. Its portability is one of its primary advantages. In the field, an inspector may readily conduct a test using aerosol penetrants. Its main disadvantage stems from the fact that penetrants can be used to detect surface deformations only; they cannot detect flaws beneath the surface. The part also must be clean and dry because the penetrant cannot enter surface defects that are already filled with dirt, oil, grease, paint, water, or other contamination. The surface condition of the part also affects whether or not it can be inspected with liquid penetrants. For example, rough or porous surfaces produce heavy background indications as the penetrant locates the small cavities responsible for the roughness or porosity. When this occurs, it is difficult to locate small defects. Mechanical surface treatments, such as wire brushing, shot peening, and sanding, also create problems for liquid penetrant inspection techniques. These surface treatments tend to smear the metal surface and cover the cracks, thus preventing the penetrant from entering the cracks (Wick and Veilleux 1987).

MAGNETIC PARTICLE TESTING

Magnetic particle testing is performed by inducing a magnetic field in ferro-magnetic materials such as iron, steel, and nickel and cobalt alloys. The test piece is then dusted with iron particles. Imperfections distort the magnetic field in such a way that the iron particles are attracted and concentrated as shown in Figure 54-15. The concentrated particles indicate not only the location of the flaw but its size, shape, and extent as well.

Magnetic particle testing is used primarily for in-process and final product inspection of castings, forgings, rollings, welded pipe, seamless pipe, and extrusions. Its main advantage is that it can detect surface flaws as well as subsurface flaws. Other than the fact that non-ferro-magnetic materials (aluminum, magnesium, copper, etc.) cannot be tested, another disadvantage is that demagnetization is often required. The size and shape of the parts inspected by this method are almost unlimited. However, abrupt changes in dimensions can cause problems with indication interpretation. In general, elaborate precleaning is not necessary; cracks

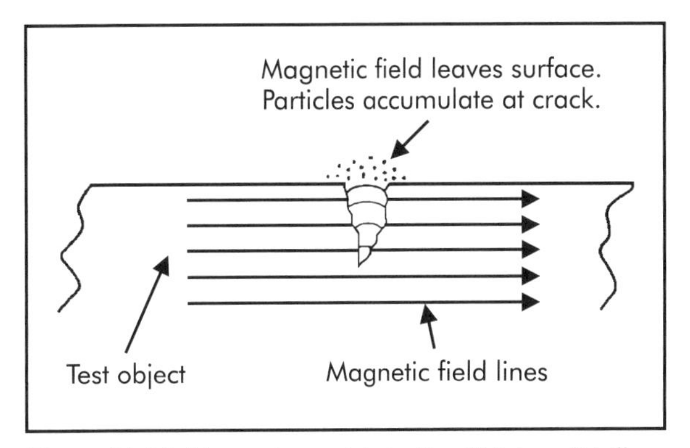

Figure 54-15. Magnetic particle testing (Wick and Veilleux 1987).

filled with nonmagnetic foreign materials can be detected (Wick and Veilleux 1987).

ULTRASONIC TESTING

With *ultrasonic testing*, beams of high-frequency sound waves are transmitted into a test piece to detect imperfections. The sound waves reflect from internal flaws and return to a receiver that analyzes their presence and location as illustrated in Figure 54-16. Materials to be tested by this method must be capable of transmitting vibrational energy, such as metals, ceramics, glass, and rubber. Ultrasonic testing is often used to inspect turbine forgings, generator rotors, pressure pipes, weldments, and nuclear reactor elements.

The major advantages of ultrasonic testing over other methods are its superior penetrating power and ability to detect extremely small defects. Ultrasonic testing is suitable for immediate interpretation, automation, rapid scanning, in-line production monitoring, and process control. The volumetric scanning ability permits inspection of a volume of material extending from the front surface to the back surface of a part.

Test pieces that are small, irregularly shaped, or rough are not recommended for ultrasonic testing. Discontinuities present in a shallow layer immediately beneath the surface may not be detectable. Couplant is needed to provide effective transfer of the ultrasonic beam between the transducer and part being tested (Wick and Veilleux 1987).

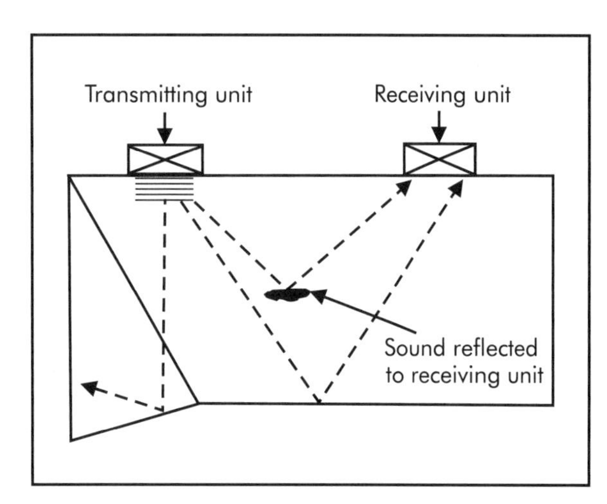

Figure 54-16. Ultrasonic testing (Wick and Veilleux 1987).

RADIOGRAPHIC TESTING

Radiographic testing involves the use of penetrating gamma- or x-radiation to examine material defects and internal flaws. Radiation is passed through a test piece and onto a film. The resulting image reveals any internal imperfections. Material thickness and density changes are indicated as lighter or darker areas on the film. Figure 54-17 illustrates the radiographic test.

The ability of x-rays to penetrate most materials is the reason radiographic testing is used extensively in industry and, in particular, on castings, weldments, solid propellants, and finished assemblies. The increasingly wide use of radiographic testing methods in the castings field is a result of the fact that most flaws and discontinuities inherent in ferrous and nonferrous

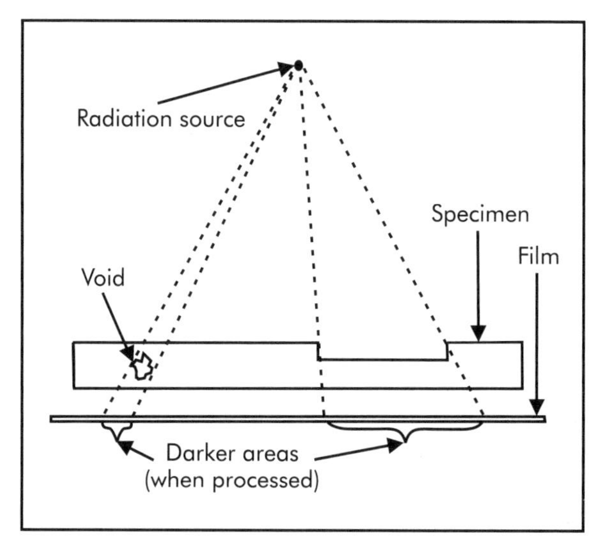

Figure 54-17. Radiographic testing (Wick and Veilleux 1987).

castings can be readily detected by this inspection medium.

The following welding imperfections or discontinuities are detectable by radiography: porosity, cracks, lack of or inadequate penetration and fusion, undercut, inclusions, and other discontinuities common in welded joints.

Interestingly, the size of the test piece (large or small) is no obstacle to radiographic testing. Test pieces of complex geometry, however, may prove to be a problem because of poor orientation of the radiation source. The major advantage of radiographic testing is its ability to provide a visual image of the test piece (Wick and Veilleux 1987).

EDDY-CURRENT TESTING

Eddy-current testing is accomplished by the occurrence of a varying magnetic field produced by a test coil, which creates small circulating currents in an electrically conductive material. Because the electrical signal of the test coil can be measured, any interruption in its magnetic field indicates a flaw in the test piece. Test piece imperfections are measured by a digital or analog meter as shown in Figure 54-18.

Eddy-current testing is used to sort for chemical and physical composition, measure conductivity, measure physical dimensions, and detect imperfections. A distinct advantage of eddy-current testing is that test results are available instantaneously; there is no waiting period for results to develop. The testing procedures are readily adaptable to go/no-go situations. Eddy-current testing is applicable for conductive materials only. The depth of penetration is restricted—approximately 0.5 in. (13 mm) in aluminum with standard probes. Testing of ferromagnetic metals can be difficult (Wick and Veilleux 1987).

LEAK TESTING

Leak testing is used to locate leaks in pressure containment parts, pressure vessels, and structures. Leaks can be detected by using bubble emission, tracer gas, electronic listening devices, pressure-change measurements, or a simple air-leak test. A leak test is typically used for inspecting engine blocks, water faucets, and sealed pressure vessels. Leak-test equipment is relatively inexpensive, simple to use, and compact (Wick and Veilleux 1987).

ACOUSTIC EMISSION TESTING

When a solid material is stressed, imperfections within the material emit sound energy, which can be detected by special receivers. *Acoustic emission testing* is typically used to detect cracking, fractures, deformations, leaks, fatigue, friction, stress, and vibration.

Various applications for acoustic emission testing include inspection of adhesive bonds, aircraft structures, bearings, ceramics, glass, concrete, plating, welds, and wood products. The main advantage of acoustic emission testing is its ability to test a complete structure in real time. The primary disadvantage is that it only detects flaws that are increas-

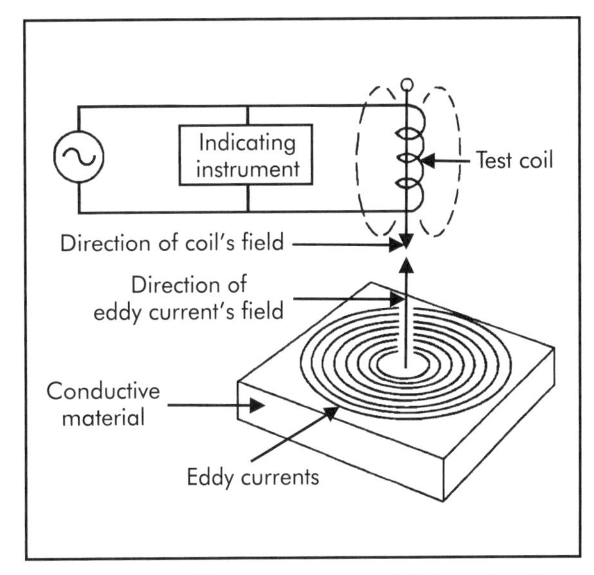

Figure 54-18. Eddy-current testing (Wick and Veilleux 1987).

ing in size; static imperfections are not uncovered (Wick and Veilleux 1987).

REVIEW QUESTIONS

54.1) Calculate the control limits for a *p* chart using the data in Table Q54-1 and the average subgroup size method.

54.2) Referring to the block diagram in Figure Q54-1, calculate the system's reliability. (The values in the boxes are the reliability factors of each component.)

54.3) Which type of gage error results in large measurement variations (on the same parts) due to differences between operators?

54.4) How many runs are necessary in a two-level DOE if four factors are being studied including all possible factor interactions?

54.5) In the signal-to-noise ratio, what does noise refer to?

54.6) What is the primary difference between ISO 9000 and QS-9000?

54.7) What title is given to a lesser trained and inexperienced six sigma practitioner?

54.8) Which NDT method uses high-frequency sound waves to detect imperfections?

Table Q54-1. Question 54.1

Subgroup	*n*	Number Defective	*p*
1	208	11	0.053
2	197	14	0.071
3	284	8	0.028
4	163	5	0.031
5	221	8	0.036
6	205	7	0.034
7	276	6	0.022
8	213	7	0.033
9	239	5	0.021
10	260	12	0.046

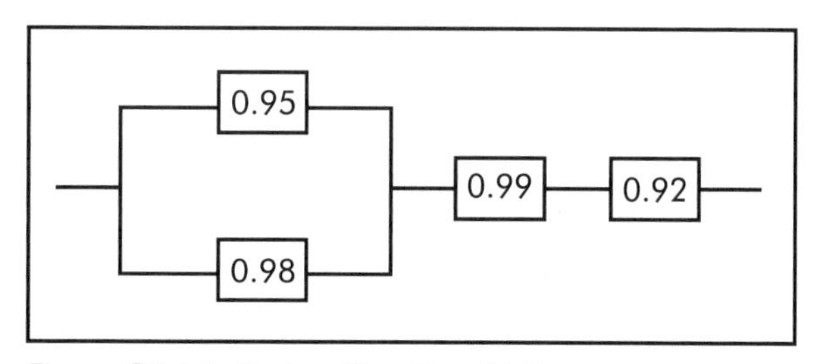

Figure Q54-1. Review Question 54.2.

REFERENCES

Barney, M. and McCarty, T. 2003. *The New Six Sigma*. Upper Saddle River, NJ: Prentice Hall.

The Black Belt Memory Jogger. 2002. Salem, NH: GOAL/QPC.

Brue, G. 2002. *Six Sigma for Managers*. New York: McGraw-Hill.

Harry, Mikel and Schroeder, Richard. 2000. *Six Sigma: the Breakthrough Management Strategy Revolutionizing the World's Top Corporations*. New York: Currency.

Maguire, Miles. 1999. "Cowboy Quality." *Quality Progress* 32, 10: 27-34.

Wick, Charles, and Veilleux, Raymond, eds. 1987. *Tool and Manufacturing Engineers Handbook*, Fourth Edition. Volume 4, *Quality Control and Assembly*. Dearborn, MI: Society of Manufacturing Engineers.

Chapter 55

Engineering Economics Analysis

55.1 DEPRECIATION METHODS

Depreciation is a decline in the value of property over the time it is used. Events that can cause property to depreciate include wear and tear, age, deterioration, and obsolescence. The cost of certain property, such as equipment used in a business or property used for the production of income, can be recovered by taking a tax deduction for depreciation.

Most types of tangible property (except land), such as buildings, machinery, vehicles, furniture, and equipment, can be depreciated. Certain intangible property (items that cannot be seen or touched), such as patents, copyrights, and computer software, either can be amortized or depreciated over time. Table 55-1 defines some common depreciation terms.

To be depreciable, the property must meet all of the following requirements (Internal Revenue Service 2003).

- It must be property you own.
- It must be used in your business or income-producing activity.
- It must have a determinable useful life.
- It must be expected to last more than one year.
- It must not be property that falls under an exception in the IRS code.

The following discussion will focus on various depreciation methods such as straight-line depreciation, sum-of-the-years' digits depreciation, declining-balance depre-

Table 55-1. Depreciation terminology (Internal Revenue Service 2003)

Term	Definition
Cost basis	Initial cost including purchase price, delivery, and installation.
Useful life	Anticipated useful life in years of an asset prior to disposal and/or replacement.
Salvage value	Estimated value of property at the end of its useful life.
Book value	The difference between the cost basis (initial cost) and the amount of depreciation charged to date.
Tangible property	Property that can be seen or touched, such as buildings, machinery, vehicles, furniture, and equipment.
Intangible property	Property that has value but cannot be seen or touched, such as good will, patents, copyrights, and computer software.
Recovery period	The number of years over which the cost basis of an item of property is recovered (not necessarily the same as the useful life).

ciation, the accelerated cost-recovery system (ACRS), and the modified accelerated cost-recovery system (MACRS). The depreciation method chosen for tax purposes is determined by the year the property or asset was placed in service, as defined in Table 55-2.

STRAIGHT-LINE DEPRECIATION

Straight-line depreciation is one of the simplest methods to apply. It assumes the decline in value is directly proportional to the age of the asset as illustrated in Figure 55-1. The annual depreciation expense, D_t, is a constant value divided by the depreciable life of the asset.

$$D_t = \frac{P - SV}{n} \qquad (55\text{-}1)$$

where:

D_t = annual depreciation expense for year t
P = cost basis
SV = salvage value
n = expected depreciable life or recovery period

If the book value of only one specific year, BV_t, is of concern, it can be calculated directly by:

$$BV_t = P - t \times D_t \qquad (55\text{-}2)$$

Table 55-2. Depreciation methods (Internal Revenue Service 1995, 2003)

Year Property was Placed in Service	Allowable Depreciation Method
Prior to 1981	Straight line Declining balance Sum-of-the-years' digits
After 1980 and before 1987	Accelerated cost-recovery system (ACRS)
After 1986 (or after July 31, 1986 if MACRS was elected)	Modified accelerated cost-recovery system (MACRS)

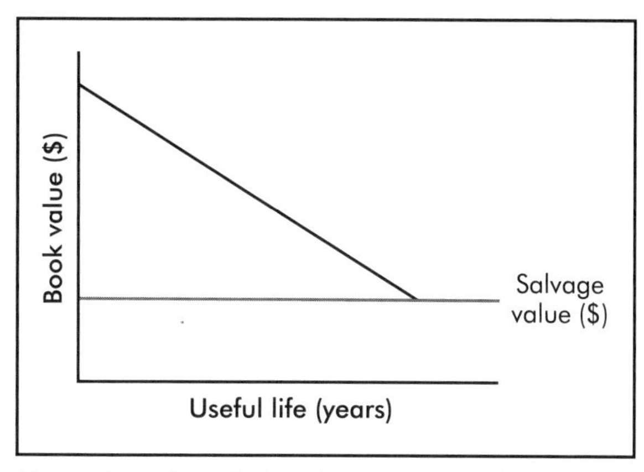

Figure 55-1. Straight-line depreciation method.

Example 55.1.1. The initial cost of a machine including shipping and installation is $40,000. The machine has a $10,000 salvage value and a 5-year useful life. Calculate the annual depreciation expense using the straight-line method.

Solution. The solution requires using Equation 55-1.

$$D_t = \frac{P - SV}{n} \qquad \text{(Eq. 55-1)}$$

where:

D_t = depreciation expense for year t
P = cost basis = $40,000
SV = salvage value = $10,000
n = expected depreciable life = 5 years

So:

$$D_t = \frac{40{,}000 - 10{,}000}{5} = \$6{,}000$$

Table 55-3 lists the annual depreciation expenses and the end-of-year book values.

Example 55.1.2. Using Example 55.1.1, what is the book value of the asset after the third year of depreciation?

Solution. The solution requires using Equation 55-2 and can be verified by Table 55-3.

$$BV_t = P - t \times D_t \qquad \text{(Eq. 55-2)}$$

Table 55-3. Examples 55.1.1 and 55.1.2

Year, t	Depreciation Expense, D_t	End-of-year Book Value, BV_t
0	$0	$40,000
1	$6,000	$34,000
2	$6,000	$28,000
3	$6,000	$22,000
4	$6,000	$16,000
5	$6,000	$10,000

where:

BV_t = book value in year 3 = BV_3
P = cost basis = $40,000
t = year = 3
D_t = depreciation expense for year 3 = $6,000

So:

$$BV_3 = 40,000 - (3 \times 6,000) = \$22,000$$

SUM-OF-THE-YEARS' DIGITS

Sum-of-the-years' digits depreciation is one of several methods used to realize depreciation at an accelerated rate. Depreciation expenses are larger in the early years and smaller in the later years as compared to straight-line depreciation, which uses a constant yearly rate.

With the sum-of-the-years' digits method, the annual depreciation expense is found by:

$$D_t = (P - SV)\left(\frac{n - t + 1}{s}\right) \quad (55\text{-}3)$$

where:

D_t = depreciation expense for year t
P = cost basis
SV = salvage value
n = expected depreciable life or recovery period
t = year
s = sum-of-the-years' digits

The sum-of-the-years' digits, s, can be expressed as:

$$s = 1 + 2 + \ldots + n = \frac{n(n+1)}{2} \quad (55\text{-}4)$$

The book value, BV_t, can be calculated directly by:

$$BV_t = P - \left(\frac{t\left(n - \frac{t}{2} + 0.5\right)(P - SV)}{s}\right) \quad (55\text{-}5)$$

Example 55.1.3. Suppose a machine has a cost basis of $40,000 and a salvage value of $10,000. Determine the annual depreciation expense and end-of-year book value during its 5-year useful life using sum-of-the-years' digits method.

Solution. The solution requires using Equation 55-3.

$$D_t = (P - SV)\left(\frac{n - t + 1}{s}\right) \quad (\text{Eq. } 55\text{-}3)$$

where:

D_t = depreciation expense for year t
P = cost basis = $40,000
SV = salvage value = $10,000
n = expected depreciable life or recovery period = 5 years
t = year
s = sum-of-the-years' digits = 1 + 2 + 3 + 4 + 5 = 15

Depreciation expense for year one ($t = 1$):

$$D_1 = (40,000 - 10,000)\left(\frac{5 - 1 + 1}{15}\right) = \$10,000$$

Book value at the end of year one:

$$BV_1 = P - D_1 = 40,000 - 10,000 = \$30,000$$

Depreciation expense for year two ($t = 2$):

$$D_2 = (40,000 - 10,000)\left(\frac{5 - 2 + 1}{15}\right) = \$8,000$$

Book value at the end of year two:

$$BV_2 = BV_1 - D_2 = 30{,}000 - 8{,}000 = \$22{,}000$$

Depreciation expense for year three ($t = 3$):

$$D_3 = (40{,}000 - 10{,}000)\left(\frac{5-3+1}{15}\right) = \$6{,}000$$

Book value at the end of year three:

$$BV_3 = BV_2 - D_3 = 22{,}000 - 6{,}000 = \$16{,}000$$

Depreciation expense for year four ($t = 4$):

$$D_4 = (40{,}000 - 10{,}000)\left(\frac{5-4+1}{15}\right) = \$4{,}000$$

Book value at the end of year four:

$$BV_4 = BV_3 - D_4 = 16{,}000 - 4{,}000 = \$12{,}000$$

Depreciation expense for year five ($t = 5$):

$$D_5 = (40{,}000 - 10{,}000)\left(\frac{5-5+1}{15}\right) = \$2{,}000$$

Book value at the end of year five:

$$BV_5 = BV_4 - D_5 = 12{,}000 - 2{,}000 = \$10{,}000$$

Table 55-4 summarizes the results.

Table 55-4. Example 55.1.3

Year, t	Depreciation Expense, D_t	End-of-year Book Value, BV_t
0	\$0	\$40,000
1	\$10,000	\$30,000
2	\$8,000	\$22,000
3	\$6,000	\$16,000
4	\$4,000	\$12,000
5	\$2,000	\$10,000

Example 55.1.4. Using Example 55.1.3, calculate the book value of the asset at the end of year three.

Solution. The solution requires the use of Equation 55-5.

$$BV_t = P - \left(\frac{t\left(n - \frac{t}{2} + 0.5\right)(P - SV)}{s}\right) \qquad \text{(Eq. 55-5)}$$

where:

$BV_t = BV_3$ = book value at the end of year three
P = cost basis = \$40,000
t = year = 3
n = expected depreciable life or recovery period = 5 years
SV = salvage value = \$10,000

$$BV_3 = 40{,}000 - \left(\frac{3\left(5 - \frac{3}{2} + 0.5\right)(40{,}000 - 10{,}000)}{15}\right)$$

$$BV_3 = \$16{,}000$$

DECLINING BALANCE DEPRECIATION

Declining balance depreciation is another accelerated depreciation method. A uniform rate of depreciation is applied to the beginning of the year book values of the asset over its useful life. The depreciation rate is a fixed percentage determined by the type of asset and the year it was purchased.

In any given year, the annual depreciation expense is defined as:

$$D_t = \left(\frac{r}{n}\right) BV_{t-1} \qquad (55\text{-}6)$$

where:

D_t = depreciation expense for year t
r = rate = 1.5, 1.75, and 2, which correspond to 150%, 175%, and 200%, respectively

n = expected depreciable life or recovery period
BV_t = book value of the asset at the end of year t

Since 200% is twice the straight-line rate, this method is also referred to as *double-declining balance depreciation*.

Example 55.1.5. Suppose a machine has an initial cost of $40,000 and a salvage value of $10,000. Determine the annual depreciation expense and end-of-year book value during its 5-year useful life using the double-declining balance depreciation method.

Solution. The solution requires using Equation 55-6.

$$D_t = \left(\frac{r}{n}\right) BV_{t-1} \qquad \text{(Eq. 55-6)}$$

where:

D_t = annual depreciation expense for year t
r = rate = 2
n = expected depreciable life or recovery period = 5 years
BV_t = book value at the end of year t

Depreciation expense for year one:

$$D_1 = \left(\frac{2}{5}\right)(40{,}000) = \$16{,}000$$

Book value at the end of year one:

$$BV_1 = P - D_1 = 40{,}000 - 16{,}000 = \$24{,}000$$

Depreciation expense for year two:

$$D_2 = \left(\frac{2}{5}\right)(24{,}000) = \$9{,}600$$

Book value at the end of year two:

$$BV_2 = BV_1 - D_2 = 24{,}000 - 9{,}600 = \$14{,}400$$

Depreciation expense for year three:

D_3 = 4,400 (Book value is never allowed to go below the salvage value.)

Book value at the end of year three:

$$BV_3 = BV_2 - D_3 = 14{,}400 - 4{,}400 = \$10{,}000$$

Depreciation expense for years four and five is 0.

Table 55-5 summarizes the results.

ACCELERATED COST-RECOVERY SYSTEM (ACRS)

The Economic Recovery Tax Act of 1981 initiated the *accelerated cost-recovery system* (ACRS) depreciation method. This method includes a recovery period that is shorter than the useful life of the investment. The shorter period for write-off was intended to stimulate the economy by encouraging capital investment.

ACRS applies to most depreciable tangible property placed in service after 1980 and before 1987. It includes new or used and real or personal property. The property must be used in a trade or business or for the production of income. With ACRS, a salvage value does not exist therefore the property or asset can be depreciated down to zero.

With ACRS all assets are grouped into 3-, 5-, 10-, 15-, 18-, and 19-year classes. The 15-, 18-, and 19-year classes include real property, such as buildings. The 3-, 5-, and 10-year classes include personal property such as vehicles, computers, equipment, etc., as illustrated in Table 55-6. Table 55-7 defines the depreciation rates for various property classes.

Table 55-5. Example 55.1.5

Year, t	Depreciation Expense, D_t	End-of-year Book Value, BV_t
0	$0	$40,000
1	$16,000	$24,000
2	$9,600	$14,400
3	$4,400	$10,000
4	$0	$10,000
5	$0	$10,000

Table 55-6. ACRS classes for property placed in service after 1980 and before 1987 (Internal Revenue Service 1995)

Personal Property	Class
Autos and light trucks, tractor units for over-the-road use, equipment used for research and development, and special tools with a life <4 years	3 year
Most other machinery and equipment, office furniture and equipment, heavy-duty trucks, computers, and copiers	5 year
Certain public utility property, railroad tank cars, and manufactured homes	10 year

Table 55-7. ACRS depreciation rates for property placed in service after 1980 and before 1987 (Internal Revenue Service 1995)

Recovery Year	3-year Property	5-year Property	10-year Property
1	25%	15%	8%
2	38%	22%	14%
3	37%	21%	12%
4		21%	10%
5		21%	10%
6			10%
7			9%
8			9%
9			9%
10			9%
11			
12			
13			
14			
15			

Example 55.1.6. A production machine is purchased and installed for $10,000. Using the ACRS method, determine the annual depreciation expenses and the end-of-year book values for the machine's useful life.

Solution. From Table 55-6 the machine is categorized in the 5-year class. The deprecation rates are found in Table 55-7 and the amount to be depreciated is the entire $10,000.

Depreciation expense for year one:

$$D_1 = 0.15 \times 10{,}000 = \$1{,}500$$

Book value at the end of year one:

$$BV_1 = P - D_1 = 10{,}000 - 1{,}500 = \$8{,}500$$

Depreciation expense for year two:

$$D_2 = 0.22 \times 10{,}000 = \$2{,}200$$

Book value at the end of year two:

$$BV_2 = BV_1 - D_2 = 8{,}500 - 2{,}200 = \$6{,}300$$

Depreciation expense for year three:

$$D_3 = 0.21 \times 10{,}000 = \$2{,}100$$

Book value at the end of year three:

$$BV_3 = BV_2 - D_3 = 6{,}300 - 2{,}100 = \$4{,}200$$

Depreciation expense for year four:

$$D_4 = 0.21 \times 10{,}000 = \$2{,}100$$

Book value at the end of year four:

$$BV_4 = BV_3 - D_4 = 4{,}200 - 2{,}100 = \$2{,}100$$

Depreciation expense for year five:

$$D_5 = 0.21 \times 10{,}000 = \$2{,}100$$

Book value at the end of year five:

$$BV_5 = BV_4 - D_5 = 2{,}100 - 2{,}100 = 0$$

Table 55-8 summarizes the results.

Table 55-8. Example 55.1.6

Year, t	Depreciation Expense, D_t	End-of-year Book Value, BV_t
0	$0	$10,000
1	$1,500	$8,500
2	$2,200	$6,300
3	$2,100	$4,200
4	$2,100	$2,100
5	$2,100	$0

MODIFIED ACCELERATED COST-RECOVERY SYSTEM (MACRS)

The modified accelerated cost-recovery system (MACRS) depreciation method was created by the Tax Reform Act of 1986. It is the mandatory method of depreciation used for tangible property placed in service after December 31, 1986. The primary differences between ACRS and MACRS are the asset class lives and the depreciation rates.

Under the MACRS method, property is grouped into 3-, 5-, 7-, 10-, 15-, 20-, 25-, 27.5- and 39-year classes. The 27.5-year class includes residential property; the 39-year class includes commercial property. The remaining classes cover personal property. Various property classes are defined in Table 55-9. Table 55-10 defines the depreciation rates for various property classes. It is based on the half-year convention, which treats all property placed in service or disposed of during a tax year as placed in service or disposed of at the midpoint of the year. So, for the year the property is placed in service or disposed of, one-half year of depreciation is allowed. For example, under MACRS the recovery period for a 5-year property is 5 years. However, the depreciation deductions are spread over six tax years (Internal Revenue Service 2003).

Example 55.1.7. A plastic injection molder is purchased and installed for $50,000 and placed in service in 1995. What is the depreciation expense in year three and

Table 55-9. MACRS property classes for property placed in service after 1986 (Internal Revenue Service 2003)

Personal Property	Class
Tractor units for over-the-road use; special tools used in the manufacture of rubber, glass, metals, boats, motor vehicles, and food and beverage products (jigs, dies, fixtures, etc.)	3 year
Computers; data handling equipment; computer-based central office switching equipment; automobiles; buses; light- and heavy-duty trucks; construction assets; equipment for manufacture of knitted goods and carpet, chemicals, metals, electronic components, and semiconductors; property used in research and experimentation	5 year
Office furniture and equipment; railroad cars and locomotives; agricultural machinery; mining equipment; equipment for manufacture of tobacco products, wood and printing products, plastic and glass products, stone and clay products, leather products, foundry and steel mill products, metal products, electrical and non-electrical machinery, motor vehicle and aerospace products, and railroad equipment; includes property not assigned to a class life	7 year
Vessels; barges and tugs; electrical power distribution systems; equipment for manufacture of grain, sugar, and vegetable-oil products	10 year
Equipment for the manufacture of cement, municipal wastewater treatment plant, telephone distribution plant	15 year

Table 55-10. MACRS depreciation rates (half-year convention) for property placed in service after 1986 (Internal Revenue Service 2003)

Recovery Year	3-year Property Rate	5-year Property Rate	7-year Property Rate	10-year Property Rate	15-year Property Rate
1	33.33%	20.00%	14.29%	10.00%	5.00%
2	44.45%	32.00%	24.49%	18.00%	9.50%
3	14.81%	19.20%	17.49%	14.40%	8.55%
4	7.41%	11.52%	12.49%	11.52%	7.70%
5		11.52%	8.93%	9.22%	6.93%
6		5.76%	8.92%	7.37%	6.23%
7			8.93%	6.55%	5.90%
8			4.46%	6.55%	5.90%
9				6.56%	5.91%
10				6.55%	5.90%
11				3.28%	5.91%
12					5.90%
13					5.91%
14					5.90%
15					5.91%
16					2.95%
17					
18					
19					
20					
21					

the asset's end-of-year book value at the end of the third year?

Solution. From Table 55-9, the machine is categorized in the 7-year class. The depreciation rates are found in Table 55-10 and the amount to be depreciated is $50,000.

From Table 55-10, the depreciation rate for a 7-year class property in year three is 17.49%. Therefore,

$$D_3 = 0.1749 \times 50{,}000 = \$8{,}745$$

The book value at the end of year three is equal to the cumulative depreciation up to and including year three.

From Table 55-10,

$$D_{1\text{-}3} = (0.1429 + 0.2449 + 0.1749) \times 50{,}000 = \$28{,}135$$

$$BV_3 = 50{,}000 - 28{,}135 = \$21{,}865$$

55.2 INVESTMENT ANALYSIS

PAYBACK PERIOD

Determining the *payback period* involves finding the number of periods (weeks, months, or years) required to recover the cost of capital, ignoring the time value of money. The payback period is among the simplest measures that can be used to evaluate the merit of investment projects. It can be calculated by:

$$\sum R = \sum C \quad (55\text{-}7)$$

where:

R = revenues
C = costs

Revenue can be generated from a number of sources, such as product or unit sales and salvage values. Revenue generated by product or unit sales is defined by:

$$R = r \times Q \times n \quad (55\text{-}8)$$

where:

R = sales revenue
r = revenue per unit
Q = units sold per period
n = number of periods (weeks, months, or years)

Cost is comprised of several components such as initial cost, fixed cost, and variable cost. First costs typically consist of equipment purchases and installation. Fixed costs, such as setup costs, are independent of the quantity produced. Variable costs, such as direct labor and material, depend upon the quantity produced. The total variable cost can be determined by:

$$V = v \times Q \times n \quad (55\text{-}9)$$

where:

V = total variable cost
v = unit variable cost
Q = units produced per period
n = number of periods (weeks, months, years)

Example 55.2.1. An automatic production machine can be purchased for $100,000. The machine produces approximately 5,000 units per year with approximately 2,000 hours of associated labor at $15 per hour. Each unit produced generates $10 in revenue. Find the payback period.

Solution. Starting with Equation 55-7,

$$\sum R = \sum C \quad \text{(Eq. 55-7)}$$

The revenues, R, are

$$R = r \times Q \times n$$

$$= \frac{\$10}{\text{unit}} \times \frac{5{,}000 \text{ units}}{\text{year}} \times n \text{ years}$$

$$= \$50{,}000n \quad \text{(Eq. 55-8)}$$

The costs, C, are

$$P = \text{purchase price} = \$100{,}000$$

$$v = \frac{2{,}000 \text{ hr}}{5{,}000 \text{ units}} \times \frac{\$15}{\text{hr}} = \$6/\text{unit}$$

$$V = v \times Q \times n$$

$$= \frac{\$6}{\text{unit}} \times \frac{5{,}000 \text{ units}}{\text{year}} \times n \text{ years}$$

$$= \$30{,}000n \quad \text{(Eq. 55-9)}$$

Inserting the information back into Equation 55-7 yields

$$50{,}000n = 100{,}000 + 30{,}000n$$

$$20{,}000n = 100{,}000$$

$$n = \frac{100{,}000}{20{,}000} = 5 \text{ years}$$

Example 55.2.2. A toy manufacturer uses approximately 80% of its capacity to break even each month. The fixed costs are $60,000 per month and the variable costs are $2.00 per unit. The revenue per unit is $8.00. What is the breakeven point (quantity Q)?

Solution. Starting with Equation 55-7,

$$\sum R = \sum C \quad \text{(Eq. 55-7)}$$

The revenues, R, are

$$R = r \times Q \times n$$

$$= \frac{\$8}{\text{unit}} \times Q \frac{\text{units}}{\text{month}} \times 1 \text{ month} = 8Q \quad \text{(Eq. 55-8)}$$

The costs, C, are

F = fixed costs = \$60,000

$$V = v \times Q \times n$$

$$= \frac{\$2}{\text{unit}} \times Q \times 1 \text{ month} = 2Q \quad \text{(Eq. 55-9)}$$

Inserting the values into Equation 55-7 yields

$$8Q = 60{,}000 + 2Q$$
$$6Q = 60{,}000$$
$$Q = 10{,}000 \text{ units per month}$$

Example 55.2.3. Using example 55.2.2, if production decreases to 8,000 units per month, what would the unit's variable cost (v) need to be to break even assuming the revenue per unit was \$8?

Solution. Starting with Equation 55-7,

$$\sum R = \sum C \quad \text{(Eq. 55-7)}$$

The revenues, R, are

$$R = r \times Q \times n$$

$$= \frac{\$8}{\text{unit}} \times 8{,}000 \frac{\text{units}}{\text{month}} \times 1 \text{ month}$$

$$= \$64{,}000 \quad \text{(Eq. 55-8)}$$

The costs, C, are

F = fixed cost = \$60,000

$$V = v \times Q \times n$$

$$= v \times 8{,}000 \frac{\text{units}}{\text{month}} \times 1 \text{ month}$$

$$= \$8{,}000v$$

Inserting the values into Equation 55-7 yields

$$64{,}000 = 60{,}000 + 8{,}000v$$
$$8{,}000v = 4{,}000$$
$$v = \$0.50 \text{ per unit}$$

Although the payback period method of analysis remains in use and is a relatively simple calculation, it has two severe shortcomings as a criterion on which to base investments. First, it ignores cash flows in the years following the payback period and, second, it ignores the time value of money.

MINIMUM ATTRACTIVE RATE OF RETURN (MARR)

Another technique for evaluating investment alternatives is the *minimum attractive rate of return* (MARR). When evaluating an investment, most companies want to recover more than just the capital invested. Most companies desire a minimum rate of return before risking money on a capital investment. This minimum rate of return is referred to as MARR.

INTERNAL AND EXTERNAL RATE OF RETURN

The *internal rate of return* (IRR) method determines the interest rate that will equate the present worth of a project's revenues with the present worth of the project's costs. IRR also can be described as the interest rate that makes the net present worth of a project zero. If a project's IRR is less than the MARR, a company may decline investing in the project based on the less than minimum attractive rate of return.

The internal rate of return method may not accurately reflect a project's true rate of return. It assumes that cash proceeds from a project can be reinvested at the IRR. If the IRR is high, such as 40%, it is unrealistic to expect proceeds will be reinvested at that rate. In response to this situation, the *external rate of return* (ERR) method can be used.

The ERR method uses a reasonable interest rate at which a project's proceeds or revenue can be reinvested. Generally, the ERR will vary slightly from the IRR. The magnitude of the variation depends on the magnitude of the difference between the IRR and the reasonable rate at which proceeds can be reinvested. If the IRR is reasonably close to the MARR, the ERR method or another rate of return method would be more appropriate.

REVIEW QUESTIONS

55.1) A company purchases a new production machine for $100,000. It will have a salvage value of $20,000 in 10 years. Calculate the book value at the end of the third year using the following methods:

- **a)** straight-line depreciation,
- **b)** sum-of-the-years' digits depreciation,
- **c)** double-declining depreciation, and the
- **d)** accelerated cost-recovery system (ACRS).

55.2) A gear manufacturer can produce 2,000 gears per month at full capacity. The variable cost is $10 per gear; overhead cost is $21,000 per month; and labor cost is $5 per gear. If the gears sell for $30 each, what is the minimum percent utilization the plant must use to break even?

REFERENCES

Internal Revenue Service. 2003. Publication 936. Washington, DC: U.S. Government Printing Office.

Internal Revenue Service. 2003. Publication 946. Washington, DC: U.S. Government Printing Office.

Internal Revenue Service. 1995. Publication 534. Washington, DC: U.S. Government Printing Office.

Chapter 56

Management Theory and Practice

56.1 ORGANIZATIONAL STRUCTURES AND STRATEGIES

Organizing is the process of grouping jobs together based on certain criteria such as job similarity or product family. The criteria used for organizing dictates the organizational structure. Organizational structures provide a framework within the organization for information flow. Some structures may enhance information flow while others inhibit it. Information flow is one factor that can determine a company's success or failure. Organizations with an appropriate structure run smoothly. Those organizations without an appropriate structure struggle in most aspects of company business. Organizations must be dynamic in nature to enable restructuring should environmental conditions dictate.

In an organization, employees generally have a combination of authority, responsibility, and accountability. Many employees often have a large amount of responsibility and accountability without much authority.

- *Authority* is power granted to individuals so they can make final decisions for others to follow.
- *Responsibility* is the obligation incurred by individuals to effectively perform assignments in their roles in the formal organization.
- *Accountability* is the state of being totally answerable for the satisfactory completion of a specific assignment.

The discussion in the following sections focuses on various organizational structures and their respective advantages and disadvantages.

LINE-AND-STAFF STRUCTURE

As discussed in Chapter 45 of *Fundamentals of Manufacturing*, Second Edition, the traditional line-and-staff structure divides the organization into functional groups or departments based on job similarity. Each group or department has a supervisor or department head who, in turn, reports to another supervisor as illustrated in Figure 56-1.

The line-and-staff structure offers several advantages, such as easier budgeting and control, and well-established communication channels. The disadvantages, however, are that decisions favor the strongest functional group and coordination can become complex. The strongest functional group in terms of number of people or budget size will probably have the largest influence on company decisions since that group has the most people or the most money or influence with money. With the departmentalized structure it is also difficult for functional groups or departments to communicate since the information flows vertically within the group as opposed to horizontally across departments. Thus the line-and-staff structure has a single communication channel and a single reporting system.

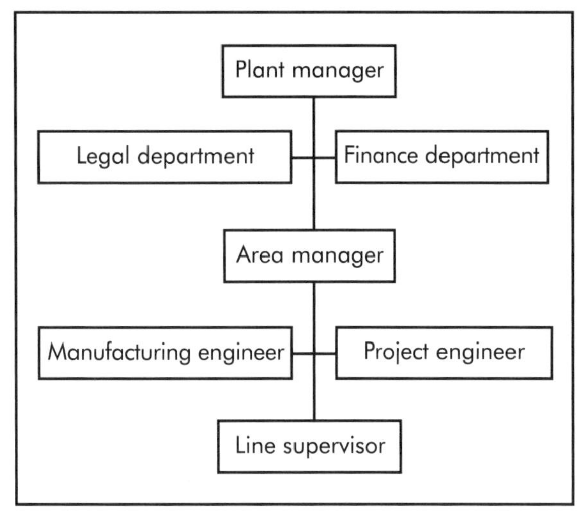

Figure 56-1. Line-and-staff organizational chart.

PRODUCT STRUCTURE

A product structure groups functions by product, such that each product line has its own unique support staff dedicated to that product only, as illustrated in Figure 56-2.

Each product line performs its own research, product development, engineering, and product planning. The product structure provides easier interaction between functions since they are in the same group. It also enables faster market and customer response. The managers who coordinate multifunctional subordinates have broader skills as opposed to most line-and-staff managers. The product structure provides strong communication channels and complete line authority over the project participants who work directly for the project manager. Thus the product structure has a single communication channel and a single reporting system.

There are drawbacks to the product structure, however. They include the cost of duplicate efforts since each product has its own engineering area. Personnel are retained on projects long after they are needed. Also, there is lack of opportunity for technical interchange between departments or groups.

MATRIX STRUCTURE

The matrix organizational structure is a combination of the line-and-staff structure and the product structure, as illustrated in Figure 56-3. It recognizes and capitalizes on the cross-functionality of an organization. In most organizations, people need to interact with various functions, such as purchasing, accounting, and quality.

The primary disadvantages of the matrix structure are its multidimensional information flow and dual reporting requirements. Multidimensional information flow can be an advantage; however, being able to achieve effective and efficient communication in many directions can be difficult. Also, report-

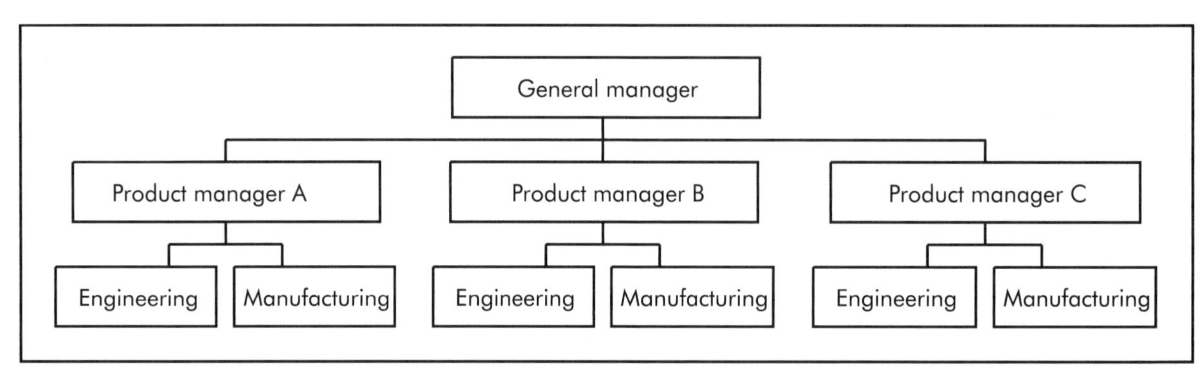

Figure 56-2. Product organizational structure.

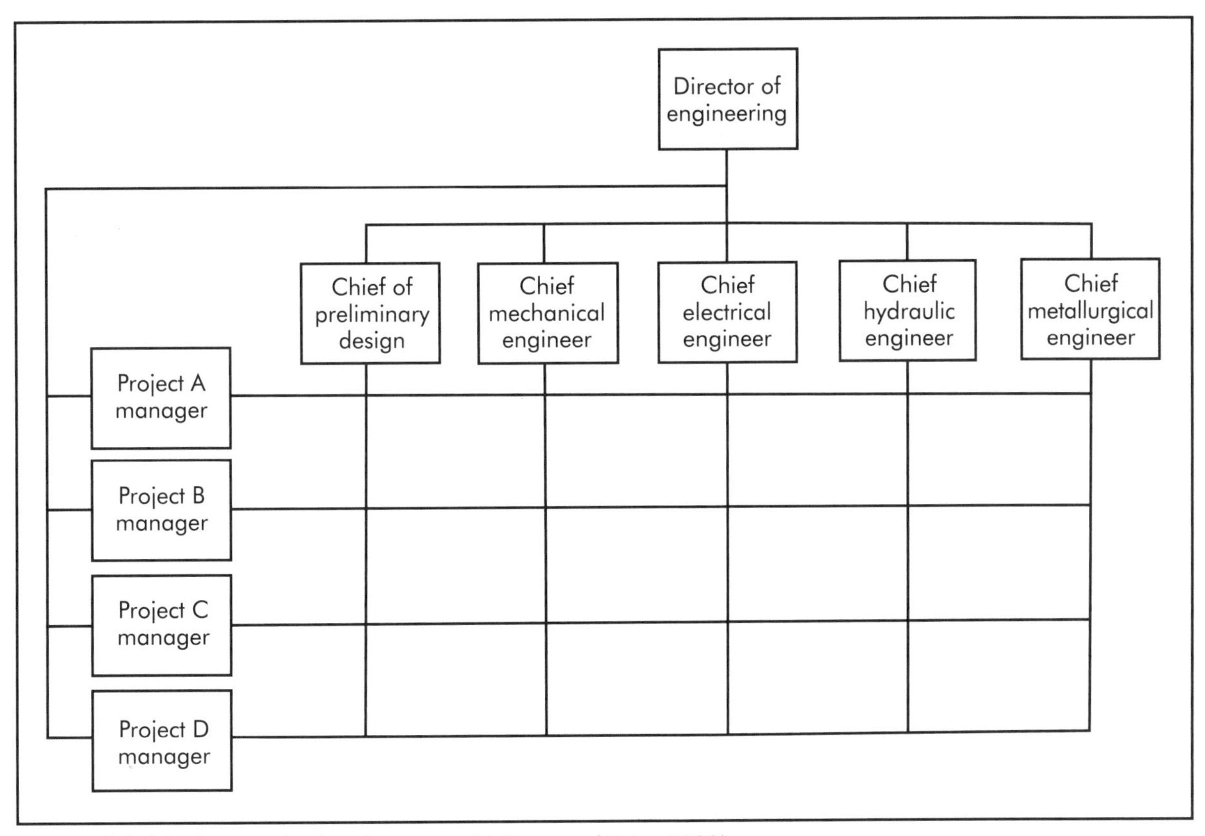

Figure 56-3. Matrix organizational structure (Veilleux and Petro 1988).

ing to two managers or team leaders can become cumbersome in some organizations.

56.2 SUPERVISION

Supervision, as discussed in Chapter 45 of *Fundamentals of Manufacturing*, Second Edition, involves the motivation and guidance of subordinates toward goals and targets. Typically, supervisors work closely with subordinates on a daily or weekly basis. Supervisors are immediately responsible for the actions of their subordinates.

There are two traditional theories of supervision know as Theory X and Theory Y. Supervisors typically prescribe to one or the other, or a combination of the two. *Theory X* assumes the average worker dislikes work and avoids it whenever possible. To induce adequate effort, the supervisor must threaten punishment and exercise careful supervision. It is assumed the average worker avoids increased responsibility and seeks to be directed. Theory X managers normally exercise authoritarian-type control and allow little participation in decision-making.

Theory Y assumes the average worker wants to be active and finds the physical and mental effort on the job satisfying. The greatest results come from willing participation, which tends to produce self-direction toward goals without coercion and control. The average worker seeks opportunity for personal improvement. Theory Y managers normally advocate participation and the management-employee relationship.

In addition to the traditional theories, there is *Theory Z*, which is a combination of

Japanese and American management styles. It emphasizes long-term employment and concern for employees themselves, not just for their work performance. Decision-making is done collectively based on input from all functional areas. Employees may not agree with every decision, but their viewpoint is always heard and considered. Theory Z encourages the bottom-up process wherein initiative, change, and problem-solving are accomplished best by those who are closest to the problems. Middle managers act as initiators and coordinate with other managers to present solutions to senior management who facilitate the process.

56.3 LEADERSHIP

Leadership is the ability to influence others toward the achievement of goals. The source of this influence may be the formal role held by the leader, such as the role of manager. However, not all leaders are managers, nor are all managers leaders. In fact, a formal position of leadership is only one of many sources of a leader's influence. The ability to use that influence goes beyond the formal organization. In addition, much of this influence depends on the perception of the followers. In fact, some would say that good leadership is a function of good (or at least willing) followers.

The following sections discuss the various leadership theories.

TRAIT THEORY

Many years ago it was thought that to be a good leader a person must possess a specific set of traits or characteristics. There is the idea that political and industrial leaders possess charismatic qualities that assure their success. These qualities certainly assure admiration and often support the willingness to follow. Leaders were thought to have these traits, which separated them from others, also known as followers. The trait theory suggested that for "natural born leaders" the traits were innate, and it would be impossible for someone to learn them and become a good leader.

Research on leadership has indicated there are no definitive predictors of leadership success. However, there are a few qualities that correlate with an effective leader. These characteristics are intelligence, dominance, self-confidence, a high energy level, and knowledge relevant to the task (Veilleux and Petro 1988).

LEADER BEHAVIOR THEORY

Given the lack of meaningful conclusions about which traits assure good leadership, researchers shifted from what good leaders are to what good leaders do to be successful. In other words, it should be possible to identify the behavior that effective leaders exhibit and, from this information, make decisions about the development of leadership skills of those in the designated roles. The emphasis shifted from finding the right people to developing a large supply of leaders through a well-ordered management training program.

Studies conducted just after World War II showed that leadership behaviors seemed to fall into one of two broad categories: job-oriented (concern for production) and employee-oriented (concern for people).

A job-centered leader practices close supervision so that subordinates may know specifically what is expected of them. The leader organizes the work, defines the relationships between group members, and tends to establish well-defined communication patterns. The job-oriented approach leads to higher productivity for the group and high ratings from superiors. The human element is not viewed in a negative light by the job-oriented leader, but rather is seen as a luxury.

The employee-centered set of behaviors includes delegating decision making, thereby creating a supportive environment that will permit subordinates to achieve and grow. Employee-centered behavior encourages the development of mutual trust and respect between the leader and his or her subordinates. These people-oriented approaches lead to more cohesive groups, greater job satisfaction, lower absenteeism, and higher group productivity. Productivity is not ignored by the leader, but rather is treated as an aspect that will take care of itself after the people concerns are addressed.

Both orientations are essential for good leadership; however, the leader must determine the appropriate ratio for specific situations. For example, if a business requires jobs that are unpleasant, a good leader may increase his or her employee-centered orientation and focus less on the job, thereby increasing employee satisfaction. Conversely, if a business lacks good housekeeping and standard work practices, employee satisfaction may increase if the leader is more job oriented rather than employee oriented. As situations change, the leader must adjust his or her orientation as well (Veilleux and Petro 1988).

CONTINGENCY THEORY

The contingency theory suggests that effective leadership occurs when there is a good match between the leader's style of interacting with subordinates and the degree to which the situation provides influence and control for the leader. Either job orientation or employee orientation is appropriate for the leader given the situation. The situational circumstances are defined by the following variables:

- the extent of the power the leader possesses due to the role assigned by the organization;
- the degree of structure of the job or task to be performed; and
- the interpersonal and psychological relationship between the leader and other employees or subordinates.

It is important to measure which approach the leader may have more of a tendency toward, job- or employee-oriented.

The *power variable* is defined as the degree of influence the leader possesses over such things as hiring, firing, discipline, pay, and promotions. The leadership position's power is expressed as weak or strong. The second variable, referred to as *task structure*, defines how routine and well defined the duties of an individual are in a specific job. Task structure is expressed as high or low. The third variable is *leader-member relations*, which is concerned with the degrees of trust, confidence, and respect subordinates have for their leader. This also is expressed as a two-sided item, either good or poor. If leader-member relations are good, the job is highly structured, and the leader possesses great position power, the situation is said to be very favorable. If the three variables are opposite, then the situation is very unfavorable. With other combinations of the variables, the situations fall somewhere between most favorable and least favorable. Figure 56-4 illustrates the combinations of variables and situations.

Given only the choice of either the job-oriented or employee-oriented style, which is appropriate in each situation? In a very favorable or very unfavorable situation, a job-oriented leader is most effective. In the moderate situation, a relationship-oriented style of leadership is likely to be more effective.

Since altering leadership styles is difficult, a good match can be obtained by altering the situation. Although not always possible, the situation can be modified by changes in the three variables: position

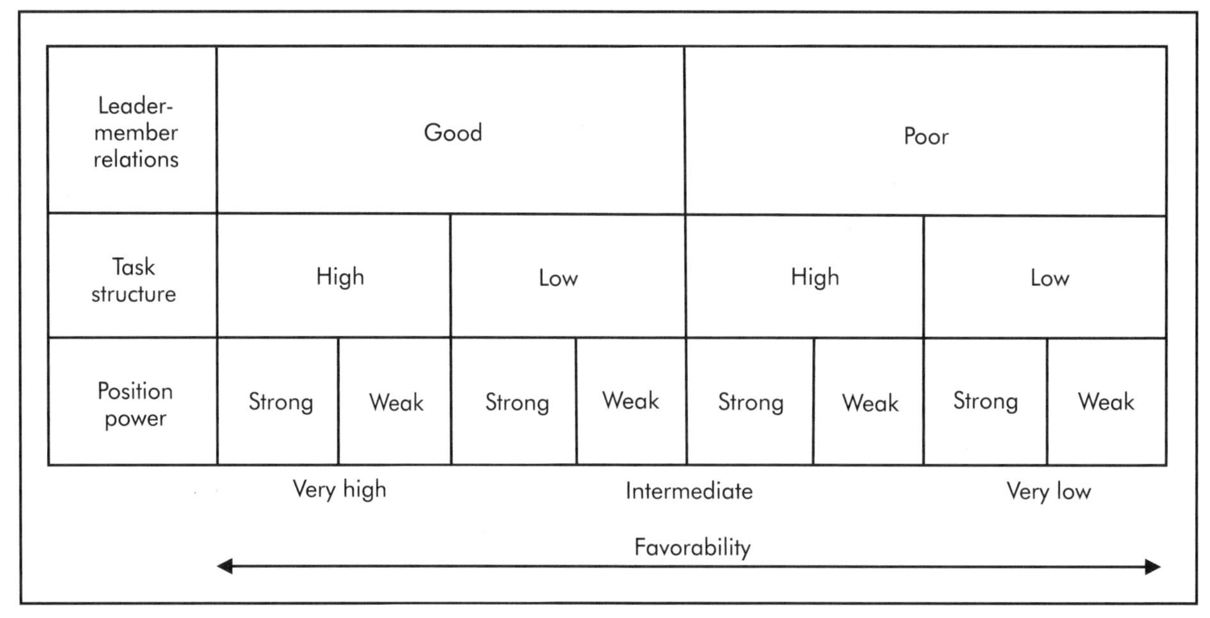

Figure 56-4. Contingency theory variables versus situational favorability (Veilleux and Petro 1988).

power, task structure, and leader-member relations. In general, it may be more practical to match a leader to a specific situation than to try to change the situation to match the leader (Veilleux and Petro 1988).

NORMATIVE THEORY

If leaders can be presumed to be flexible in their general style, a more useful situational theory is one that gives leaders a set of rules (norms) to guide their behavior. This decision-making theory uses two criteria for measuring the effectiveness of a decision: quality and acceptance.

Decision quality refers to the effect a decision may have on job performance. The quality of some decisions will result in variable degrees of job effectiveness. Acceptance refers to the need for followers to commit to or accept a decision. Some leader decisions can be implemented without group acceptance, while others would be unsuccessful unless the followers clearly commit to the decision.

Leaders can choose from among a set of decision styles that range from a clear autocratic style to an almost entirely group decision process, with varying degrees of those styles available between the two ends of the spectrum. The definitions are listed as follows (A = autocratic, C = combined, G = group):

- A1—The leader makes the decision alone using the information available at the time.
- A2—Subordinates provide information requested by the leader. They may or may not be aware of the nature of the problem or decision. The leader solves the problem or makes the decision.
- C1—The problem is shared by the leader with the subordinates individually. Each may provide suggestions or ideas. Subordinates do not function as a group. The leader's decision may reflect the subordinates' influence.
- C2—The problem or decision is shared with the subordinates as a group. Ideas

and suggestions are obtained, but the leader makes the decisions.

- G1—Problems are shared with the subordinates as a group. Information, alternatives, and consequences are analyzed. The group reaches consensus on a solution. The leader's primary role is facilitator of the group process. The decision made or solution chosen is one supported by the leader and his or her subordinates.

To select a decision style, the leader follows a set of decision rules (norms). The norms are simply questions with yes-no alternatives. By answering the questions in sequence, a decision tree is formed, which indicates the most appropriate style (Veilleux and Petro 1988).

PATH-GOAL THEORY

In the *path-goal theory* the leader emphasizes the relationship between the employee's goals and the organization's goals. The leader also identifies paths that will allow employees to reach their personal goals while achieving organizational goals at the same time. The leader can utilize an appropriate set of behaviors based on consideration of the personal characteristics of each subordinate, especially each person's perception of their own abilities and experience in a given situation, and analysis of environmental demands, such as the organization's authority system, the tasks, and the nature of the work group.

The path-goal theory is very similar to the expectancy theory of motivation discussed later, whereby employees are motivated if the expectancy of a desired outcome is high. The choice of the leader's behavior style is similar to other theories. The leader may be:

- directive—the leader lets subordinates know what is expected of them;
- supportive—the leader treats subordinates as equals while providing encouragement;
- participative—the leader consults with subordinates and uses their suggestions and ideas; or
- achievement-oriented—the leader sets challenging goals, expects subordinates to perform at a high level, and seeks continuous improvement (Veilleux and Petro 1988).

56.4 MOTIVATION

An essential element of leadership is determination of the forces that cause people to act in the ways they do. If what causes action is known, then perhaps the leader can utilize these forces for the benefit of the organization.

Motivation is defined simply as the force, the cause, the internal reason why people act in a certain way. A person decides to eat, sleep, talk, work, or play based on certain causes—including some not easily rejected ones, such as when a person feels hungry. Motivation also can be the result of a learning experience. Certain connections are made in a person's mind between pleasant or unpleasant results of actions that have been committed. Based on these chains of actions and results, a person can choose to behave similarly again or not.

For convenience, most of the theories of motivation are divided into two broad categories, content theories and process theories.

Content theories focus on the immediate triggers for the way a person behaves. For example, hunger causes eating and, in the context of manufacturing, individuals work because they wish to earn a living. Content theories include those proposed by Maslow, McClelland, and Herzberg.

Process theories focus on the things leaders provide to followers and the learning and interpreting that individuals do in the context of work. By studying the process that individuals go through in coming to a decision about whether they wish to behave in a certain way, organizations can find ways to assist them in behaving in productive ways.

The two most complete and practical theories of motivation are the equity theory and expectancy theory.

MASLOW'S APPROACH

Abraham Maslow clustered motivating factors into five now well-known groups he believed had a rank ordering for most people as illustrated in Figure 56-5. At the base level is a set of physiological needs that drive people to act, such as food and water. As this group of needs is generally satisfied, individuals next seek a level of safety, such as personal safety or economic security. After safety, people seek to fulfill social needs, such as group affiliation. Next he suggests that individuals have a need for self-esteem, such as building a reputation and self-confidence. Finally, the last need is self-actualization whereby a person realizes his or her own potential. Unless lower-order needs are fulfilled first, people do not seek out the next level, and once a need is satisfied it no longer remains a motivator.

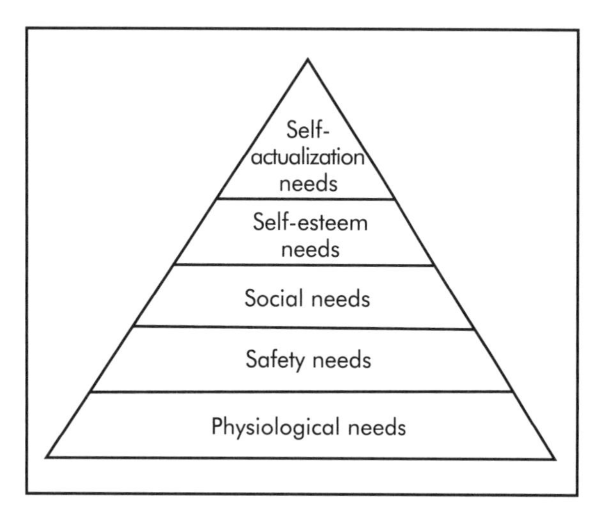

Figure 56-5. Maslow's hierarchy of needs.

To be effective, the leader identifies the level of need of the follower and then by providing an opportunity to fulfill that need, the leader obtains work output from the follower.

Maslow's idea appears to be very intuitive. However, variations exist between individuals as to the magnitude and duration of their respective needs. From the leadership viewpoint, it seems inefficient to try to provide a measure of satisfaction for each individual. Motivation is usually determined by a combination of needs rather than one at a time. Leaders should identify the needs of individual employees and fulfill their respective needs (Veilleux and Petro 1988).

MCCLELLAND'S APPROACH

Another content approach to motivation was developed by David McClelland. He suggested a model that relates the needs of leaders and followers in terms of power, achievement, and affiliation. Each individual may have a need to give or receive each of these factors to some degree. Again, by recognizing the extent of these needs in each individual, organizations can provide satisfaction and, consequently, increased productivity.

McClelland's theory suggests that candidates for leadership roles be knowledgeable of the motivational forces to assist the organization in doing more effective selection of employees. If an individual possesses a high need for power, he or she may be suited for a leadership role. Of course, the individual must have a need for power and control that is not strictly a personal need, but rather an organizational one. An individual with a high need for achievement could be provided with opportunities to be successful in a job, and the individual with a high need for affiliation could be given chances to work with others to accomplish goals.

It should be recognized that almost everyone in the organization is to some extent a follower regardless of needs or formal position (Veilleux and Petro 1988).

HERZBERG'S APPROACH

The last of the content theories of motivation is that of Frederick Herzberg, who suggests in more practical ways how a leader (and the organization) can provide for fulfillment of motivational forces. He suggests a two-factor approach, with one set of factors called motivators or satisfiers and the other called the hygiene factors or dissatisfiers. Motivators create job satisfaction; however, their absence does not create job dissatisfaction. The presence of hygiene factors does not create job satisfaction; however, their absence creates job dissatisfaction.

In Herzberg's view, many intangible items provide sources of motivation. These include an individual's autonomy, job responsibility, and the work itself. The organization has a normal duty to prevent dissatisfaction by providing hygiene factors such as a quality work environment, reasonable supervision, and adequate pay and rewards. Without these, individuals become dissatisfied and less productive.

Herzberg suggests that traditional needs as usually provided by leaders will only serve to make individuals unhappy when they are not provided. These motivating forces, usually thought of as necessary, are also those that are easily recognized and must be weighed from a financial point of view. Intangible items may cost less, but are more difficult to understand, define, and provide (Veilleux and Petro 1988).

EQUITY THEORY

The *equity theory* looks at how individuals weigh the relationship between the outcome, such as various rewards and returns from the organization, against their input such as effort, experience, and education.

In addition to the ratio of outcome received to input given, the individual compares his or her own ratio to that of others. Others may be fellow workers in the same organization, those in other related organizations, or persons totally unrelated to the organization's immediate environment. The theory is expressed as:

$$\frac{\text{Outcome (individual)}}{\text{Input (individual)}} = \frac{\text{Outcome (other)}}{\text{Input (other)}} \qquad (56\text{-}1)$$

The thrust of the theory is that each person is motivated to maintain a balanced equation. The organization, through its leader, provides some of the rewards (outcome) for the individual, some sense of the required demands on the individual (input), and information used by the individual to fill in the "other" side of the equation.

In the most direct example, consider the individual who perceives that his or her outcome to input ratio is less than that of another person. The theory suggests that to rebalance the equation, he or she would seek additional rewards or reduce his or her efforts. Either of these would tend to raise the "individual" side of the equation and therefore achieve balance.

An alternative in this same setting is for the individual to work harder in the hope that the organization will respond with more than sufficient reward, thereby reducing the inequity. If the individual, however, sees that he or she is over-rewarded or under-worked, the reverse imbalance can cause an increase in efforts to deserve the outcome. Leaders must then provide equitable rewards, an adequate definition of what is expected of the worker, and facts in regard to the outcome to input ratio of others. If all of these elements are provided, the individual will maintain the correct outcome to input ratio.

Problems arise in large part with the equity theory because the leader cannot be sure which intangibles exist in the individual's outcome to input ratio, or which "other" is the individual's reference, or indeed how the individual will adjust. The theory is based on individual perceptions that cannot always be well understood by the leader. The perspective is the individual's, which could cause the organization to always be making estimations of the various inputs and outcomes (Veilleux and Petro 1988).

EXPECTANCY THEORY

The *expectancy theory* of motivation is based on these factors:

- Effort will lead to success.
- Success will be rewarded.
- The reward is of personal value to the employee.

For example, suppose a person studies to take the Certified Manufacturing Engineer (CMfgE) exam. The person believes that if he or she studies hard there is a high probability of passing the exam. Secondly, there is a good chance of receiving a better job assignment at work and a raise if he or she becomes a CMfgE. Finally, the raise is personally valuable because it will help pay for his or her daughter's college tuition.

If any of the three conditions were not true, there would be no motivation to put forth effort to become a CMfgE. If the person thought preparing for the exam was futile or passing it would do nothing for them at work, or the reward would only be a pat on the back, then it is unlikely the person would ever begin the process of becoming certified.

The organization has many opportunities to motivate. Naturally, it usually provides the reward or outcomes. It sets the standard for performance or success. By providing information, tools, and training, it increases the probability of success after effort. In the administration of the reward system, it develops a strong link between success and receipt of the reward. Only with regard to the value of given rewards to the individual does the organization have limited influence. However, the organization may establish a culture wherein certain rewards are seen to be more valuable.

Overall, the expectancy theory includes many of the concepts contained in other motivation theories. It provides the mechanisms for management intervention and allows for the interpretation of changes in either the individual's or the organization's approach to motivation (Veilleux and Petro 1988).

56.5 JOB DESIGN

Often, jobs are designed to meet the needs of the machinery, without regard for the humans who must do the work. In fact, the quality of the human-machine interface, both physical and psychological, can make a critical difference in the productivity of a plant.

INTERDEPENDENCE

Jobs are more motivating if they require workers to interact with others. People are motivated when they know how what they do fits in with what other people do. Most people want to spend at least part of their work time interacting with others about work-related matters. Some jobs cannot be completed without the help of others. Those jobs that require people to work in isolation with little social interaction are less motivating than those that require a sense of interdependence.

JOB ENLARGEMENT

Skill variety, task identity, and task significance are three variables that can increase the meaningfulness of work. Jobs can be more motivating if a person has responsibility for doing a larger portion of the work. This is what is meant by *job enlargement*.

Skill variety refers to the degree to which the job includes a number of different activities that require the individual to use a range of talents and abilities. When a job includes skill variety, the person is likely to experience it as challenging and, therefore, personally meaningful. *Task identity* refers to the employee doing something that is an identifiable piece of work. A job that requires an employee to create a product from start to finish would be a job with high task identity. Low task identity refers to a job requiring moderate effort and minimal work structuring.

The third variable, *task significance*, refers to the degree that employees perceive the job as having an important effect on their lives or work, or on other people. When people feel their product or service is important to others, they tend to be more motivated to do the best they can (Veilleux and Petro 1988).

JOB ENRICHMENT

Autonomy is the extent to which a person has authority to make decisions about the work they do. Building autonomy into the job is what is meant by *job enrichment*, where the person's decision-making authority is increased. Theoretically, as autonomy increases, so does motivation (Veilleux and Petro 1988).

FEEDBACK

Jobs that give people immediate feedback on results are more motivating than jobs that do not supply such information.

INCENTIVE PLANS

The most basic production incentive plans are built on the observation that workers paid by the day or hour are usually not exerting as much effort as those who are salaried. Manufacturing engineers use the term "low task" to refer to the basic state of moderate effort and minimal structuring of work. Effort can be improved by methods analysis and standardization to attain the level of "medium task." Then it can be increased by appropriate incentives to attain "high task" output.

One of the biggest reasons incentive plans fail is the mismatch between the incentive and the employees' value system. Motivating employees is more complicated than an incentive of money, gift certificates, etc. What works for one person may not work for others. For example:

- Is withholding punishment an incentive?
- Is praise an incentive?
- Is job enlargement an incentive?
- Is money or a gift certificate an incentive?

56.6 PROJECT MANAGEMENT

A project encompasses all the activities associated with achieving a set of specific objectives regarded as important and worthy of financial support. By its very nature, a project has a finite life; it requires specific resources and has a clear definition of when the job is complete (closure). A project's goals require accomplishing something that has not been done before, thereby making the project unique.

Project management is comprised of many different aspects such as planning, budgeting, controlling, analyzing, and managing people and teams. A project should have a well-defined set of goals and a finite lifetime. The responsibility of the project manager is to integrate all of the functions and oversee the project to completion.

The appropriate person to head a project is a matter of major importance. Successful project managers are persons who have distinguished themselves as good managers of time, assets, and people, and who have highly developed communication skills. The intangible, but all-important quality of leadership is also vital to the project management function. In most cases, the project manager is one who has been responsible for, or close to, the formulation and marketing of the proposal that resulted in project funding.

Successful project management requires careful attention to meeting the technical objectives, meeting the project's time requirements, and meeting the budget. These three factors characterize any project and compete with one another. To be successful, the project manager needs to understand this inherent conflict and manage it effectively within a competitive environment.

Project management can be divided into two distinct phases: a) project planning and b) project execution and control. Project planning can be broken down into several components: setting objectives and goals, task planning, budgeting, and scheduling (Veilleux and Petro 1988).

OBJECTIVES AND GOALS

The statement of objectives is the foundation on which all project planning and execution is built. Objectives must be agreed to mutually by upper management (or the client) and the project manager. They should define the desired outcome of the project. Each objective should be written with understanding and appreciation for its feasibility. If consideration is not given, the objective may prove to be unachievable within the resources available to the project. The objectives themselves may or may not be quantifiable, but they should be supported by measurable goals. Goals are specific statements intended to quantify the project's objectives. They should be as precise as possible.

Next, the feasibility of the project needs to be determined. Three things must be taken into consideration. First is technical feasibility. Is the requisite technology available to enable the objectives to be met? If not, can it be adapted from another field or application? Projects that rely on the development of technology are better regarded as exploratory research initiatives, not technical projects. The second consideration is operational feasibility. Is the resulting system or process likely to operate as expected? What are the most likely problems and approaches to their solution? Finally, the project objectives should address the issue of economic feasibility. Are the resources adequate for successful project completion? What are the anticipated financial results? Upon project completion, financial analysis will be used as a measure of success.

Timing, a basic factor involved in any project, should be addressed in the statement of objectives. Timing includes the overall period of performance, as well as the major project milestones. As with objectives, timing should be supported by measurable goals. Network techniques such as the program evaluation and review technique (PERT) or critical path method (CPM) and Gantt charts can be used to set measurable goals.

TASK PLANNING

The objectives and the goals developed in support of them are used to break the project into a set of work packages. Work packages define specific tasks that contribute to the clearly defined goals.

A task has an identifiable beginning and end. The collection of tasks should be necessary and sufficient to satisfy the work package requirements. Tasks may be arranged in series and in parallel, according to their mutual dependencies. They should be linked directly to milestones. *Milestones* are events of significant accomplishment such as the start or completion of tasks and jobs, achievement of objectives and goals, completion of customer reviews and approvals, or the demonstration of prototype performance. They are convenient points at which to report status or measure and evaluate progress.

Many approaches are used in task planning. At one extreme, a project that is small and of short duration can be broken down easily and obviously into tasks without sophisticated planning tools. At the other extreme, a large, complex project can profit from systematic planning using modeling techniques and computer software.

The development of a work breakdown structure (WBS) can be of great help in task planning. The WBS represents the project functionally; much like an organization chart describes the functions of a business unit. It gives the hierarchical relationships between work elements (design, hardware, software, and services) (Veilleux and Petro 1988). An example of a transceiver project WBS is shown in Figure 56-6.

RESOURCE REQUIREMENTS

The project manager and the project team must be wise architects of the project during the planning phase. The two main items that must be taken into consideration at this phase are human resource requirements and capital investment. In addition, the project manager should decide whether any other supporting services will be required during the life of the project.

The project manager and the project planning team should develop detailed, realistic plans for project staffing. Typically, workers are not all under the line control of the project manager. Therefore, human resources must be drawn out from other units of the firm to staff the project. In addition, there may be a need for outside subcontractors or consultants. All these anticipated resources need to be defined in the project planning phase.

If the project requires either the acquisition or delivery of equipment, plans must be made to accommodate this need. A detailed set of equipment performance specifications should be drawn up. These specifications should be rigorous enough so conforming equipment will perform adequately. Another important factor to evaluate is the setting in which the equipment is to operate. Consideration should be given to space and special power needs, contaminants that may have an environmental impact, and other support needs such as specially trained or dedicated operators and software development.

BUDGETING

The next step in project planning is to develop a detailed budget. The budget is based on the task descriptions and resource requirements. It should reflect the most re-

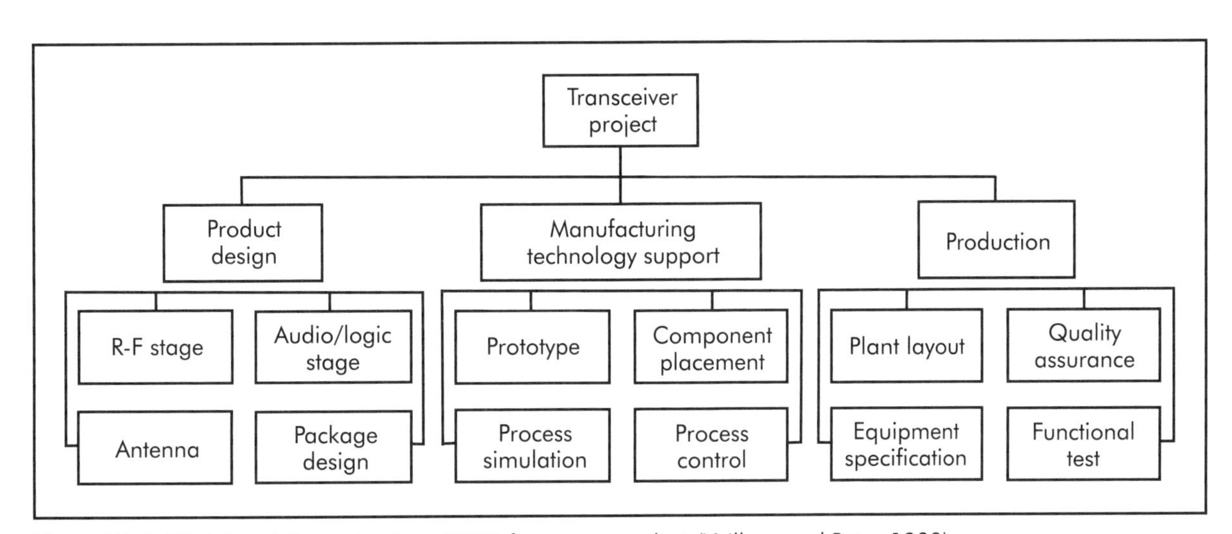

Figure 56-6. Work breakdown structure (WBS) for a new product (Veilleux and Petro 1988).

alistic estimate of all expected costs without being overly conservative (high) or optimistic (low). Those costs that can be predicted accurately should be budgeted as such. Elements of the project whose costs cannot be reliably predicted should be budgeted at expected levels, with contingency amounts posted in the budget to provide for costs that may exceed expected values. Such an approach to budgeting builds integrity into the project and makes it defensible to management. It also provides a means to manage excessive costs through effective project control measures.

Many approaches exist for constructing the project budget. The project budgeting process described here represents the "bottom-up" approach. The project is broken down into a fine structure of tasks. Each task is broken out into all the cost elements involved, and the total project cost is the sum of all the cost components. All project budgets must account for the same elements of cost, one way or another. The five main elements are labor, expense, capital, overhead, and profit.

Labor costs may be direct and/or indirect. Direct labor may consist of exempt and nonexempt employees. Exempt personnel would include managers and other professional employees who do not receive overtime compensation. Direct labor costs should be estimated for each task of the project.

Indirect labor includes administrative and general labor costs. These costs are common to many or all activities of the organization and, therefore, cannot be directly allocated to individual projects. Typical of these are the costs of senior management, finance, legal, personnel, and health care. Indirect labor costs are often computed as a percentage of direct labor.

For accounting purposes, expense costs encompass all other non-labor costs. Expenses may include travel, expendable supplies, project materials, and consultant fees.

Capital costs are the fixed assets owned by the company, which can be depreciated over their useful lives as determined by standard accounting practices. Overhead costs include all non-salary indirect costs, such as fringe benefits, taxes, insurance, and heating and lighting. Overhead rates are set by cost accounting practices and usually figured as a percentage of direct labor.

The project manager must evaluate the cost estimate. Does it appear realistic? Does it seem too high or too low? Often, "top-down" budget pressures will suggest revisions to bring the costs in line with available resources or with what is regarded as fair and reasonable. The challenge, of course, in such budget tuning is not to cut the cost estimate without reducing the cost elements commensurately, that is, by adjusting the task descriptions (Veilleux and Petro 1988).

SCHEDULING

Scheduling is the process of converting the objectives, goals, and tasks into a time-phased execution plan. It is the project manager's responsibility to ensure all time commitments to a client or upper management are met. The time-phased schedule must simultaneously consider the required tasks, resource requirements, and budget.

Scheduling includes the time to do the work, the people responsible for doing it, where the work will be done, resources required, and monitoring and reporting on the work.

There are many scheduling techniques and software packages available. However, the basic concept behind scheduling is the development of a task network. The task network defines the sequential and parallel relationships between tasks.

This section discusses two scheduling techniques: the program evaluation and review technique (PERT) and critical path method (CPM).

Program Evaluation and Review Technique (PERT)

The program evaluation and review technique (PERT) was first applied in 1958 to manage the Navy's Polaris submarine program (Veilleux and Petro 1988). It is well suited for project planning and control.

The heart of PERT is the construction of a network comprising all the tasks and milestones of the project. Tasks and milestones are linked together in a flow network represented by circles called *events*; successive events are connected by arrows indicating the direction and content of the activity. Each activity has a beginning and end point. These activities, or tasks, may be related in series or in parallel relationships, depending on the requirements of the project plan.

PERT requires that three time estimates be associated with the completion of each activity. The unit for time is generally in weeks, but days and months are also used. Based on these time estimates, the expected time to complete an activity between two successive events can be calculated.

Time estimates are often difficult to determine. In cases where such work has not been done before, the estimates may be mere guesses. Nonetheless, they must be based on the best experience available. Such estimates can be modified as the project progresses.

The network plan can then be analyzed by computerized linear programming methods to determine the most time-consuming path that connects project initiation and completion. This information serves as a management flag to suggest reallocation of resources to improve overall project performance.

Figure 56-7 illustrates a PERT network for a project consisting of nine events. The bold arrows highlight the critical path over which the total estimated time for the project is calculated. Other paths, as can be shown by calculation, require less time. These are termed *semi-critical paths* or *slack paths*, depending on how close they are to the critical path in terms of total time required. All other possible paths lie between the critical and slack paths. By reallocating some of the human resources from the slack to the critical and semi-critical paths, it may be possible to reduce the overall project completion time and project costs. Unless such adjustments are made, the duration of the project will be governed by the critical path.

Critical Path Method (CPM)

The critical path method (CPM), as discussed in Chapter 45 of *Fundamentals of Manufacturing*, Second Edition, has adaptations that are quite similar to PERT. In CPM, only a single number, the expected time to completion, is associated with each activity arrow. In this sense, CPM is merely a special adaptation of PERT.

PROJECT MONITORING

Project monitoring is the process of collecting data regarding all facets of the project and reporting it to the people involved such as coworkers, superiors, and the client, if applicable. Monitoring the project supplies data for control decisions; therefore, collecting accurate, meaningful, and timely data is essential. Project control, as discussed later, uses monitored data to compare the current state with the project plan and schedule. Control techniques are used to make the project performance coincide with the project plan.

The monitoring system must be designed to monitor key project characteristics such as cost, time, labor hours, and milestones. The project manager must ensure that items such as cost and labor hours are assigned to the correct work package. After data collection, reports involving status, earned value, variance, and productivity are generated. Reports provide a snapshot of project status and goals for everyone involved with the

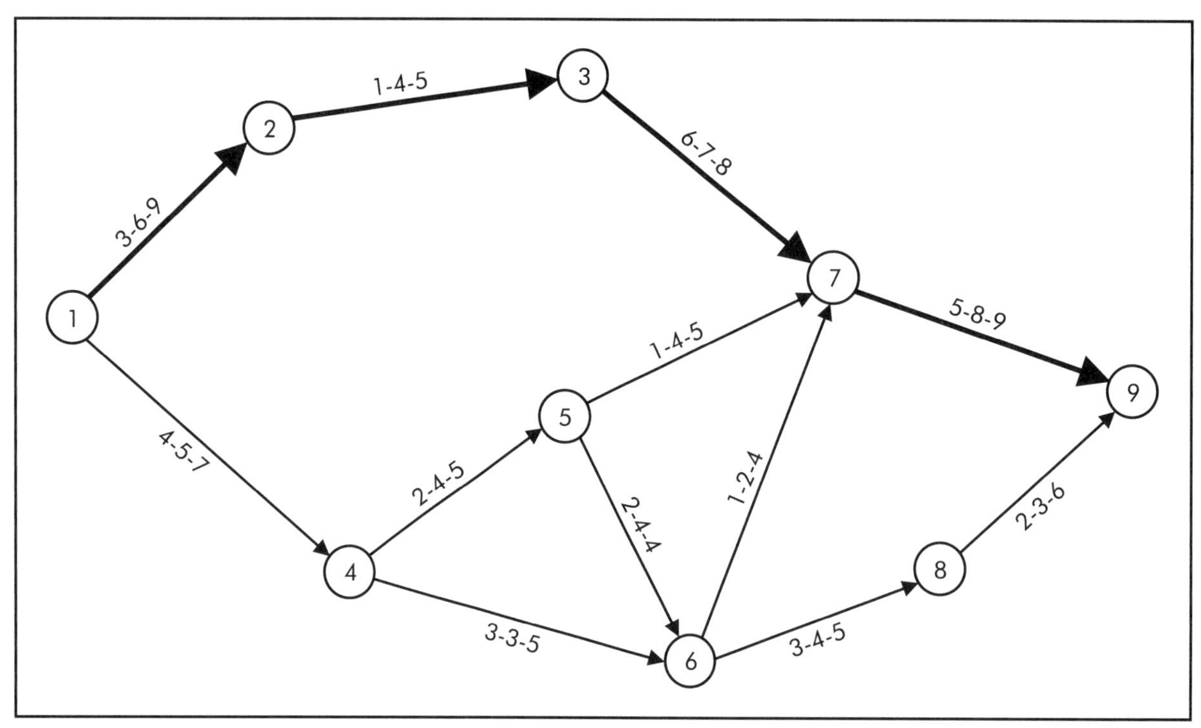

Figure 56-7. PERT network for a project containing nine events. The three time estimates included on the linkages between the events are given in weeks. The bold line denotes the critical path (Veilleux and Petro 1988).

project so there is a common understanding. Additionally, they identify the relationship of tasks to each other and the total project (Meredith and Mantel 1989).

One source of reporting information is individual or work package status. Tasks must have unambiguous measures of progress or business measurables. In addition to a task due date, other expectations must be quantified. For example, if part of the project involves software development, then the details of what the software must do need to be articulated and measurable. Project managers can use this information to help bring individuals back on schedule when they are running late.

Earned value is a measurement of how much work has been completed on a project in terms of dollars or labor hours. It is the percent complete of a task or a group of tasks multiplied by the planned or budgeted value. A scenario can exist where the actual labor expended matches the planned labor expenditure; however, the earned value is lower than the actual labor. For example, the planned labor for Task A is 10 hours and the actual labor expended is 10 hours. However, if the earned value for Task A is 50%, this equates to five earned labor hours. This indicates that very little is being accomplished despite the fact that the actual labor hours are on target. Using earned value, managers can easily compare what has been accomplished in a project and what remains to be done.

Variance analysis reports describe variances from the project plan in terms of schedule and cost. There are predetermined threshold values for the planned project variables such as labor, materials, capital, etc. A variance analysis report is generated when the actual value exceeds the threshold value.

Schedule variance is the variation between the earned value in terms of dollars or labor hours and the planned value. As work is performed, it is considered to be earned in the same units it was planned (dollars or labor hours). *Cost variance* is the variation between the earned value and the actual cost for the work performed.

Productivity is the ratio of earned labor hours to actual labor hours. For example, if 100 hours are planned for Task A and 50% of the task is completed, this would equate to 50 earned labor hours. If 100 actual labor hours are expended, the productivity is 50%. The productivity report can help project managers detect wasted or nonproductive efforts.

PROJECT CONTROL

Careful project planning can greatly enhance the likelihood of a project's success. However, whether or not a project will meet all its stated objectives and goals will depend entirely on the resources allocated and how well the project is executed and controlled.

It takes the coordinated effort of all participants to execute the project according to the plan. Problems are identified and solved along the way. Comparing the project status (technical, financial, and schedule) with the plan and making suitable adjustments when the two differ brings the project under control. The project plan itself must be viewed as dynamic and updated as appropriate in light of changing requirements.

One of the simplest and probably the most commonly used of all project control tools is the Gantt chart. Gantt charts are simple to construct and easy to maintain. They show at a glance the relative staging of all tasks and how each measures up to progress expected by the current date (see Figure 56-8).

The disadvantages of using Gantt charts stem precisely from their advantage—simplicity. For example, Gantt charts only provide gross information about project status. Detail is obscured or lost, such as reasons for schedule deviation, technical and financial performance, and what is likely to happen in the near future. It is also often difficult to determine how the degree of shading relates to task performance. Does the length of shading indicate the proportion of task resources already expended or the degree of technical accomplishment or some subjective measurement of both? Another drawback is that the Gantt chart does not reveal mutual interactions or dependencies among project tasks. Rather, the chart treats tasks as mutually independent activities.

The program evaluation and review technique (PERT), which was described earlier for scheduling, also can be used for project control. Unlike Gantt charts, PERT charts do not lend themselves to easy visual identification of project status. On the other hand, they are richer in information content and demand a more detailed analysis of the task status for periodic updates of the network. PERT networks are primarily used for major projects comprised of many tasks that are mutually interconnected. Small projects

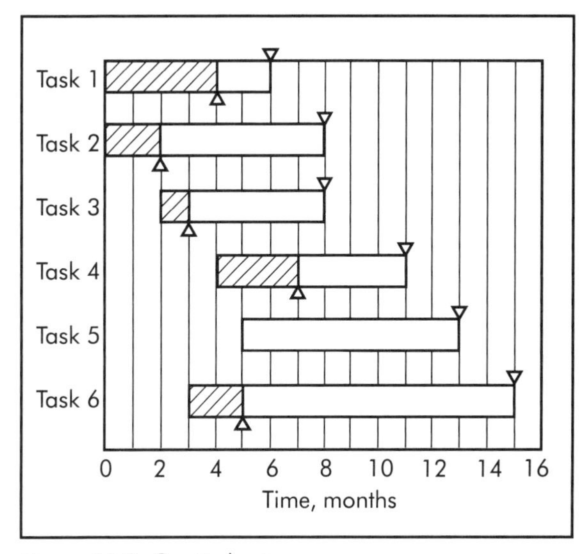

Figure 56-8. Gantt chart.

comprised of about 10 tasks are more usefully represented by Gantt charts, especially when there is a high degree of task independence.

The critical path method (CPM), which was described as a scheduling technique, also can be used for project control. As with PERT, CPM enables the critical path (longest time) and slack paths to be calculated. CPM networks are used in the same way as PERT charts.

56.7 STRATEGIC PLANNING

Strategic planning is a methodology that defines the mission and objectives of an organization in the long term. It identifies the core business and customer base of the organization, along with the business or businesses that it plans to pursue in the future. Strategic planning provides a framework within which an organization can make appropriate decisions that will fulfill its objectives. Decisions regarding capital investments and project proposals, for example, will be based on the strategic plan.

Several components, such as the organization's vision, mission, values, objectives, goals, environmental scan, and programs are included in the strategic plan.

An organization's *vision* indicates the desired future state of the organization. Items such as the anticipated size of the organization and the type of business it expects to participate in are included. In effect, the vision is an idealistic representation of the organization's future state.

The *mission* of an organization or business is its purpose. For example, in higher education, the mission of a university is to provide educational opportunities for students seeking to further their education. Following the mission statement, objectives, goals, and values are defined. Objectives and goals further define the vision and mission of the organization. Objectives are fairly broad and goals are more specific.

Values define what is important to the organization or business. For example, a business may value its employees and their respective job satisfaction. Hopefully, all businesses value their customers.

Next, a scan of the organization's external and internal operating environment is performed. Economics, in addition to social, political and technological factors, are examined.

A common method of evaluating the environment is through strengths, weaknesses, opportunities, and threats (SWOT) analysis. Addressing these factors can help an organization develop a strategic plan.

A second tool for analyzing the environment is the *five-forces model* developed by Michael Porter. The five environmental forces influencing business success are as follows (Harrison and St. John 2002).

- Potential entrants—new entrants are typically a threat to existing businesses.
- Suppliers—in an ever-increasing supplier-dependent manufacturing environment, suppliers can have a significant effect on the success or failure of a business.
- Customers—all customers are important to a business; however, customers that purchase in large volumes can have a significant impact.
- Substitutes—other businesses may provide a substitute product. For example, a money market fund is a substitute for a bank savings account.
- Existing competition—the direct competitors currently in the marketplace.

Another common tool for evaluating the environment is benchmarking. *Benchmarking* compares the operations and performance of an organization or business with those of an apparently better or more successful organization or business. The businesses used for benchmarking purposes do not have to be similar depending on the criteria being used. For example, the best method of soliciting customer input is not necessarily dependent on the type of business.

After the vision, mission, objectives, goals, and environmental scan are complete, strategic programs can be developed. If, for example, customer satisfaction is valued, and one of the organization's objectives is to increase customer satisfaction to meet a goal set at 100% satisfaction, what is the plan or program for achieving that? Other programs would be developed in a similar fashion.

Although strategic planning can be very beneficial, there are hurdles or limitations to overcome. Strategic planning requires the commitment of top level managers and executives. As with any organizational initiative, strategic planning will fail if not supported at the top of the organization. Second, everyone in the organization must "buy-in" to the strategic planning process and the eventual plan. Without organizational buy-in, the plan will be unsuccessful. Third, an organization must be disciplined to complete the strategic plan. Often, the planning process is lengthy and it is easy for people to lose interest. Also, the organization needs discipline to adhere to the strategic plan after it is developed. It can be easy to forget about the plan a year later. Finally, managers can get trapped in a "plan for the sake of planning" cycle. Planning is part of a manager's duties; however, the plan must lead to a desired outcome or business measurable.

56.8 GROUP DISCUSSION TECHNIQUES

BRAINSTORMING

Brainstorming is a group discussion process that emphasizes creativity and free thinking. It is a technique used for generating a broad range of ideas and solutions. Given the non-critical environment brainstorming provides, participants' creativity is increased and inhibitions about participating in a group environment are decreased.

The brainstorming session begins with an explanation of the rules given by the group facilitator or leader. The facilitator or leader focuses on the processes of the group. The facilitator remains neutral, helps the team operate smoothly, prepares for the meeting, creates an agenda, plans for completing the meeting work, and informs team members of the meeting time. To promote a successful brainstorming process, there is no criticism of ideas during the idea generation phase. After all the ideas are generated and printed on a board or flip chart, the leader eliminates repetitive ideas. The group eliminates those that are impractical and decides on the best idea(s) from the remaining ones.

NOMINAL GROUP TECHNIQUE

The *nominal group technique* is appropriate for problems that are difficult to quantify and for groups whose members do not know each other. Each person in the group is given an equal voice in making decisions. Group members silently generate solutions to a given problem after which the group facilitator records the ideas on a board or flip chart. The group discusses the ideas and makes clarifications where necessary, after which the group votes on the ideas and ranks them. The idea ranking is open for discussion. After ranking the group typically formulates one solution.

DELPHI GROUP TECHNIQUE

The *delphi group technique* begins with group members silently writing possible solutions to the problem in question. The group facilitator collects the anonymous suggestions and distributes them to everyone in the group for another round of suggestions. After several iterations a common suggestion will emerge. This technique allows group members to express their ideas without intimidation or the fear of rejection. With the delphi technique group members can be separated geographically and communicate via mail or e-mail.

FOCUS GROUPS

Focus groups generally consist of a group of participants and a moderator. The participants can be selected randomly from the general public or they may be familiar with the topic in question. The participants, however, do not know each other. A session begins with an introduction by the focus group moderator, who explains the purpose of the discussion and discussion rules. The moderator controls the flow of the discussion by probing, summarizing, paraphrasing, and direct questioning if necessary. The group discussion progresses and consensus usually develops. Focus groups can provide more information than impersonal surveys or questionnaires.

56.9 INTELLECTUAL PROPERTY

In the United States, intellectual property is protected by patents, licenses, copyrights, trademarks, and in industry as trade secrets.

PATENTS

A *patent* for an invention is the grant of a property right to the inventor issued by the United States Patent and Trademark Office (USPTO). The right conferred by the patent grant is, in the language of the statute and of the grant itself, "the right to exclude others from making, using, offering for sale, or selling" the invention in the United States or importing the invention into the United States. The extent of that right is limited by the invention specifications and claims of the patent. What is granted is not the right to make, use, offer for sale, sell, or import, but the right to exclude others from making, using, offering for sale, selling, or importing the invention. Once a patent is issued, the patentee must enforce the patent without aid of the USPTO. For an invention to be patentable it must be novel, non-obvious, and useful.

There are three types of patents:

- *Utility patents* may be granted to anyone who invents or discovers a new and useful process, machine, article of manufacture, composition of matters, or any new useful improvement thereof. Generally, the term of a new utility patent is 20 years from the date on which the application for the patent was filed in the United States. Under certain circumstances, patent term extensions or adjustments may be available.
- *Design patents* may be granted to anyone who invents a new, original, and ornamental design for an article of manufacture. Design patents have a term of 14 years from the date granted.
- *Plant patents* may be granted to anyone who invents or discovers and asexually reproduces any distinct and new variety of plant. Plant patents have a term of 20 years from the filing date.

Infringement of a patent is the unauthorized exercise of any rights granted to the inventor, such as unauthorized manufacture, use, or sale. When these exclusive rights have been infringed, the inventor may file suit in federal court to recover damages and obtain an injunction prohibiting future infringement. In deciding an infringement suit, the court may not only adjudicate the infringement question, but also may reconsider the validity of the patent itself. For example, if it was found that the USPTO was intentionally misled, a federal court could invalidate the patent based on fraud.

To make or sell patented articles, a patentee or his agent is required to mark the articles with the word "Patent" and the number of the patent. The penalty for failure to mark is that the patentee may not recover damages from an infringer unless the infringer was duly notified of the infringement and continued to infringe after the notice.

The marking of an article as patented when it is not in fact patented is against the law and subjects the offender to a penalty. Some persons mark articles sold with the terms "Patent Applied for" or "Patent Pending." These phrases have no legal effect, but give information that an application for patent has been filed with the USPTO. The protection afforded by a patent does not start until the actual grant of the patent. False use of these phrases or their equivalent is prohibited.

Treaties and Foreign Patents

The rights granted by a U.S. patent extend only throughout the territory of the United States and have no effect in a foreign country. An inventor who wants patent protection in other countries must apply for a patent in each of the other countries or in regional patent offices. Almost every country has its own patent law. A person desiring a patent in a particular country must make an application in that country, in accordance with its particular requirements.

The Patent Cooperation Treaty, which came into force on January 24, 1978, facilitates the filing of patent applications for the same invention in member countries. It provides, among other things, centralized filing procedures and a standardized application format.

The timely filing of an international application affords applicants an international filing date in each country. It provides a search of the invention and an extended time period within which the national applications for patent must be filed (Department of Commerce 2003).

LICENSING

In many cases, an organization or individual that owns all rights and title to a patent may not be in a position to capitalize on it. An organization may not be in a position to manufacture and sell the invention because of limited manufacturing or marketing capabilities or because the invention may not be compatible with existing product lines. If this is the case, several options may be available to the organization to capitalize on the invention. Two such options are: 1) manufacture the invention and license other organizations to market it, or 2) license other organizations to manufacture, use, and market the invention. Either of these options may be chosen with the hope of realizing immediate gain, depending on the licensee's manufacturing and marketing capabilities.

COPYRIGHT

A *copyright* is a form of protection provided under the laws of the United States to authors of "original works of authorship." This includes literary, dramatic, musical, artistic, and certain other intellectual works. Protection is provided for published and unpublished works.

Though protection extends to written, pictorial, audio, and theatrical works, it does not to extend to ideas, processes, principles, and the like. The owner of a copyright has exclusive rights to do (or authorize) reproduction, preparation of derivative works, or distribution of the copyrighted work for sale, lease, rental, or transfer of ownership.

Copyright protects original works of authorship that are fixed in a tangible form of expression. The fixation need not be directly perceptible so long as it may be communicated with the aid of a machine or device.

The way in which copyright protection is secured is frequently misunderstood. Publication, registration, or other action with the Copyright Office is not required to secure copyright. There are, however, definite advantages to registering the copyright, particularly when it comes to enforcement.

Copyright is secured automatically when the work is created. A work is "created" when it is fixed in a copy or phono-record for the first time. A work created (fixed in tangible form for the first time) on or after January 1, 1978 is automatically protected from the moment of its creation. It is ordinarily given a term enduring for the author's life plus an additional 70 years after the author's death.

The use of a copyright notice is no longer required under U. S. law, although it is often beneficial. Because prior law did contain such a requirement, however, the use of notice is still relevant to the copyright status of older works. Use of a notice may be important because it informs the public that the work is protected by copyright; it identifies the copyright owner; and it shows the year of first publication. Furthermore, in the event that a work is infringed upon, if a proper notice of copyright appears on the published copy or copies to which a defendant in a copyright infringement suit had access, then no weight shall be given to interposition of a defense based on innocent infringement in mitigation of actual or statutory damages (Department of Commerce 2003).

TRADEMARK

A trademark is a distinctive mark used to distinguish the products of one producer from those of another. As defined in the Trademark Act of 1946, a trademark includes any word, name, symbol, device, or any combination thereof, adopted and used by a manufacturer or merchant to identify goods and distinguish them from those manufactured or sold by others. Its basic function is to identify the origin of the product to which it is affixed.

As with patents and copyrights, trademarks enjoy legal protection. A trademark offers a monopoly of sorts, because it provides the owner with the exclusive right to use the mark. Its value is derived from its function as a visual assurance of the source or manufacture of the product bearing the mark, thereby creating and maintaining demand for the product.

Rights can be established for a mark based on its legitimate use. However, owning a federal trademark registration on the Principal Register provides several advantages, such as:

- constructive notice to the public of the registrant's claim of ownership to the mark;
- a legal presumption of the registrant's ownership of the mark and the registrant's exclusive right to use the mark nationwide on or in connection with the goods and/or services listed in the registration;
- the ability to bring an action concerning the mark in federal court;
- the use of the U.S registration as a basis to obtain registration in foreign countries; and
- the ability to file the U.S. registration with the U.S. Customs Service to prevent importation of infringing foreign goods.

Any time rights are claimed in a mark, the "TM" (trademark) or "SM" (service mark) designation should be used to alert the public to the claim, regardless of whether there is an application with the USPTO. However, the federal registration symbol "®" may be used only after the USPTO actually registers a mark, and not while an application is pending. Also, the registration symbol may be used with the mark only on or in connection with the goods and/or services listed in the federal trademark registration.

Rights in a federally registered trademark can last indefinitely if the owner continues to use the mark on or in connection with the goods and/or services in the registration and files all necessary documentation in the USPTO at the appropriate times (Department of Commerce 2003).

TRADE SECRETS

In industry, a trade secret offers a legitimate and effective means of market protection. A *trade secret* is simply a means of restricting information on formulas, designs, systems, or compiled information, thus giving the organization an opportunity to obtain an advantage over competitors who do not know or use the secret.

Courts have held that a bona-fide trade secret must involve information that exhibits a measurable amount of novelty and originality, provides a competitive advantage over competitors who do not use it, and it is generally unpublished.

Not withstanding the legal aspects of trade secret law, trade secrets offer an attractive and practical alternative to patents and copyrights as a means of protecting proprietary information. In many cases, an organization would rather avoid the time, expense, and risk involved in the more common patent procedures. It may choose therefore to enshroud the project with secrecy and capitalize on it immediately. Once a patent has been issued, the entire file becomes public information, making it easy for competitors to analyze and improve the product or process, thereby weakening the advantages the patent is supposed to provide.

Generally, a trade secret is considered to be a property right afforded protection under law from fraudulent access by outside parties. However, it is legitimate to uncover a trade secret through the practice of reverse engineering, that is, starting with a finished product or process and working backward in logical fashion to discover the underlying new technology (Department of Commerce 2003).

56.10 WARRANTIES AND LIABILITY

A *warranty* is an assurance or guarantee that a product or service will provide a specified level of quality. There are two general types of warranties, express warranties and implied warranties. An *express warranty* is a statement regarding the satisfactory performance of a product or service after purchase. In the past, courts have indicated that it is the buyer's responsibility to inspect goods prior to purchasing and the consumer does not have to buy anything if they find defects or are uncertain about the product. More recently, however, the courts have placed more responsibility on the seller. Claims made by the seller reduce or relieve the buyer's responsibility for inspecting the goods prior to purchasing.

There are several types of *implied warranties*, such as warranty of merchantability and warranty of fitness, which are defined as follows.

- *Warranty of merchantability* is an assurance by the seller that the goods or services in question are fit for their ordinary or intended purpose.
- *Warranty of fitness* applies when a buyer describes his or her intended use for a product and then relies on the skill or judgment of the seller to select the appropriate product.

LIABILITY

There are several types of *liability* including negligence, breach of warranty, and strict liability.

Product liability is an issue for any company in the supply chain regardless of the role they play in the process. It is important to note that there are no federal liability laws. States form their own liability statutes and generally rely on common law, the Uniform Commercial Code, and tort law.

Negligence is indicated or proven when a product's producer has not demonstrated careful design responsibility and this negligence causes a buyer's injury.

Breach of warranty refers to a breach of express or implied warranty. It is similar to breach of contract, except it can typically extend beyond the seller to include the manufacturer.

Under *strict liability*, any manufacturer or supplier can be found guilty if their product is defective and causes harm. Manufacturers and suppliers can be found liable regardless of the amount of care exercised.

Product defects consist of three basic types: design defects, manufacturing defects, and marketing defects. *Design defects* arise from flaws in the design and engineering of a product. *Manufacturing defects* are created during the product's manufacture. Even if the design is good, errors in the manufacturing sequence can cause field failures. Finally, *marketing defects* are a result of inadequately warning or instructing buyers. Marketing defects may consist of incomplete or misleading directions for using the product or lack of warning of potential danger when using the product.

REVIEW QUESTIONS

56.1) Which organizational structure focuses on cross-functional and multidirectional information flow?

56.2) Which leadership theory suggests that leaders should choose from a set of decision styles according to a set of rules?

56.3) According to Maslow, if someone has just eaten, will offering that person more food be a motivator?

56.4) Which strategic planning tool is used to compare one company with a more successful company for the purpose of identifying best practices?

56.5) Which group discussion technique requires the group members to be strangers?

56.6) Patenting a screwdriver today would be impossible because a screwdriver is ___________?

56.7) Which type of warranty assures the buyer that the goods or services in question are fit for their ordinary purpose?

REFERENCES

Department of Commerce. 2003. *General Information Concerning Patents*. Washington, DC: United States Patent and Trademark Office.

Harrison, J. and St. John, C. 2002. *Foundations in Strategic Management*, Second Edition. Cincinnati, OH: South-Western College Publishing.

Meredith, J. and Mantel, S. 1989. *Project Management: A Managerial Approach*, Second Edition. New York: John Wiley and Sons.

Veilleux, R. and Petro, L., eds. 1988. *Tool and Manufacturing Engineers Handbook*, Fourth Edition. Volume 5: *Manufacturing Management*. Dearborn, MI: Society of Manufacturing Engineers.

Chapter 57

Industrial Safety, Health, and Environmental Management

57.1 ACCIDENT THEORY

Accidents are unplanned occurrences that result in injury, death, lost production time, and/or damage to property (Raouf 2002). Each year, workplace accidents exact a heavy price on businesses in the form of medical payments, insurance premiums, reduced productivity, and lost work time. Protecting employees from hazards in the workplace is not only required, but it can save costs. Understanding why accidents happen helps to prevent future incidents, reduces costs to businesses, and reduces worker injuries and fatalities.

Theories of accident causation widely accepted by occupational health and safety professionals include the domino theory, the human factors theory, and the multiple causation theory. These three are discussed here since they cover the broad range of accident theory. However, there are many others that have merit.

DOMINO THEORY

Herbert Heinrich worked for the Travelers Insurance Company. In the late 1920s and early 1930s, he observed that accidents followed certain patterns. His study of accidents resulted in the observation that 88% of industrial accidents are caused by unsafe acts; 10% are caused by unsafe conditions; and 2% were just unavoidable or "acts of God" (Goetsch 2002). Heinrich came to the conclusion that accidents were caused by a series of factors in sequence (dominos) that resulted in the accident.

1. Ancestry and social environment—certain persons are more prone to accidents, for example, a person with a tendency toward risk-taking or anti-social behavior (ancestry) or who reacts to peer pressure or workplace expectations (social environment) in an unsafe way.
2. Fault of person—character flaws or poor choices by workers contribute to hazardous situations.
3. Unsafe act and mechanical or physical hazard—an unsafe act by a person compounded by a workplace setting that has an inherent mechanical or physical hazard present leads to the accident.
4. Accident—an unexpected event occurs.
5. Injury—as a result of the accident, minor or serious injuries are usually the unfortunate outcome of the chain of events.

If it were possible to prevent the unsafe act or remove the unsafe conditions, the chain of events would be broken and the accident avoided. Removing the key "domino"—the unsafe act that occurs in unsafe conditions—will prevent the accident (Goetsch 2002).

HUMAN FACTORS THEORY

Realizing that humans are error prone, the *human factors theory* associates accident causation to a chain of events put in motion by human error. There are three factors that lead to this error.

- Overload—individuals have limitations as to the amount of stress they can handle. Stress can take the form of job responsibilities, personal commitments or problems, overall health and physical condition, or the work environment (for example, lighting, temperature, noise levels, other workers). When a person reaches the limit of internal and/or external stress he or she can tolerate, errors can occur.
- Inappropriate response—experience, training (or the lack of it), and demeanor determine how a person responds to a situation. An inappropriate response would be failure to recognize a hazard, choosing to ignore warnings or clues, or electing not to use safety equipment or follow safe procedures. The result is an accident.
- Inappropriate activities—accidents can occur when people perform a task without the necessary training or when they incorrectly judge the degree of risk involved and continue based on that incorrect judgment (Goetsch 2002).

MULTIPLE CAUSATION THEORY

The cause of an accident often has a variety of contributory factors that do not always fit neatly into a unique theory or explanation. For this reason, the multiple causation theory is valuable in accident causation analysis. An extension of the domino theory, multiple causation groups the factors that lead to accidents as either behavioral or environmental.

- Behavioral—these are the human elements that contribute to accidents such as attitude, inadequate training or skill, inattentiveness, or physical limitations such as fatigue. When the behavior of the worker is contrary to safe work practice or common-sense thinking, the result is typically a near-miss or, in some cases, an accident.
- Environmental—the conditions at the time or scene of the accident contribute to the unfortunate result. Unsafe conditions that can cause accidents include mechanical or electrical failure, poor housekeeping, dim lighting, excessive noise, inadequate safeguards, or improper personal protective equipment (PPE) or apparel.

The combination of unsafe behavior and unsafe environment is the cause of accidents. It is not possible to have an accident with only one element and, conversely, it is possible to prevent accidents by removal of one element (Raouf 2002).

In summary, safe work practices and adherence to work rules and applicable safety and health standards can successfully reduce injury and illness in the workplace. Manufacturing managers and their employees have a responsibility to promote and maintain a safe work environment.

57.2 OCCUPATIONAL SAFETY AND HEALTH ADMINISTRATION (OSHA)

The Williams-Steiger Occupational Safety and Health Act of 1970 (OSHAct) has had a significant impact on workplace safety. Since its enactment, there has been a reduction in the number of worker deaths, injuries, and illnesses. It has also reduced employer liability. The OSHAct created the Occupational Safety and Health Administration (OSHA), which enforces standards to assure safe and healthful working conditions. The OSHAct also created the National Institute for Occupational Safety and Health (NIOSH), which is an agency that develops standards and performs ongoing research, education, and training in occupational health and safety. The OSHAct established separate but dependent responsibilities for employers and employees. Employer duties are explained in Section 5(a)(1) of the OSHAct in the general duty clause:

> "Each employer shall furnish to each of his or her employees employment and a place of employment free from recognized hazards that are causing or are likely to cause death or serious physical harm to his employees and shall comply with occupational safety and health standards promulgated under this Act."

Thus, the employer has the obligation to proactively maintain a safe work environment, rather than reacting to situations as they develop. To facilitate the employer in his or her charge, the OSHAct states that employees have the responsibility to "...comply with occupational safety and health standards and all rules, regulations, and orders issued pursuant to this Act that are applicable to his or her own actions and conduct." However, OSHA can not penalize employees who do not comply. Only employers can receive citations for violating standards.

It should be noted that, although the safety standards created by OSHA are fairly complete, there are instances where a regulation may not exist. When this situation occurs and the employer fails to take corrective action when the problem was recognized, and there was a feasible method to solve the problem, a citation under the "catch-all" general-duty clause may be issued.

The OSHAct covers any employer engaged in business who affects commerce and who has employees. It is important to note that it covers every employer, whether there is one employee or 1,000. The exception is the applicability of recording and reporting guidelines, which exempts employers with fewer than 11 employees from keeping a log and summary of occupational illnesses and injuries. All other standards and rules apply regardless of the number of employees. It is important to note that self-employed individuals with no employees are not considered employers and thus are not subject to OSHA requirements. Also, OSHA does not have jurisdiction over federal and state employees, mining, and facilities covered by the Atomic Energy Act of 1954.

To facilitate understanding of the rights and responsibilities of employees and employers, the OSHAct requires the employer to post the OSHA notice or poster in a conspicuous place. The poster, provided free by the Department of Labor, informs employees of protections and obligations covered in the OSHAct. It also includes contact information for the nearest office of the Department of Labor.

GOVERNMENT AGENCIES AND REGULATIONS

The Code of Federal Regulations (CFR) is published every year by the U.S. Government and independent book publishers. It contains the rules and regulations made by federal agencies and executive departments. For example, Title 29 of the CFR contains standards put forth by the Department of Labor. Occupational Safety and Health "Standards for General Industry" are contained in 29 CFR Part 1910; Part 1926 is construction; and Parts 1915 through 1918 apply to shipbuilding, ship repair, and longshoring. For most manufacturers, general industry standards (29 CFR 1910) apply and should be reviewed on a regular basis. The Department of Labor also maintains an extensive database of workplace-specific information and guides for employers and employees at the OSHA website (http://www.osha.gov).

The information presented in this chapter is accurate at the time of writing. It is advisable for the reader to review the specific titles of the CFR relevant to environmental and industrial safety as they are updated annually.

The OSHAct allows individual states to establish their own safety and health programs. The minimum requirement is that they must have a standard that is identical to, or at least as effective as, the federal standard.

OSHA has 10 regional offices that can provide off-site and on-site inspection and consultation services. These are the offices of the area directors who are the direct supervisors of the field compliance officers. The offices also can provide contact information for the states and territories with individual programs.

INSPECTIONS, CITATIONS, AND PENALTIES

OSHA inspections are, with few exceptions, conducted without prior notice (29 CFR 1903). Upon presenting appropriate credentials to the owner, operator, or agent in charge, an OSHA compliance officer may:

- enter without delay;
- inspect and investigate at reasonable times all pertinent conditions, structures, machines, apparatus, devices, equipment, and materials;
- question privately any employer, owner, operator, agent or employee; and
- review records required by the OSHAct and any other records directly related to the inspection.

This does not mean that OSHA's right to inspect is unlimited, rather the inspection must be reasonable and the agent should provide the cause for the inspection, enabling the employer to give access to the particular area. The OSHA compliance safety and health officer (CSHO) must sign a confidentiality agreement, if requested, to protect any trade secrets that might be revealed during an inspection. In addition, the officer usually initiates opening and closing inspection conferences.

Employer and employee representatives may participate in the inspection and conferences, but not in any way that hampers the process. It is highly recommended that an employee or representative of the employer walk along with the compliance officer during the workplace inspection. The employer should use the same media that the compliance officer is using to document conditions in the facility. For example, if the officer brings a camera or video recorder, it is appropriate for the employer to also take photos of or videotape the same area in the plant from the best angle to portray the situation or possible problem. Photos and/or videos document workplace conditions as they existed at the time of the inspection and can serve as an objective observer of the situation. If the compliance officer takes written notes, it is acceptable to ask what they are writing down at that point in the inspection and also to make your own notes about the situation. This does not imply that the employer should take an aggressive or disrespectful stance with the officer. Rather, the employer should be inquisitive and assure that he or she has good documentation of the conditions the OSHA officer observed.

Reasons for Inspections

The priority for workplace inspections and investigations is as follows.

1. Imminent danger condition—an imminent danger is a situation that requires immediate correction. It is a condition that will cause or is likely to cause death or serious physical harm before standard enforcement procedures under the OSHAct can take effect. Health hazards, although not commonly considered imminent danger, can fall under this category if toxic chemicals are present or serious harm could be caused by improper use. An imminent danger condition or allegation usually will provoke an inspection within 24 hours. Employees should first inform the supervisor if they detect or suspect an

imminent danger condition. If corrective action is not taken, employees have the right to contact OSHA, and the compliance office will determine if the situation warrants immediate attention. The employee may refuse to be exposed to imminent danger by asking "What would a reasonable person do?"

2. Fatalities/catastrophes—if there is a fatality or serious accident that requires hospitalization of three or more employees, the incident must be reported to OSHA within 8 hours. If there is a death at the workplace, even if it is not apparently related to work conditions (for example, a worker dies of heart attack), it is better to report the incident to OSHA rather than risk a citation and large penalty for failure to report. Should the investigation that follows prove the death was not work-related, the incident will not be recorded as a workplace injury or illness.
3. Employee complaints—these are characterized as alleged standard violations—unsafe or unhealthful working conditions. They may be reported from other sources such as a visitor to a facility, another governmental agency officer, or other personnel. If the situation is an imminent danger complaint, an immediate investigation will occur. Otherwise, the OSHA compliance office makes a decision about the seriousness of the complaint.
4. Programmed high-hazard inspections—some industries, such as construction or foundry operations, have a high rate of incidents. These industries and those with high injury and illness rates due to chemical or environmental conditions warrant a higher priority level for inspection.
5. Follow-up inspections—inspections that resulted in an employer being cited in the previous year may require reinspection to determine whether the problem has been corrected. In some instances, a letter and/or photographs from the employer are acceptable to prove compliance.

Violations, Citations, and Penalties

There are a variety of violations and resulting citations and penalties that an OSHA compliance officer can levy against an employer, depending on the seriousness of the situation, the history of the problem, and the actions of the employer to correct the situation.

- *De minimus violations* are violations of standards that have no direct relationship to safety or health and are not included in citations. These are usually deviations of a standard that do not endanger the employee. For example, 29CFR 1910.28 (b)(15) and (c)(14) require guarding on all open sides of scaffolds more than 10 ft (0.3 m) above the ground. Where employees are tied off with safety belts in lieu of guarding, the intent of the standard is met, and the absence of guarding may be de minimus.
- An *other than serious violation* is cited when the hazardous condition would not normally cause death or serious injury, but does have a direct relationship to employee safety and health. This violation may be assessed a civil penalty of up to $7,000 for each violation. This penalty may be reduced if the employer's good faith efforts, history of positive workplace safety and health, and size of business warrant reconsideration by OSHA.
- A *serious violation* is deemed to exist if there is a substantial probability that death or serious injury could result from a condition or from one or more practices, operations, or processes adopted or in use unless the employer did not know and should not have

known of the presence of the violation. A serious violation has a mandatory penalty of up to $7,000 for each violation.

- The *willful violation* is cited where evidence shows either an intentional violation of the OSHAct or plain indifference to its requirements. A *repeated violation* results when reinspection finds the same or substantially similar violation. A repeated violation must be based on a citation that is not under appeal and has become final. The penalty for repeated violation is the same as for willful violations. Each willful or repeated violation may be assessed a civil penalty of up to $70,000, but not less than $5,000.
- A *criminal/willful violation* of a standard causes the death of an employee and there is evidence the violation was the cause of death. A fine of up to $250,000 for an individual, $500,000 for a corporation, and/or imprisonment for not more than six months may be ordered.
- *Failure to abate* is cited when an employer fails to correct a violation for which a citation has been issued within the period permitted for its correction. A civil penalty of not more than $7,000 may be assessed for each day the violation continues.

The employer, employee, or representative of employees may request an informal conference with an OSHA assistant regional director for the purpose of discussing issues raised by an inspection, citation, notice of proposed penalty, or notice of intention to contest. If the employer receives a citation, an unedited copy must be posted at or near the place where the violation occurred. The citation must remain posted until the violation has been abated, or for 3 working days, whichever is longer. The employer may also post a notice in the same location indicating the citation is being contested and the reasons for such contest. The notice of intention to contest must be postmarked within 15 working days of the receipt of the notice of the proposed penalty.

In many cases, the site or employer's safety officer is key to successful preparation of an OSHA case should the employer choose to contest a proposed penalty or alleged serious violation. Many cases are based on the question of feasibility—how the workplace should be made safer, how much safer, at what cost, and by what timetable. Usually, the safety professional has the training and experience to identify the relevant facts and their importance to the workplace under question, and evaluate the possible alternatives. Employees familiar with the operation may offer significant insight into the details of the problem as well as possible solutions. If the OSHA agent and/or the employer electronically recorded the site visit, this evidence can be combined with other data, charts, prints, or drawings and professional photographs to build the defense case. Attorneys, consultants, and management must work together as a team to bring a well-prepared case before the review commission. The review commission is an independent federal agency that decides contests of citations.

RECORD-KEEPING AND REPORTING

Each employer is required to maintain in each establishment a log and summary (OSHA Form No. 300 or equivalent) of all recordable occupational injuries and illnesses for that facility. Exceptions to the maintenance of the annual log and summary are employers who had no more than 10 employees at any time during the preceding calendar year. Each recordable incident must be entered into the log and summary within 7 calendar days after receiving information that an incident has occurred. Form 300 covers an entire year on a calendar basis. A summary of the OSHA 300 log from the previous year must be posted for employees no later than February 1 and remain posted until April 30 of the same year. In addition to the log and summary, an illness and incident report, such as OSHA Form 301, workmans' compensation, insurance, or other reports are kept for each recordable incident.

57.3 GENERAL INDUSTRY STANDARDS (29 CFR 1910)

MEANS OF EGRESS (SUBPART E)

A *means of egress* is a continuous and unobstructed way of exit travel from any point in a building or structure to a public way. It consists of three separate and distinct parts: the way of exit access, the exit, and the way of exit discharge. Essentially, an egress is a path or way out of a building.

"CFR for General Industry, Subpart E—Means of Egress" has the following main requirements for any establishment.

- The exits will be designed for prompt escape of occupants in an emergency and shall be such that there is more than one safeguard in case of human or mechanical failure. In each case, there must be a backup safeguard (for example, an additional exit or override for automatic doors) for employees to escape, especially if the building has more than one floor.
- Building structures will be constructed, maintained, and operated to avoid danger to occupants from fire, smoke, fumes, or resulting panic during the period of time reasonably necessary to escape from the building during an emergency.
- There shall be an adequate number, location, and kinds of exits to allow everyone to escape. There are documented cases where people have been trampled during a panic because the exits were too few or too narrow to allow orderly evacuation during a fire.
- No lock or device to prevent emergency egress is allowed (exceptions are mental, penal, or corrective institutions where supervisors are continually on duty). Doors that automatically close and lock from the outside must have a release and swing to the outside easily. Allowing boxes, barrels, or other objects to be stacked against an outside exit door is a disaster waiting to happen. Workplace rules that forbid such practice are one method of prevention.
- Every exit shall be clearly visible or the route to reach it shall be conspicuously marked so that every occupant who is capable will readily know the direction of escape. Any passageway or door not leading to an exit but that could be mistaken for an exit must be marked or arranged to minimize its possible confusion with an exit. It is best to label doors or passages as "Not an Exit" or "Storeroom" or "To Basement." Every exit sign must have the word "Exit" in plainly visible letters not less than 6 in. (152 mm) high. Signs also must be illuminated by a light source that has a backup power system.

EMERGENCY ACTION AND FIRE PREVENTION PLANS (SUBPART E)

Emergency Action Plan

One of the most common violations cited is failure to plan for, document, and implement an employee emergency action plan and fire prevention plan. The OSHA standard requires the following minimum elements to be included in an emergency action plan:

- emergency escape procedures and route assignments,
- procedures for employees who must remain to operate critical equipment prior to evacuation,

- a way to account for all employees after the evacuation has been completed,
- rescue and medical duty assignments for those employees who are to perform them,
- a preferred means to report fires and other emergencies, and
- names or job titles of responsible persons or departments.

An employee alarm system must be provided and be distinctly different for each type of situation (for example, fire, tornado, end of shift, or call for fire brigade). Employees must be trained on the meaning of each alarm, how to report emergencies, where emergency phone numbers are posted, and special provisions for visual or hearing-impaired employees.

Fire Prevention Plan

A written fire prevention plan must be developed. All employees should be trained on the fire hazards of the materials and processes to which they are exposed as well as the correct procedures in the event of a fire. Good housekeeping can prevent many fires. The employer must control accumulations of flammable and combustible waste materials so they do not contribute to a fire emergency. No place in the standard requires employees to fight fires or otherwise endanger themselves in the event of a fire.

The minimum requirements for a written fire prevention plan are a list of the major workplace fire hazards and their proper handling and storage procedures, the names of those persons responsible for maintenance of equipment and systems in place to prevent or control fires, and the names of persons responsible for control of fuel source hazards. Plant equipment or heat-producing equipment must be properly maintained to prevent accidental ignition of combustible materials. Maintenance procedures must be included in the fire prevention plan.

Fire Safety (Subpart L)

Most small- to medium-sized manufacturers do not have the resources to support a company fire brigade. The OSHA standards applying to industrial fire departments or fire brigades require extensive documentation, training, and equipment. For those companies without internal fire-fighting crews, most must rely on portable fire extinguishers, standpipe and hose systems, automated sprinkler systems, and/or fixed extinguishing systems.

Portable fire extinguishers are the most common equipment available for fire protection. The placement, use, maintenance, and testing of portable extinguishers must follow specific guidelines and requirements, depending on the situation and the fire protection plan. It is important to note that where an employer has established and implemented a written fire safety policy requiring the immediate and total evacuation of employees from the workplace upon the sounding of a fire alarm signal, and where an employer has an emergency action plan and a fire prevention plan that meet OSHA requirements, portable fire extinguishers are not required unless a specific standard requires them. Most establishments have fire extinguishers installed at the insistence of insurance companies to mitigate property damage.

When portable fire extinguishers are required, they must be located, identified and made readily accessible to employees. The extinguishers must be maintained in a fully charged and operable condition and kept in their designated place at all times, except during use. One of the most common violations of this part of the standard is when fire extinguishers are blocked by stacked boxes, pallets, or other misplaced materials. It is easy to forget where the fire extinguisher is located when its location is not well marked, access is not kept clear, or it is moved without replacement.

Fire extinguishers must be selected according to the hazards present in the workplace. Figure 57-1 illustrates identifiers for different types of fire extinguishers and the types of fires on which they should be used.

Portable fire extinguishers must be inspected at least once annually and a record of annual maintenance data kept for the life of the equipment. When maintenance or recharging requires the extinguisher to be removed, a replacement must be provided. In addition to maintenance, hydrostatic testing of extinguishers must be performed every 5 or 12 years depending on the type of extinguisher. Training for employees on the general principles of fire extinguisher use and beginning stage firefighting is required if employees are expected to use fire extinguishers.

Automatic sprinkler systems are another layer of fire protection common in industrial settings. The main concerns regarding these systems are that they be properly installed using standard components, maintained by performing a main drain flow test each year, and have an adequate water supply to provide water flow for at least 30 minutes. Fixed extinguishing systems using gas, dry chemical, water spray, or foam as the extinguishing agent must be used properly. Most of these systems should have a pre-discharge employee alarm, particularly those that displace oxygen in the air as the method of fire suppression.

57.4 WELDING, CUTTING AND BRAZING (SUBPART Q)

Basic precautions for fire prevention in welding, cutting, and brazing have been developed from the National Fire Protection Association (NFPA) standards. The OSHA standard relevant to welding, cutting, and brazing addresses fire hazards, fire watch requirements, personal welding protection, oxygen fuel welding and cutting safety, and welding or brazing in confined spaces.

FIRE HAZARDS

The object to be cut or welded should be removed from any fire hazards in the area. If this is not possible, all movable fire hazards in the vicinity must be taken to a safe place. At that point, guards must be used to contain the sparks, heat, and slag and protect immovable fire hazards. Wherever there are cracks in floors or openings that cannot be closed, precautions shall be taken so readily combustible materials are not exposed to sparks or slag. Fire extinguishers must be kept in the area and ready for instant use.

FIRE WATCH

After welding or cutting in an area where appreciable combustible materials are closer than 35 ft (10.7 m) to the point of operation, or where sparks from the job could ignite materials more than 35 ft (10.7 m) away, a

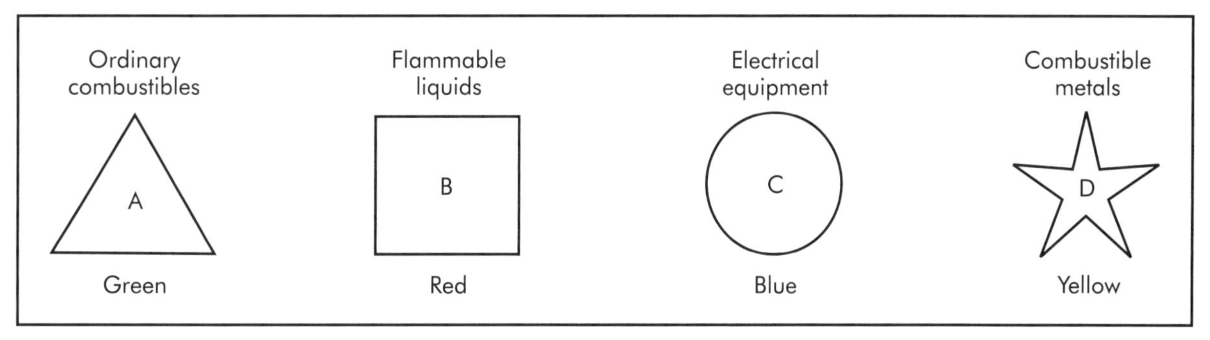

Figure 57-1. Portable fire extinguisher classes (Drozda and Wick 1983).

fire watch must be maintained for at least 30 minutes to detect and extinguish possible smoldering fires. Paper, wood shavings, and/or textile fibers must be swept from a radius of 35 ft (10.7 m). Combustible floors should be kept wet, covered with damp sand, or protected by fire-resistant shields. It is important that personnel operating arc welding or cutting equipment be protected from shock on damp or wet floors.

PERSONAL WELDING PROTECTION

Welders and helpers must have eye, hand, foot, and body personal protective equipment (PPE) available depending on the type, size, nature, and location of the work to be performed. Helmets and hand shields must be used during arc welding or arc cutting operations. Resistance welding or brazing equipment operators must use face shields or goggles, depending on the particular job, to protect their faces and eyes. Goggles or other suitable eye protection shall be used for all gas/oxygen welding and cutting operations. Safety glasses without side shields may be used if the user has suitable filter lenses for performing gas welding on light work, torch brazing, or inspection.

The specifications for helmets and hand shields are stringent, requiring them to be made of material that insulates against heat and electricity and protects the face, neck, and ears from direct radiant energy from the arc. Helmets, shields, and goggles must not be readily flammable and must be able to be sterilized. Goggles should be ventilated to prevent fogging as much as possible.

Welding operations have the capability to damage an employee's eyesight. Table 57-1 is a guide for the selection of proper shade numbers.

Where work permits, the welder should be located in an individual booth painted with a low-reflectivity finish such as zinc oxide and lamp black (to absorb ultraviolet radiation). If this is not possible, portable noncombustible screens can be used. It is important to allow air circulation at floor level to prevent the buildup of noxious fumes. Workers in nearby areas should be protected from the welding rays by screens or shields, or be required to wear appropriate goggles.

OXYGEN FUEL WELDING AND CUTTING SAFETY

Mixtures of fuel gases and air or oxygen may be explosive and must be guarded against combustion. Acetylene fuel gas should not be generated, piped, or utilized at a pressure in excess of 15 psi (103 kPa). When using acetylene cylinders, the cylinder valve should not be opened more than one and one-half turns of the spindle, and preferably no more than three-fourths of a turn, thereby allowing quick shut down in an emergency.

Compressed gas cylinders must be legibly marked with either the chemical or trade name of the gas. Cylinders must be kept away from radiators and other sources of heat. Inside of buildings, cylinders must be

Table 57-1. Welding shade selection guide (29CFR 1910.252)

Welding Operation	Shade Number
Shielded metal-arc welding: 1/16, 3/32, 1/8, 5/32 in. electrodes	10
Shielded metal-arc welding: 3/16, 7/32, 1/4 in. electrodes	12
Shielded metal-arc welding: 5/16, 3/8 in. electrodes	14
Torch brazing	3 or 4
Cutting, up to 1 in.	3 or 4
Cutting, 1–6 in.	4 or 5
Gas welding, up to 1/8 in.	4 or 5
Gas welding, 1/8–1/2 in.	5 or 6

stored in a well-protected, well-ventilated, dry location, at least 20 ft (6.1 m) from highly combustible materials. Cylinders should be stored in assigned places away from elevators, stairs, or gangways and where they will not be knocked over or damaged by passing or falling objects, or subjected to tampering by unauthorized persons. Cylinders must not be kept in unventilated enclosures such as lockers and cupboards.

Valve protection caps must always be in place and hand-tight, except when cylinders are in use or connected for use. Oxygen cylinders in storage should be separated from fuel-gas cylinders or combustible materials (especially oil or grease) by a minimum distance of 20 ft (6.1 m), or by a noncombustible barrier at least 5 ft (1.5 m) high with a fire-resistance rating of at least 30 minutes. Empty cylinders must have their valves closed.

Unless cylinders are secured on a special truck, regulators shall be removed and valve-protection caps put in place before cylinders are moved. Before a regulator is removed from a cylinder valve, the cylinder valve shall be closed and the gas released from the regulator.

WELDING OR BRAZING IN CONFINED SPACES (SUBPART J)

Confined space refers to a space that:

- is large enough and so configured that an employee can bodily enter and perform assigned work,
- has limited or restricted means for entry or exit (for example, tanks, vessels, silos, storage bins, hoppers, vaults, and pits), and/or
- is not designed for continuous employee occupancy.

A *non-permit confined space* (NPCS) is a confined space that does not contain or, with respect to atmospheric hazards, have the potential to contain, any hazard capable of causing death or serious physical harm. A *permit-required confined space* (PRCS) is a confined space that contains or has a potential to contain a hazardous atmosphere, or contains a material that has the potential for engulfing an entrant, or has an internal configuration such that an entrant could be trapped or asphyxiated by inwardly converging walls or by a floor that slopes downward and tapers to a smaller cross-section, or contains any other recognized serious safety or health hazard.

When welding or brazing is to be done in a confined space, there are a series of safeguards for personnel that must be obeyed, as the consequences can be tragic if ignored. Subpart J 1910.146 covers the requirements for a confined space program and the permit system.

Ventilation is key to protecting welders and helpers from accumulation of toxic gases and fumes or oxygen deficiency. Air-replacing equipment that provides air exchange for the space or airline respirators/hoods are solutions for areas not immediately hazardous to life. In the case where the area is immediately hazardous to life, full-face pressure demand, self-contained breathing apparatus or a combination approved by the National Institute for Occupational Safety and Health (NIOSH) under 42 CFR Part 84 must be used. Regardless of the situation, an attendant must be stationed outside of the confined space to ensure the safety of those persons working inside.

Other regulations for confined space welding operations include leaving the gas cylinders and welding machines outside and, if they are on wheels, they must be securely blocked to prevent accidental movement; and a retrieval line must be used if the welder descends over 5 ft (1.5 m) vertically to remove the worker in case of emergency. An attendant outside the space must be able to perform a preplanned non-entry rescue procedure and summon the rescue team.

When welding is to be suspended for any substantial period of time (such as during a break), the electrode must be removed from the holder and the holder positioned so that accidental contact cannot occur; when practical, gas welding or cutting torches should be removed from the space during breaks.

57.5 PERSONAL PROTECTIVE EQUIPMENT (SUBPART I)

Personal protective equipment (PPE), when carefully selected and consistently used in the workplace, has been able to prevent many lost workdays and fatalities. The American National Standards Institute (ANSI) has provided codes, standards, and recommended practices that have been adopted by OSHA, particularly in the area of PPE. In recent years, updates to OSHA's general requirements for PPE have evolved to incorporate the most recent ANSI standards. They provide guidance for selection and use of PPE based on performance-oriented requirements.

In general, the employer is required to assess the workplace to determine if there are hazards or conditions present, or likely to be present, which warrant the use of personal protective equipment. Training of employees who are required to use PPE must include (at minimum):

- when PPE is necessary;
- which PPE is necessary;
- how to properly don, doff, adjust, and wear PPE;
- limitations of the PPE; and
- proper care, maintenance, useful life, and disposal of PPE.

HEARING PROTECTION (SUBPART G)

Hearing loss from overexposure to noise could be considered a chronic rather than a critical problem for managers. Workers can incur significant hearing loss from the gradual effects of exposure to excessive noise levels over time at work or at home, rather than as the result of a single catastrophic incident. The technology to measure workplace sound levels via a sound level meter and test worker hearing function by audiometric devices has been available for decades; thus hearing conservation programs have had a long history of success.

When employees are subjected to sound levels exceeding those listed in Table 57-2, engineering controls such as sound dampening devices, design and use of quieter equipment, and changes in the workplace environment should be considered first. Administrative controls such as limiting worker exposure by workplace rules, scheduling or rotation of workers, or other management initiatives are the second line of defense against excessive noise.

It should be noted that people must be exposed to noise before the regulations are applicable. A machine producing 130 dBA (130 decibels on the "A" scale of a sound meter at slow response) is not in violation of the OSHA standard if no one is exposed to it. Sounds produced by equipment (noise emissions) should not be confused with sounds received by workers (noise exposure). Figure 57-2 illustrates typical sound levels encountered by machine operators.

Table 57-2. Noise exposure limits permitted by OSHA

Duration per Day (hours)	Sound Level (dBA)
8	90
6	92
4	95
3	97
2	100
1.5	102
1	105
0.5	110
0.25 or less	115

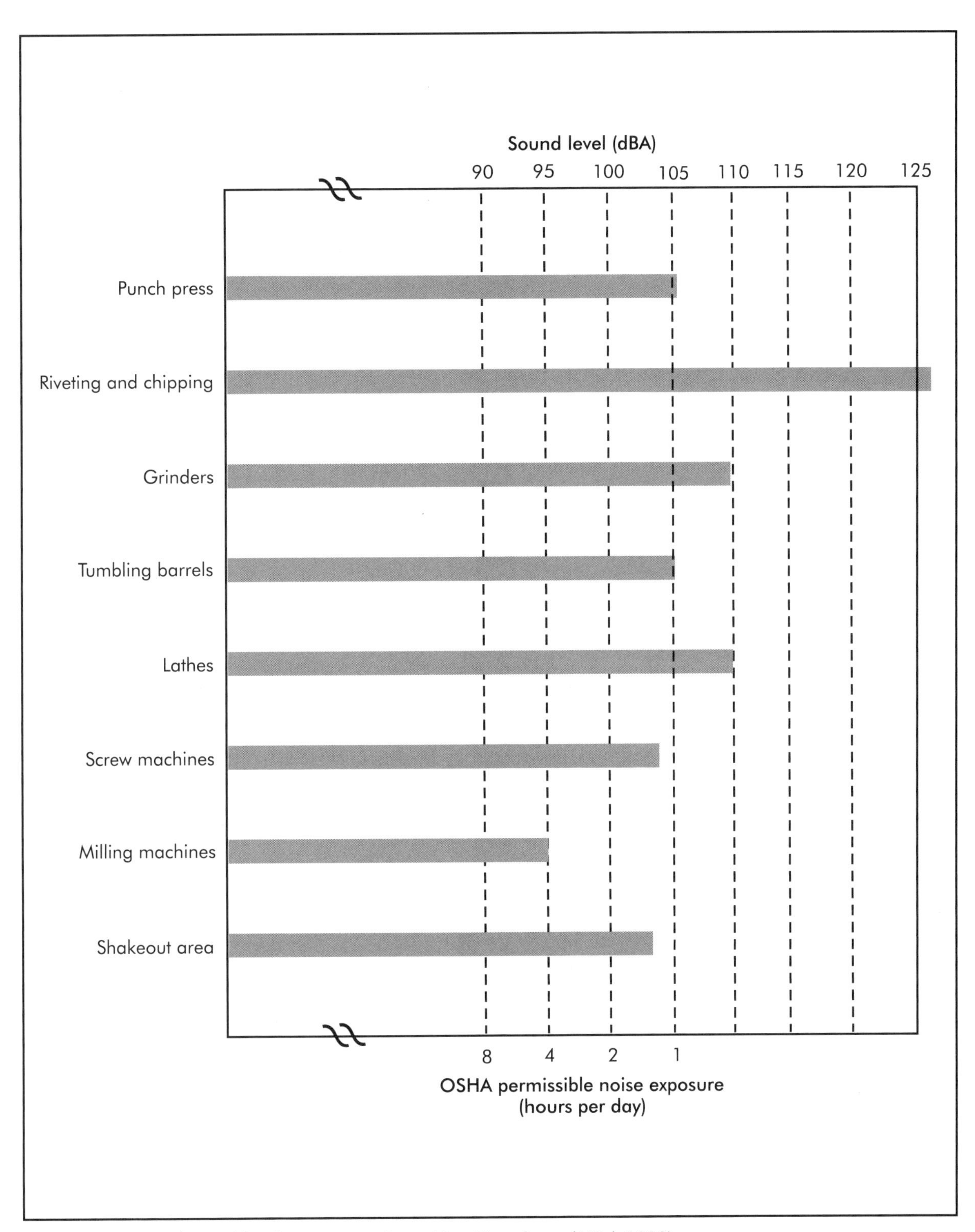

Figure 57-2. Typical sound levels at operator's position (Drozda and Wick 1983).

In most industrial settings, noise levels vary during the course of the average workday. Machines may operate at different loads, employees may move between departments, and some equipment may be used irregularly; all generate what is called a "noise dose." The time-weighted-average (TWA) noise dose, composed of noise exposures at varying levels, is calculated by:

$$T_A = 16.61\log_{10}\left(\frac{D}{100}\right) + 90 \quad (57\text{-}1)$$

$$D = 100\left(\frac{C_1}{T_1} + \frac{C_2}{T_2} + \frac{C_3}{T_3} + \ldots \frac{C_n}{T_n}\right) \quad (57\text{-}2)$$

where:

T_A = time-weighted-average noise dose (dBA)
D = noise dose (%)
C_n = total time of exposure at a specific noise level (hours)
T_n = reference duration as given in Table 57-3 for the specific noise level (hours)

If the noise levels rise above the established criteria, the employer must use suitable engineering and administrative controls to reduce noise levels. Employers must implement a hearing conservation program and make hearing protectors available to all employees exposed to 8 hours TWA of 85 dB or greater at no cost. Regardless of the type used, if the noise level is above 90 dB TWA, the hearing protection must be able to attenuate (reduce) noise levels back to the TWA of 90 dB at minimum. Proper fitting and requiring employees to wear hearing protection are elements of a successful hearing conservation program. Offering employees a variety of hearing protection devices will improve compliance.

In addition to a documented and implemented hearing conservation program, the company must perform baseline and annual audiograms on all employees exposed to elevated noise doses. The baseline measurement is important as it establishes the employee's capacity to hear at the time of employment. Annual audiograms track changes in hearing capacity. The employer must maintain these records for the duration of the employee's employment. It is recommended, however, that they be held for several years after the employee leaves the company should there be a question later about the nature or cause of an employee's hearing loss. In addition to these records, employers are required to maintain noise exposure measurement records for at least two years.

EYE AND FACE PROTECTION

Some of the associated eye and face hazards in manufacturing include flying par-

Table 57-3. Reference durations for time-weighted-average calculation of sound level (29 CFR 1910.95 App. A)

A-weighted Sound Level, dB	Reference Duration, *T* (hours)
80	32
81	27.9
82	24.3
83	21.1
84	18.4
85	16
90	8
95	4
100	2
105	1
110	0.5
115	0.25
120	0.125
125	0.063
130	0.031

ticles, molten metal, liquid chemicals, chemical gases or vapors, and exposure to light radiation from welding or cutting operations. Eye and face protection must be in accordance with ANSI Z87.1-1989. Most eyewear manufacturers stamp Z87.1 on their safety glasses to denote compliance. Side protectors (permanent, slide-on, or clip-on) are required when there is a possibility employees will be exposed to flying objects. Depending on the particulate matter and its hazard, full-face shields may be needed in grinding operations. Splash goggles, rather than glasses, should be used when working with liquids.

HEAD, HAND, LEG, AND FOOT PROTECTION

When there is potential for injury to employees' heads from falling objects, or bumping their heads on pipes or beams, employees are required to wear protective helmets. In operations where materials are stored on tall racks or where equipment is running overhead, hardhats are required. All hardhats must comply with ANSI Z89.1-1997 and resist penetration by objects, absorb the shock of a blow, be water-resistant and slow burning, and come with instructions for proper adjustment of the suspensions and headbands.

Another hazard in the workplace is exposure to energized electrical conductors. The standard requires headgear that can protect the head from accidental contact. Electricians and other maintenance workers should be fitted with class A or B helmets that offer protection from low-voltage conductors (tested to 2,000 volts) or high-voltage conductors (tested to 20,000 volts), respectively. Class C helmets only provide impact and penetration resistance and should not be used around electrical hazards.

When the workplace presents the danger of foot and leg injuries due to falling or rolling objects, objects piercing the sole of a boot, hot or wet surfaces, or where feet are exposed to electrical hazards, protective footwear is required. Safety shoes or boots must comply with ANSI Z41-1991, thus providing impact and compression protection. In certain hazardous environments, safety shoes with puncture protection, metatarsal protection, heat-resistant soles, or those that are electrically conductive or insulated must be worn. Metal insoles protect against puncture, and can have metatarsal guards built into the top of the shoe. Shoes that are designed to be electrically conductive prevent the buildup of static electricity and are necessary in explosive atmospheres. Nonconductive shoes protect workers from electrical hazards.

The choices for foot and leg PPE include:

- leggings that protect the lower legs and feet from heat hazards like molten metal or welding sparks;
- metatarsal guards that can be strapped to the shoe to protect the instep area from impact or crushing;
- toe guards that fit over the toes of regular shoes; or
- safety shoes.

One of the more recent updates to the general industry standards is the requirement for hand protection in the workplace. When employees are exposed to hazards such as skin absorption of chemicals, severe cuts or lacerations, severe abrasions, chemical or thermal burns, or harmful temperature extremes, the employer must require employees to use hand and arm protection. Dexterity, frequency and degree of exposure, and physical stresses that will be applied should be considered when selecting hand and arm PPE.

RESPIRATORY PROTECTION

Breathing harmful dust, fumes, mists, gases, smoke, sprays, or vapors can cause occupational diseases and injuries. Very hazardous atmospheres can be classified as "immediately dangerous to life or health (IDLH)"—where there is a direct threat to life, irreversible health effects from exposure, or impairment of a person attempting to escape from a dangerous situation.

In general, it is more common to find contaminants that fall within permissible exposure limits (PEL) as determined by 1910.1000 Table Z-1 in the general industry standard. It should be noted that as research and testing of new chemical compounds and air contaminants progresses, new limits are established. It is best to resolve exposure issues by using engineering controls, since a respirator program requires extensive documentation, administration, and maintenance costs.

If engineering controls will not adequately protect workers from air contaminants, a respiratory protection program must be developed, documented, and implemented. A respirator program must include worksite-specific procedures for:

- selecting respirators;
- medical evaluations of employees required to use respirators;
- fit testing of respirators;
- proper use of respirators;
- ensuring adequate quality, quantity, and flow of breathing air for atmosphere-supplying respirators;
- training of employees on the respiratory hazards they are potentially exposed to in routine and emergency situations; and
- regularly evaluating the effectiveness of the program.

Employers are annually required to provide respirators, training, and medical evaluations at no cost to employees when respirators are deemed necessary.

57.6 ELECTRICAL SAFETY (SUBPART S)

Electricity is the key energy source in manufacturing for machine tools and related equipment, communication and information systems, illumination, ventilation, and security systems. The effects of uncontrolled electrical energy can threaten workers' lives and health. Industrial injuries that result from contact with electrical sources are of four primary types: electrocution, electric shock, burns, and falls from elevated places caused when a person contacts a live circuit (Casini 1998).

One of the most serious myths about electricity is that ordinary 110-volt circuits can not hurt people. This fallacy is based on the assumption that low amperage or voltage is not harmful. However, contact with current as low as 20 mA can be fatal if the current flows through vital organs, causing paralysis of respiratory muscles and subsequent death.

ELECTRICAL STANDARDS

OSHA Subpart S addresses electrical safety requirements for protection of employees in the workplace. The requirements fall under two main categories: design of electrical systems and safety-related work practices. Much of the content of the standard is based on the National Electric Code and ANSI safety standards for equipment.

One of the main provisions of the electrical standard is the general requirement: "Electrical equipment shall be free from recognized hazards that are causing or are likely to cause death or serious physical harm to employees" (1910.303 [b] [1]). The standard also requires all listed or labeled equipment to be used or installed in accordance with the instructions in the listing.

Misuse or misapplication of electrical equipment can be a serious hazard, particularly when the problem is not evident through visual inspection. An example of this would be a situation where the grounded or neutral conductor is incorrectly connected to the ungrounded or "hot" terminal of a plug, receptacle, or other conductor. This potentially dangerous situation is called *re-*

verse polarity. Many pieces of equipment will operate properly even when the conductors are reversed. However, the equipment may not stop when the switch is released or may start unexpectedly when the unit is plugged in. This problem is more commonly encountered with standard 120-volt outlets, lights, and cord and plug-connected equipment, as these circuits and equipment are more likely to be wired or modified by unqualified persons.

The path to ground from circuits, equipment, panels, and enclosures must be permanent and continuous. Ground-fault-circuit interrupters (GFCIs) are used to protect people where the path to ground could be bypassed, particularly in damp or wet conditions. The GFCI contains a sensor that monitors current flow in wiring and, when an imbalance as little as 5 mA is detected, the circuit is rapidly opened, interrupting the flow of current.

Other requirements of the electrical standard include:

- electrical equipment must be durably identified with the manufacturer's name or trademark, voltage, current, and wattage;
- circuits for motors, appliances, tools, and branch circuits must have the disconnecting means clearly identified or located and arranged so its purpose is evident to any person; and
- circuit breakers must indicate the on (closed) and off (open) position on the box, panel, or breaker.

Live parts of electrical equipment of 50 volts or more must be guarded against accidental contact by approved enclosures or cabinets. Electrical panels must be accessible and free from obstructions within a 3 ft (0.9 m) radius.

Plug- and cord-connected equipment must be grounded if operated in damp or wet conditions, used outdoors (for example, hedge clippers or snow blowers), or if it may come in contact with water (for example, washers/dryers or sump pumps). Equipment need not be grounded if it is double-insulated or supplied by a secondary transformer of not over 50 volts. Refrigerators, freezers, and air conditioners must be grounded.

Another important aspect of the standard relates to work on energized electrical equipment. Only qualified persons are allowed to work on or in the vicinity of energized parts, overhead lines, or live circuits. The standard defines a qualified person as:

- trained in safety-related work practices for electrical systems,
- having the skill to distinguish exposed live parts from other pieces of equipment,
- able to determine the nominal voltage of live parts, and
- having knowledge of the clearance distances required for the voltages to which they will be exposed (1910.332 [b]).

57.7 MACHINE GUARDING (SUBPART O)

Much of the OSHA standard on machine guarding focuses on the mechanical hazards at the point of operation. Hazards include persons making contact with the drive system of the machine, parts or objects flying from the machine, and machine malfunction. Guards, safety devices, or feeders/ejectors can be very effective in protecting employees from these hazards.

A basic rule is any machine part, function, or process that may cause injury must be safeguarded. When the operation of a machine or accidental contact with it can injure the operator or others in the vicinity, the hazard must be either controlled or eliminated. To protect workers against mechanical hazards, safeguards must meet the following minimum general requirements.

- The guard must prevent hands, arms, or any other part of the worker's body from making contact with dangerous moving parts.

- Workers should not be able to easily remove or tamper with the safeguards. Guards and safety devices should be made of durable material to withstand normal use. Safeguards must be firmly secured to the machine or, in some instances, to the tooling.
- The guard must ensure that no objects can fall into moving parts. A small tool or loose machine part dropped into a cycling machine could become a projectile that may strike and injure someone.
- A safeguard defeats its purpose if it creates a hazard of its own, such as a shear point, a jagged edge, or an unfinished surface that can cause a laceration. The edges of guards, for instance, should be rolled or deburred and smoothed to eliminate sharp edges and pinch points.
- Safeguards should not interfere with work. Any safeguard that impedes a worker from performing the job quickly and comfortably might soon be overridden or disregarded.
- Employees should be able to lubricate the machine without removing the safeguards. Locating oil reservoirs outside the guard, with a line leading to the lubrication point, reduces the need for the operator or maintenance worker to enter the hazardous area.

METHODS OF SAFEGUARDING

A thorough hazard analysis of each machine and particular situation is essential before the principle of safeguarding by location/distance can be applied. To safeguard a machine by location, a machine or its dangerous moving parts must be so positioned that hazardous areas are not accessible and do not present a hazard to a worker during normal operation or during other interactions with the machine such as walking past it. This may be accomplished by locating the machine so that a plant design feature, such as a wall, protects the worker and other personnel. Additionally, enclosure walls or fences can restrict access to machines. Another possible solution is to locate dangerous parts high enough to be out of the normal reach of any worker.

A feeding process can be safeguarded by location if a safe distance can be maintained to protect the worker's hands. The dimensions of the stock being worked on may provide adequate safety. For instance, if the stock is several feet long and only one end of the stock is being worked on, the need for the operator to hold the opposite end while the work is being performed may provide built-in protection. However, depending upon the machine, protection still might be required for other personnel. The positioning of the operator's control station provides another potential approach to safeguarding by location. Operator controls may be located a safe distance from the machine if there is no reason for the operator to be in the vicinity of working parts.

The method of loading or unloading a machine should be carefully evaluated, and training should be provided when necessary. Often there is more than one way to manually load a part into a machine. By loading in the incorrect manner, the operator's hand is subjected to a hazard; by loading in the proper way, no hazard exists or the hazard is greatly reduced. When the point of operation cannot in any way be guarded, adequate well-placed lighting is an important contribution to safe operation.

For safety reasons, many feeding and ejection methods do not require the operator to place hands in the danger area. Operator training can aid in avoiding a hazard. In some cases, no operator involvement is necessary after the machine is set up. In other situations, operators can manually feed the stock with the assistance of a feed mechanism. Properly designed ejection processes do not require operator involvement after the machine starts to function.

Some feeding and ejection methods may create hazards. For instance, a robot may eliminate the need for an operator to be near the machine, but may create a new hazard by the movement of its arm. Using robots does not eliminate the need for guards and safety devices.

Machine Guards

Several types of machine guards exist.

Fixed guards. Guards may be fixed as a permanent part of the machine and not dependent on moving parts to perform their safety function. Fixed guards can be constructed of any material that is substantial enough to withstand long use or abuse. The fixed guard is usually preferred because of its permanence and simplicity. Figure 57-3 illustrates an example of a fixed guard.

Interlocking guards. When opened or removed, an interlocking guard trips a mechanism that automatically shuts off or disengages the power and the machine cannot cycle or be started until the guard is back in place. Replacing the guard, however, should not automatically start the machine.

Adjustable guards. Adjustable guards accommodate for different material being processed or for varying stock sizes.

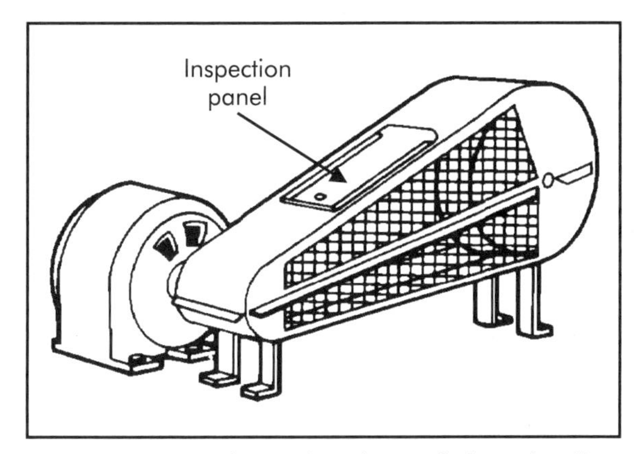

Figure 57-3. Fixed guard enclosing belt and pulleys (Drozda and Wick 1983).

Self-adjusting guards. The movement of stock determines the openings of self-adjusting guards. As the operator moves the stock into the danger area, the guard is pushed away, providing an opening that is just large enough for the workpiece. After removal of the work, the guard returns to its rest position.

Safety Devices

A safety device may perform one of several functions. It may:

- stop the machine if a hand or any part of the body is inadvertently placed in the danger area;
- restrain or withdraw the operator's hands from the danger area during operation;
- require the operator to use both hands on machine controls, thus keeping hands and body out of danger; or
- provide a barrier synchronized with the operating cycle of the machine, thereby preventing entry into the danger area during a hazardous part of the cycle.

Safety Aids

Consideration may be given to a number of miscellaneous safety aids. While these aids do not give complete protection from machine hazards, they may provide the operator with an extra margin of safety. Sound judgment is needed in their application.

Awareness barriers. The awareness barrier does not provide physical protection, but serves to remind a person that he or she is approaching the danger area. Generally, awareness barriers are not considered adequate when continual exposure to the hazard exists. A rope may be used as an awareness barrier on the rear of a machine. Although the barrier does not physically prevent a person from entering the danger area, it calls attention to the potential danger.

Shields. Shields are barriers used to provide protection from flying particles and splashing cutting oils or coolants. Figure 57-4 shows two applications.

Signs. Signs notify employees of certain types of hazards. Warning signs are placed on tanks containing various chemical solutions and on potentially hazardous machines.

Signals. Audio or visual signals are used to warn of potential hazards. For example, they are used prior to starting a long conveyor. Employees are able to avoid injury because they are aware of its starting. Signals also are used to warn of an approaching overhead crane or powered industrial truck.

Color. Color is used as notification of a hazard. For instance, orange paint on the side of guards, tools, or other areas may indicate a mechanical hazard exists.

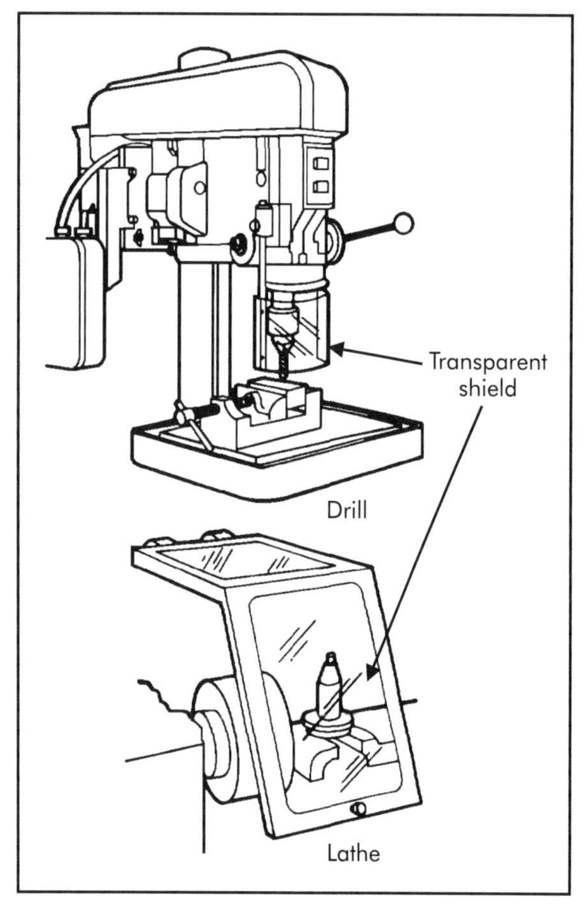

Figure 57-4. Transparent shields protect against flying particles (Drozda and Wick 1983).

Personal protection. Personal protective equipment must be depended upon to protect employees. For example, safety glasses or face shields protect the eyes and face from flying objects.

Expendable tools. Expendable tools frequently are used to avoid a point-of-operation hazard. A typical use is for reaching into the danger area of a machine to place or remove stock or workpieces. An expendable tool may be a rod, pliers, tongs, magnet, suction cup, or other item that serves as an extension of an employee's hands. Such tools should not be used instead of other machine safeguards; they are merely a supplement to the protection that other guards provide.

57.8 EQUIPMENT LOCK OUT AND TAG OUT (SUBPART J)

Maintenance workers and production employees required to perform adjustments or servicing of machines must be able to lock the machines out in a zero energy state. This condition means a machine has been isolated from all sources of hazardous energy (for example, electrical, hydraulic, or pneumatic energy). Also, all stored energy, which could cause the machine to start up or move without warning, has been restrained or dissipated (for example, the pressure in a hydraulic accumulator). Tragic accidents in the workplace have resulted from a machine automatically cycling or another worker starting it unaware that someone is performing maintenance without a lock or tag applied.

The lockout/tagout (LOTO) section of the standard is 1910 Subpart J—"General Environmental Controls." It covers:

- servicing and/or maintenance of machines and equipment;
- any time a production worker must bypass a guard or other safety device; and

- any time an employee is required to place any part of their body into a danger zone or too near the point of operation during the machine cycle.

The standard does not apply to minor tool changes and adjustments that take place during normal production if they are routine, repetitive, integral to the operation, and alternate protection (for example, interlocks) is provided. It also does not apply to plug- and cord-operated equipment where unplugging it will eliminate hazardous energy or to special circumstances regarding distribution systems for gas, water, or steam.

The difference between locking out and tagging out a piece of equipment lies in the use of a padlock versus a tag, respectively, on the energy-isolating device. Lockout is the preferred method, as the lockout device cannot be removed by anyone other than the person who applied it. A tag fastened to the energy-isolating device (for example, an electrical panel) warns that the equipment may not be operated until it is removed by the worker who placed it. The tag is not completely foolproof as it can be purposely or accidentally removed. An employer can use a tagout system if certain conditions can be met, but again, lockout systems are preferred. All machines purchased, renovated, or having undergone major repairs since 1990 must allow a lock(s) to be applied to the energy-isolation device(s).

LOCKOUT/TAGOUT PROGRAM

The LOTO program must have established procedures for energy isolation. In addition, training on how to verify the zero energy state has been reached and how to properly apply lockouts/tagouts must be provided for employees. Employers must perform periodic inspections to assure employees are following the program. The hardware used for LOTO purposes must be supplied by the employer. It also must meet the following criteria.

- Singularly identifiable: the tags, locks, and other equipment will not be used for any other purpose. They should be easily recognized as LOTO devices.
- Durable: the devices must hold up under adverse conditions such as wetness, exposure to caustic chemicals, or against becoming unidentifiable after prolonged exposure.
- Standardized: all devices used in the plant must be the same shape, color, or size, and the print and format on tagout devices must be standardized.
- Substantial: lockout devices must prevent removal by anything except bolt cutters or other metal-cutting tools. Tagout devices must withstand accidental or inadvertent removal. The tagout attachment must be non-reusable and self-locking such as a nylon cable tie.
- Identifiable: the LOTO device must identify the employee who applied the device. The tag must contain a warning, such as “Do Not Start” or “Do Not Operate,” or other warnings that indicate a hazardous condition is present.

LOCKOUT/TAGOUT PROCEDURE

Locking out equipment in a prescribed, logical, and safe fashion serves to protect employees from the hazards of unexpected machine movement or released energy. The procedure for lockout/tagout of equipment is as follows.

1. Prepare for shutdown—authorized employee(s) need knowledge of the energy to be controlled and the means for controlling it.
2. Machine or equipment shutdown—machine or equipment should be shut down according to the manufacturer’s recommended procedure.

3. Machine or equipment isolation—isolate the machine or equipment from the energy source(s).
4. Application of lockout or tagout device—lockout or tagout devices should be applied to each energy-isolating device. Lockout devices should be applied so they will hold the energy-isolating device in the "off" or safe position.
5. Release stored energy—all potentially hazardous stored or remaining energy should be released.
6. Verify isolation—before starting to work, each authorized employee should verify the isolation and de-energization of the machine or equipment.
7. Release from lockout or tagout—the workplace should be inspected to ensure all persons are safely out of the way and all nonessential items have been removed. Each lockout or tagout device must be removed by the employee who applied the device.

57.9 TOXIC AND HAZARDOUS SUBSTANCES (SUBPART Z)

On November 25, 1985, the OSHA Hazard Communication (HazCom) standard (29 CFR 1910.1200) took effect. This standard is also known as the employee "Right to Know" standard. It covers all hazardous chemicals in the workplace. The standard requires chemical manufacturers and importers to assess the hazards of chemicals they sell and affix warning labels to containers they ship. They must also provide material safety data sheets (MSDS) to their clients in the manufacturing sector. The purpose of the standard is to facilitate the flow of information from chemical manufacturers and importers to distributors, to industrial firms that use the chemicals and, ultimately, on to the employee exposed to the chemical in the workplace.

There are many misconceptions about the HazCom standard. Because of this, violations of it are the most frequently cited by OSHA. The standard dictates stringent requirements for documentation of the chemicals used in the workplace and training of employees. Many firms make mistakes involving the HazCom standard for a variety of reasons. These range from the belief that they are too small to require compliance, to poor documentation practices, to failure to provide access for employees to material safety data sheets at all times. Errors in judgment can be expensive if an exposure incident occurs and the company is found not to be in compliance. It is important to note the standard does not cover non-chemical hazards, hazardous waste, tobacco, wood, personal articles, labeling on pesticides, drugs, or liquor. Also, common chemicals, such as window cleaner or whiteout correction fluid, do not require material safety data sheets when the duration and frequency of exposure is no greater than the typical consumer would experience.

The manufacturing manager who wants to know how best to comply with the standard should read it in its entirety—the preamble, the summary, and the guidelines prepared for compliance officers or consultants in the field. OSHA also defines key terms within the standard.

To comply with the standard, manufacturers must compile a list of hazardous chemicals in the workplace and cross-reference each to its individual MSDS. The MSDS must include:

- the identity used on the label, including the chemical and its common name(s);
- the physical and chemical characteristics of the hazardous chemical such as vapor pressure or flash point;
- the physical hazards of the chemical including potential for fire, explosion, and reactivity;
- the health hazards of the chemical including signs and symptoms of expo-

sure, and any medical conditions generally recognized as being aggravated by exposure;
- the primary routes of entry into the human body;
- OSHA's permissible exposure limit, the American Conference of Governmental Industrial Hygienists' (ACGIH) threshold limit value, and/or any other exposure limit used or recommended by the manufacturer;
- whether the chemical is listed in the National Toxicology Program (NTP) Annual Report on Carcinogens, the International Agency for Research on Cancer (IARC), or has been found to be a potential carcinogen by OSHA or others;
- any generally applicable precautions for safe handling and use known to the chemical manufacturer, importer, or employer preparing the MSDS;
- emergency and first-aid procedures;
- the date of preparation of the MSDS or last change to it; and
- the name, address, and telephone number of the chemical manufacturer, importer, employer, or other responsible party preparing or distributing the MSDS.

A material safety data sheet must be kept for each chemical. Generally, the MSDS is sent with the first shipment of any chemical to the facility. It must be available at all times to employees in their work area. There can be no barriers to immediate access. For example, a MSDS should not be kept in a supervisor's office if there is a night shift operation and the office is locked during this time. It is best to have a folder or binder kept in a prominent place in the department so it can be accessed at any time.

Another element of the HazCom standard specifies training for employees on how to read container labels and material safety data sheets. New employees should be trained initially and all employees trained on any new chemicals brought into the workplace. Minimum training includes:

- methods to detect the presence or release of a hazardous chemical;
- physical or health hazards of chemicals in the work area; and
- protective measures.

BLOOD-BORNE PATHOGENS

With the spread of human immunodeficiency virus (HIV), hepatitis, and other viruses, protecting employees from exposure to potentially infectious diseases has become critical for health and safety. Although much press has been duly generated about HIV and its ultimate result, AIDS, there is also great risk from the hepatitis B virus (HBV) for several reasons. First, hepatitis B is quite hardy; it is able to live on surfaces for up to a week. Second, HBV is much more concentrated than HIV, thus exposure to small quantities of bodily fluid can be contagious. Third, it is widespread in the population, causing nearly 300,000 new cases in the United States each year. The symptoms are varied ranging from no effects to jaundice, joint pain, internal bleeding, and liver cancer (Goetsch 2002).

The good news is there is a hepatitis B vaccine that can be administered over the course of several months. OSHA requires employers to offer HBV vaccinations free to employees who are exposed to blood or other potentially infectious materials as part of their regular job duties. In the manufacturing setting, this may include persons who administer first-aid or medical care in the plant, supervisors who may need to treat minor injuries such as abrasions or cuts on employees, janitorial or housekeeping staff, or emergency responders employed within the facility. In general, it is best to err on the side of caution and avoid all contact with bodily fluids by maintaining distance, using protective clothing, or allowing only persons who have been adequately trained in universal precautions to administer aid or follow-up after an incident.

Subpart Z 1910.1030 covers the methods of compliance, including exposure control and personal protective equipment use when employees may have risk of occupational exposure. A written exposure control plan is needed if there is anticipated contact with infectious materials during the course of an employee's duties. Job classifications that may encounter blood-borne pathogens or other infectious materials such as nursing, housekeeping, or laboratory work must be identified and exposure determination made without regard to the use of PPE. In the development of the exposure control plan, engineering and work practices used to reduce or mitigate exposure must be solicited from affected employees. When there is occupational exposure, the employer must provide PPE at no cost to the employee.

57.10 HAZARDOUS WASTE MANAGEMENT (TITLE 40 CFR)

Nearly everything people do leaves behind some kind of waste. Households create ordinary garbage. Industrial and manufacturing processes create solid and hazardous waste. To manage wastes more effectively, Congress passed the Resource Conservation and Recovery Act (RCRA) in 1976. Although it actually amends the Solid Waste Disposal Act, the legislation is so comprehensive it is generally referred to simply as "RCRA," without reference to the original act (U. S. Environmental Protection Agency 2002).

RCRA's goals are to:

- Protect the public from hazardous waste.
- Conserve energy and natural resources by recycling and recovery.
- Reduce or eliminate waste.
- Clean up waste that may have spilled or leaked.

RCRA gave the Environmental Protection Agency (EPA) the authority to control hazardous waste from the "cradle to grave." This includes the generation, transportation, treatment, storage, and disposal of hazardous waste.

IDENTIFYING HAZARDOUS WASTE

A *hazardous waste* has properties that make it dangerous or capable of having a harmful effect on human health or the environment (U.S. Environmental Protection Agency 2002). It is generated from many sources, ranging from industrial manufacturing processes, to batteries, to fluorescent light bulbs. Hazardous waste may come in many forms, including liquids, solids, gases, and sludges.

Answers to the following questions can help a facility determine if it is producing hazardous waste:

- Is the material in question a solid waste?
- Is the material excluded from the definition of solid waste or hazardous waste?
- Is the waste a listed or characteristic hazardous waste?
- Is the waste delisted?

The statutory definition points out that whether or not a material is a solid waste is not based on the physical form of the material (that is, whether or not it is a solid opposed to a liquid or gas), but rather that it is a waste. The regulations further define *solid waste* as any material discarded by being either abandoned, inherently waste-like, or recycled.

After determining a waste is a solid waste, and that it is not either excluded from the definitions of solid or hazardous waste or exempt from the hazardous waste regulation, the owner and operator must determine if it is hazardous. The first step in this process is determining if it is a listed hazardous waste.

The hazardous waste listings consist of:

- the F list,
- the K list,

- the P list, and
- the U list.

The F list includes wastes from certain common industrial and manufacturing processes. Because the processes generating these wastes can occur in different sectors of industry, the F list wastes are known as wastes from nonspecific sources.

The K list includes wastes from specific industries, thus they are known as wastes from specific sources.

The P list and U list include pure or commercial grade formulations of specific unused chemicals (those spilled, no longer needed for their intended purpose, etc.). Chemicals are included on the P list if they are acutely toxic. A chemical is acutely toxic if it is fatal to humans in low doses, if scientific studies have shown that it has lethal effects on experimental organisms, or if it causes serious irreversible or incapacitating illness. The U list is generally comprised of chemicals that are toxic, but also includes those that display other characteristics, such as ignitability or reactivity.

A *characteristic waste* exhibits measurable properties that indicate it poses enough of a threat to deserve regulation as a hazardous waste. As a result, a plant is required to determine whether its waste possesses a hazardous property (that is, exhibits a hazardous waste characteristic). The characteristics are applicable to any RCRA solid waste from any industry. The EPA has established the following hazardous waste characteristics:

- ignitability,
- corrosivity,
- reactivity, and
- toxicity.

HAZARDOUS WASTE GENERATORS

Under RCRA, hazardous waste generators are the first link in the cradle-to-grave management system. All generators must determine if their waste is hazardous and must oversee the ultimate fate of it. RCRA Subtitle C requires generators to ensure and fully document that the hazardous waste they produce is properly identified, managed, and treated prior to recycling or disposal (U.S. Environmental Protection Agency 2002).

The degree of regulation to which each generator is subject depends to a large extent on how much waste is produced in a calendar month. The term *generator* includes any entity, by site, that first creates or produces a hazardous waste or first brings a hazardous waste into the country (for example, imports a hazardous waste into the United States). Generators are further defined as:

- Large quantity generators (LQGs)—facilities that generate greater than 2,200 lb (1,000 kg) of hazardous waste per calendar month or greater than 2.2 lb (1 kg) of acutely hazardous waste per calendar month.
- Small quantity generators (SQGs)—facilities that generate between 220 lb (100 kg) and 2,200 lb (1,000 kg) of hazardous waste per calendar month and accumulate less than 13,200 lb (6,000 kg) of hazardous waste at any time.
- Conditionally exempt small quantity generators (CESQGs)—facilities that produce less than 220 lb (100 kg) of hazardous waste per calendar month or less than 2.2 lb (1 kg) of acutely hazardous waste per calendar month. The CESQG requirements additionally limit a facility's waste accumulation quantities to less than 2,200 lb (1,000 kg) of hazardous waste, 2.2 lb (1 kg) of acutely hazardous waste, or 220 lb (100 kg) of any residue from the cleanup of a spill of acutely hazardous waste at any time.

A particular state's classification of generators may be different from those outlined above. Some states regulate all generators of hazardous waste (that is, there is no exempt category), while others classify generators by waste type rather than by generated volume.

LQGs and SQGs are subject to regulations that require each to:

- identify and count waste;
- obtain an Environmental Protection Agency (EPA) identification number;
- comply with accumulation and storage requirements (including requirements for training, contingency planning, and emergency arrangements);
- prepare the waste for transportation;
- track the shipment and receipt of waste; and
- meet recordkeeping and reporting requirements.

To determine which standards a facility must comply with, generators are required to identify each waste and determine all applicable listings and characteristics. After determining which wastes are hazardous each month, generators are responsible for totaling the weight of each waste to determine if it will be regulated as a LQG, SQG, or CESQG for that particular month.

If an entity generates, treats, stores, disposes of, transports, or offers for transportation any hazardous waste, it must have an identification number. The generator is forbidden from transferring hazardous waste to any transporter who does not also have an EPA identification number.

A LQG may accumulate hazardous waste on site for 90 days or less. Under temporary, unforeseen, and uncontrollable circumstances, this 90-day period may be extended for up to 30 days by the state or EPA on a case-by-case basis. LQGs storing wastewater treatment sludges from electroplating operations (F006) may store that waste for 180 or 270 days if the waste is to be recycled.

LQGs must comply with the following requirements.

- Proper management—the waste must be properly accumulated in containers, tanks, drip pads, or containment buildings. Hazardous waste containers must be kept closed and marked with the date on which accumulation began. Tanks and containers are required to be marked with the words "Hazardous Waste." The generator must ensure and document that the waste is shipped off site within the allowable 90-day period.
- Emergency plan—formal written contingency plans and emergency procedures are required to prepare for the event of a spill or release.
- Personnel training—there must be an established training program to instruct facility personnel on the proper handling of hazardous waste.

SQGs must comply with the following requirements.

- Hazardous waste can be accumulated on site for 180 days or less. It may be accumulated for up to 270 days if it must be transported off-site for treatment, storage, or disposal over distances greater than 200 miles (322 km).
- Proper management—the waste must be properly accumulated in either tanks or containers marked with the words "Hazardous Waste." Containers also must be marked with the date on which accumulation began.
- Emergency plan—emergency response procedures must be specified. However, written contingency plans are not required. An emergency coordinator is required to be on the premises or on-call at all times. Basic facility safety information must be readily accessible.
- Personnel training—a training program is not required, but the generator must ensure that employees handling hazardous waste are familiar with proper handling and emergency procedures.

PREPARATION FOR TRANSPORT

Pre-transport regulations are designed to ensure safe transportation of hazardous waste from its point of origin to the ultimate disposal site. In developing hazardous waste pre-transport regulations, the EPA adopted the Department of Transportation's (DOT's) regulations for packaging, labeling, marking, and placarding (U.S. Environmental Protection Agency 2002). DOT regulations require:

- proper packaging to prevent leakage of hazardous waste during normal transport conditions and potentially dangerous situations (for example, if a drum falls off of a truck); and
- labeling, marking, and placarding of the packaged waste to identify the characteristics and dangers associated with its transport.

Pre-transport regulations only apply to generators who ship waste off site for treatment, storage, or disposal. Transportation on site is not subject to these regulations.

Uniform Hazardous Waste Manifest

The Uniform Hazardous Waste Manifest (Form 8700-22), or *manifest*, allows all parties involved to track hazardous waste movement from the point of generation to the point of ultimate treatment, storage, or disposal (U.S. Environmental Protection Agency 2002). An RCRA manifest contains the following federally required information:

- name, address, and EPA identification number of the hazardous waste generator, transporter(s), and designated facility;
- DOT shipping name including hazard class and identification number;
- EPA hazardous waste code(s); and
- quantities of the wastes transported and the container type.

Each time a waste is transferred (for example, from a transporter to the designated facility or from a transporter to another transporter) the manifest must be signed to acknowledge receipt.

RECORD-KEEPING

The record-keeping and reporting requirements for LQGs and SQGs provide the EPA and states with methods to track the quantities of hazardous waste generated and transported (U.S. Environmental Protection Agency 2002).

Generators must keep a copy of each biennial report and any exception reports for at least three years from the date of the report. Copies of all manifests must be kept for three years or until a signed and dated copy is received from the designated facility. The manifest received from the designated facility must be kept for at least three years from the date the hazardous waste was accepted by the initial transporter. Finally, the generator's records of waste analyses and determinations must be kept for at least three years from the date the waste was last sent to an on-site or off-site treatment, storage, or disposal facility.

REVIEW QUESTIONS

57.1) OSHA standards apply to employers with how many employees?

57.2) What are the minimum requirements for the design of an exit sign?

57.3) A Type D fire extinguisher is capable of putting out what kind of fire?

57.4) At what time-weighted-average decibel level are employers required to provide hearing protection?

57.5) Why are awareness barriers not considered a suitable method of machine guarding?

57.6) Where should material safety data sheets (MSDS) be stored?

57.7) In the multiple causation accident theory, what two key factors cause most accidents?

57.8) To qualify as large quantity generator, a generator must produce how much non-acute hazardous waste per calendar month?

57.9) What is the purpose of the Uniform Hazardous Waste Manifest?

REFERENCES

Casini, V. 1998. "Overview of Electrical Hazards." Atlanta, GA: National Institute for Occupational Safety and Health/Centers for Disease Control & Prevention. Available at http://www.cdc.gov/hiosh/elecovrv.html.

Code of Federal Regulations. 2001. 29 CFR Part 1910–"General Industry," First Edition. Davenport, IA: Mangan Communications.

Drozda, T. and Wick, C., eds. 1983. *Tool and Manufacturing Engineers Handbook*, Fourth Edition. Volume 1: *Machining*. Dearborn, MI: Society of Manufacturing Engineers.

Goetsch, D. L. 2002. *Occupational Safety and Health for Technologists, Engineers and Managers*, Fourth Edition. Upper Saddle River, NJ: Prentice-Hall.

Raouf, A. 2002. "Theory of Accident Causes." Geneva, Switzerland: International Labour Organization/Tampere University of Technology (Finland). Available at http://turva.me.tut.fi/iloagri/natu/tac.htm.

U.S. Environmental Protection Agency, Office of Solid Waste/Communications, Information, and Resources Management Division. 2002. *RCRA Orientation Manual*. Washington, DC: U.S. Environmental Protection Agency.

Appendix E

Calculating the Time Value of Money

The tables included here give the discrete compounding interest factors for a variety of interest rates. The interest factors are used when performing calculations involving the time value of money.

Table E-1. Discrete compounding factors: $i = 0.25\%$

Year, n	To Find F Given P F/P	To Find P Given F P/F	To Find A Given P A/P	To Find P Given A P/A	To Find A Given F A/F	To Find F Given A F/A
1	1.0025	0.9975	1.0025	0.9975	1.0000	1.0000
2	1.0050	0.9950	0.5019	1.9925	0.4994	2.0025
3	1.0075	0.9925	0.3350	2.9851	0.3325	3.0075
4	1.0100	0.9901	0.2516	3.9751	0.2491	4.0150
5	1.0126	0.9876	0.2015	4.9627	0.1990	5.0251
6	1.0151	0.9851	0.1681	5.9478	0.1656	6.0376
7	1.0176	0.9827	0.1443	6.9305	0.1418	7.0527
8	1.0202	0.9802	0.1264	7.9107	0.1239	8.0704
9	1.0227	0.9778	0.1125	8.8885	0.1100	9.0905
10	1.0253	0.9753	0.1014	9.8639	0.0989	10.1133
11	1.0278	0.9729	0.0923	10.8368	0.0898	11.1385
12	1.0304	0.9705	0.0847	11.8073	0.0822	12.1664
13	1.0330	0.9681	0.0783	12.7753	0.0758	13.1968
14	1.0356	0.9656	0.0728	13.7410	0.0703	14.2298
15	1.0382	0.9632	0.0680	14.7042	0.0655	15.2654
16	1.0408	0.9608	0.0638	15.6650	0.0613	16.3035
17	1.0434	0.9584	0.0602	16.6235	0.0577	17.3443
18	1.0460	0.9561	0.0569	17.5795	0.0544	18.3876
19	1.0486	0.9537	0.0540	18.5332	0.0515	19.4336
20	1.0512	0.9513	0.0513	19.4845	0.0488	20.4822
21	1.0538	0.9489	0.0489	20.4334	0.0464	21.5334
22	1.0565	0.9466	0.0468	21.3800	0.0443	22.5872
23	1.0591	0.9442	0.0448	22.3241	0.0423	23.6437
24	1.0618	0.9418	0.0430	23.2660	0.0405	24.7028
25	1.0644	0.9395	0.0413	24.2055	0.0388	25.7646
26	1.0671	0.9371	0.0398	25.1426	0.0373	26.8290
27	1.0697	0.9348	0.0383	26.0774	0.0358	27.8961
28	1.0724	0.9325	0.0370	27.0099	0.0345	28.9658
29	1.0751	0.9301	0.0358	27.9400	0.0333	30.0382
30	1.0778	0.9278	0.0346	28.8679	0.0321	31.1133
35	1.0913	0.9163	0.0299	33.4724	0.0274	36.5292
40	1.1050	0.9050	0.0263	38.0199	0.0238	42.0132
45	1.1189	0.8937	0.0235	42.5109	0.0210	47.5661
50	1.1330	0.8826	0.0213	46.9462	0.0188	53.1887
55	1.1472	0.8717	0.0195	51.3264	0.0170	58.8819
60	1.1616	0.8609	0.0180	55.6524	0.0155	64.6467
65	1.1762	0.8502	0.0167	59.9246	0.0142	70.4839
70	1.1910	0.8396	0.0156	64.1439	0.0131	76.3944
75	1.2059	0.8292	0.0146	68.3108	0.0121	82.3792
80	1.2211	0.8189	0.0138	72.4260	0.0113	88.4392
85	1.2364	0.8088	0.0131	76.4901	0.0106	94.5753
90	1.2520	0.7987	0.0124	80.5038	0.0099	100.7885
95	1.2677	0.7888	0.0118	84.4677	0.0093	107.0797
100	1.2836	0.7790	0.0113	88.3825	0.0088	113.4500

Table E-2. Discrete compounding factors: $i = 0.5\%$

Year, *n*	To Find *F* Given *P* *F/P*	To Find *P* Given *F* *P/F*	To Find *A* Given *P* *A/P*	To Find *P* Given *A* *P/A*	To Find *A* Given *F* *A/F*	To Find *F* Given *A* *F/A*
1	1.0050	0.9950	1.0050	0.9950	1.0000	1.0000
2	1.0100	0.9901	0.5038	1.9851	0.4988	2.0050
3	1.0151	0.9851	0.3367	2.9702	0.3317	3.0150
4	1.0202	0.9802	0.2531	3.9505	0.2481	4.0301
5	1.0253	0.9754	0.2030	4.9259	0.1980	5.0503
6	1.0304	0.9705	0.1696	5.8964	0.1646	6.0755
7	1.0355	0.9657	0.1457	6.8621	0.1407	7.1059
8	1.0407	0.9609	0.1278	7.8230	0.1228	8.1414
9	1.0459	0.9561	0.1139	8.7791	0.1089	9.1821
10	1.0511	0.9513	0.1028	9.7304	0.0978	10.2280
11	1.0564	0.9466	0.0937	10.6770	0.0887	11.2792
12	1.0617	0.9419	0.0861	11.6189	0.0811	12.3356
13	1.0670	0.9372	0.0796	12.5562	0.0746	13.3972
14	1.0723	0.9326	0.0741	13.4887	0.0691	14.4642
15	1.0777	0.9279	0.0694	14.4166	0.0644	15.5365
16	1.0831	0.9233	0.0652	15.3399	0.0602	16.6142
17	1.0885	0.9187	0.0615	16.2586	0.0565	17.6973
18	1.0939	0.9141	0.0582	17.1728	0.0532	18.7858
19	1.0994	0.9096	0.0553	18.0824	0.0503	19.8797
20	1.1049	0.9051	0.0527	18.9874	0.0477	20.9791
21	1.1104	0.9006	0.0503	19.8880	0.0453	22.0840
22	1.1160	0.8961	0.0481	20.7841	0.0431	23.1944
23	1.1216	0.8916	0.0461	21.6757	0.0411	24.3104
24	1.1272	0.8872	0.0443	22.5629	0.0393	25.4320
25	1.1328	0.8828	0.0427	23.4456	0.0377	26.5591
26	1.1385	0.8784	0.0411	24.3240	0.0361	27.6919
27	1.1442	0.8740	0.0397	25.1980	0.0347	28.8304
28	1.1499	0.8697	0.0384	26.0677	0.0334	29.9745
29	1.1556	0.8653	0.0371	26.9330	0.0321	31.1244
30	1.1614	0.8610	0.0360	27.7941	0.0310	32.2800
35	1.1907	0.8398	0.0312	32.0354	0.0262	38.1454
40	1.2208	0.8191	0.0276	36.1722	0.0226	44.1588
45	1.2516	0.7990	0.0249	40.2072	0.0199	50.3242
50	1.2832	0.7793	0.0227	44.1428	0.0177	56.6452
55	1.3156	0.7601	0.0208	47.9814	0.0158	63.1258
60	1.3489	0.7414	0.0193	51.7256	0.0143	69.7700
65	1.3829	0.7231	0.0181	55.3775	0.0131	76.5821
70	1.4178	0.7053	0.0170	58.9394	0.0120	83.5661
75	1.4536	0.6879	0.0160	62.4136	0.0110	90.7265
80	1.4903	0.6710	0.0152	65.8023	0.0102	98.0677
85	1.5280	0.6545	0.0145	69.1075	0.0095	105.5943
90	1.5666	0.6383	0.0138	72.3313	0.0088	113.3109
95	1.6061	0.6226	0.0132	75.4757	0.0082	121.2224
100	1.6467	0.6073	0.0127	78.5426	0.0077	129.3337

Table E-3. Discrete compounding factors: $i = 0.75\%$

Year, n	To Find F Given P F/P	To Find P Given F P/F	To Find A Given P A/P	To Find P Given A P/A	To Find A Given F A/F	To Find F Given A F/A
1	1.0075	0.9926	1.0075	0.9926	1.0000	1.0000
2	1.0151	0.9852	0.5056	1.9777	0.4981	2.0075
3	1.0227	0.9778	0.3383	2.9556	0.3308	3.0226
4	1.0303	0.9706	0.2547	3.9261	0.2472	4.0452
5	1.0381	0.9633	0.2045	4.8894	0.1970	5.0756
6	1.0459	0.9562	0.1711	5.8456	0.1636	6.1136
7	1.0537	0.9490	0.1472	6.7946	0.1397	7.1595
8	1.0616	0.9420	0.1293	7.7366	0.1218	8.2132
9	1.0696	0.9350	0.1153	8.6716	0.1078	9.2748
10	1.0776	0.9280	0.1042	9.5996	0.0967	10.3443
11	1.0857	0.9211	0.0951	10.5207	0.0876	11.4219
12	1.0938	0.9142	0.0875	11.4349	0.0800	12.5076
13	1.1020	0.9074	0.0810	12.3423	0.0735	13.6014
14	1.1103	0.9007	0.0755	13.2430	0.0680	14.7034
15	1.1186	0.8940	0.0707	14.1370	0.0632	15.8137
16	1.1270	0.8873	0.0666	15.0243	0.0591	16.9323
17	1.1354	0.8807	0.0629	15.9050	0.0554	18.0593
18	1.1440	0.8742	0.0596	16.7792	0.0521	19.1947
19	1.1525	0.8676	0.0567	17.6468	0.0492	20.3387
20	1.1612	0.8612	0.0540	18.5080	0.0465	21.4912
21	1.1699	0.8548	0.0516	19.3628	0.0441	22.6524
22	1.1787	0.8484	0.0495	20.2112	0.0420	23.8223
23	1.1875	0.8421	0.0475	21.0533	0.0400	25.0010
24	1.1964	0.8358	0.0457	21.8891	0.0382	26.1885
25	1.2054	0.8296	0.0440	22.7188	0.0365	27.3849
26	1.2144	0.8234	0.0425	23.5422	0.0350	28.5903
27	1.2235	0.8173	0.0411	24.3595	0.0336	29.8047
28	1.2327	0.8112	0.0397	25.1707	0.0322	31.0282
29	1.2420	0.8052	0.0385	25.9759	0.0310	32.2609
30	1.2513	0.7992	0.0373	26.7751	0.0298	33.5029
35	1.2989	0.7699	0.0326	30.6827	0.0251	39.8538
40	1.3483	0.7416	0.0290	34.4469	0.0215	46.4465
45	1.3997	0.7145	0.0263	38.0732	0.0188	53.2901
50	1.4530	0.6883	0.0241	41.5664	0.0166	60.3943
55	1.5083	0.6630	0.0223	44.9316	0.0148	67.7688
60	1.5657	0.6387	0.0208	48.1734	0.0133	75.4241
65	1.6253	0.6153	0.0195	51.2963	0.0120	83.3709
70	1.6872	0.5927	0.0184	54.3046	0.0109	91.6201
75	1.7514	0.5710	0.0175	57.2027	0.0100	100.1833
80	1.8180	0.5500	0.0167	59.9944	0.0092	109.0725
85	1.8873	0.5299	0.0160	62.6838	0.0085	118.3001
90	1.9591	0.5104	0.0153	65.2746	0.0078	127.8790
95	2.0337	0.4917	0.0148	67.7704	0.0073	137.8225
100	2.1111	0.4737	0.0143	70.1746	0.0068	148.1445

Table E-4. Discrete compounding factors: i = 1.00%

Year, n	To Find F Given P F/P	To Find P Given F P/F	To Find A Given P A/P	To Find P Given A P/A	To Find A Given F A/F	To Find F Given A F/A
1	1.0100	0.9901	1.0100	0.9901	1.0000	1.0000
2	1.0201	0.9803	0.5075	1.9704	0.4975	2.0100
3	1.0303	0.9706	0.3400	2.9410	0.3300	3.0301
4	1.0406	0.9610	0.2563	3.9020	0.2463	4.0604
5	1.0510	0.9515	0.2060	4.8534	0.1960	5.1010
6	1.0615	0.9420	0.1725	5.7955	0.1625	6.1520
7	1.0721	0.9327	0.1486	6.7282	0.1386	7.2135
8	1.0829	0.9235	0.1307	7.6517	0.1207	8.2857
9	1.0937	0.9143	0.1167	8.5660	0.1067	9.3685
10	1.1046	0.9053	0.1056	9.4713	0.0956	10.4622
11	1.1157	0.8963	0.0965	10.3676	0.0865	11.5668
12	1.1268	0.8874	0.0888	11.2551	0.0788	12.6825
13	1.1381	0.8787	0.0824	12.1337	0.0724	13.8093
14	1.1495	0.8700	0.0769	13.0037	0.0669	14.9474
15	1.1610	0.8613	0.0721	13.8651	0.0621	16.0969
16	1.1726	0.8528	0.0679	14.7179	0.0579	17.2579
17	1.1843	0.8444	0.0643	15.5623	0.0543	18.4304
18	1.1961	0.8360	0.0610	16.3983	0.0510	19.6147
19	1.2081	0.8277	0.0581	17.2260	0.0481	20.8109
20	1.2202	0.8195	0.0554	18.0456	0.0454	22.0190
21	1.2324	0.8114	0.0530	18.8570	0.0430	23.2392
22	1.2447	0.8034	0.0509	19.6604	0.0409	24.4716
23	1.2572	0.7954	0.0489	20.4558	0.0389	25.7163
24	1.2697	0.7876	0.0471	21.2434	0.0371	26.9735
25	1.2824	0.7798	0.0454	22.0232	0.0354	28.2432
26	1.2953	0.7720	0.0439	22.7952	0.0339	29.5256
27	1.3082	0.7644	0.0424	23.5596	0.0324	30.8209
28	1.3213	0.7568	0.0411	24.3164	0.0311	32.1291
29	1.3345	0.7493	0.0399	25.0658	0.0299	33.4504
30	1.3478	0.7419	0.0387	25.8077	0.0287	34.7849
35	1.4166	0.7059	0.0340	29.4086	0.0240	41.6603
40	1.4889	0.6717	0.0305	32.8347	0.0205	48.8864
45	1.5648	0.6391	0.0277	36.0945	0.0177	56.4811
50	1.6446	0.6080	0.0255	39.1961	0.0155	64.4632
55	1.7285	0.5785	0.0237	42.1472	0.0137	72.8525
60	1.8167	0.5504	0.0222	44.9550	0.0122	81.6697
65	1.9094	0.5237	0.0210	47.6266	0.0110	90.9366
70	2.0068	0.4983	0.0199	50.1685	0.0099	100.6763
75	2.1091	0.4741	0.0190	52.5871	0.0090	110.9128
80	2.2167	0.4511	0.0182	54.8882	0.0082	121.6715
85	2.3298	0.4292	0.0175	57.0777	0.0075	132.9790
90	2.4486	0.4084	0.0169	59.1609	0.0069	144.8633
95	2.5735	0.3886	0.0164	61.1430	0.0064	157.3538
100	2.7048	0.3697	0.0159	63.0289	0.0059	170.4814

Table E-5. Discrete compounding factors: $i = 1.25\%$

Year, n	To Find F Given P F/P	To Find P Given F P/F	To Find A Given P A/P	To Find P Given A P/A	To Find A Given F A/F	To Find F Given A F/A
1	1.0125	0.9877	1.0125	0.9877	1.0000	1.0000
2	1.0252	0.9755	0.5094	1.9631	0.4969	2.0125
3	1.0380	0.9634	0.3417	2.9265	0.3292	3.0377
4	1.0509	0.9515	0.2579	3.8781	0.2454	4.0756
5	1.0641	0.9398	0.2076	4.8178	0.1951	5.1266
6	1.0774	0.9282	0.1740	5.7460	0.1615	6.1907
7	1.0909	0.9167	0.1501	6.6627	0.1376	7.2680
8	1.1045	0.9054	0.1321	7.5681	0.1196	8.3589
9	1.1183	0.8942	0.1182	8.4623	0.1057	9.4634
10	1.1323	0.8832	0.1070	9.3455	0.0945	10.5817
11	1.1464	0.8723	0.0979	10.2178	0.0854	11.7139
12	1.1608	0.8615	0.0903	11.0793	0.0778	12.8604
13	1.1753	0.8509	0.0838	11.9302	0.0713	14.0211
14	1.1900	0.8404	0.0783	12.7706	0.0658	15.1964
15	1.2048	0.8300	0.0735	13.6005	0.0610	16.3863
16	1.2199	0.8197	0.0693	14.4203	0.0568	17.5912
17	1.2351	0.8096	0.0657	15.2299	0.0532	18.8111
18	1.2506	0.7996	0.0624	16.0295	0.0499	20.0462
19	1.2662	0.7898	0.0595	16.8193	0.0470	21.2968
20	1.2820	0.7800	0.0568	17.5993	0.0443	22.5630
21	1.2981	0.7704	0.0544	18.3697	0.0419	23.8450
22	1.3143	0.7609	0.0523	19.1306	0.0398	25.1431
23	1.3307	0.7515	0.0503	19.8820	0.0378	26.4574
24	1.3474	0.7422	0.0485	20.6242	0.0360	27.7881
25	1.3642	0.7330	0.0468	21.3573	0.0343	29.1354
26	1.3812	0.7240	0.0453	22.0813	0.0328	30.4996
27	1.3985	0.7150	0.0439	22.7963	0.0314	31.8809
28	1.4160	0.7062	0.0425	23.5025	0.0300	33.2794
29	1.4337	0.6975	0.0413	24.2000	0.0288	34.6954
30	1.4516	0.6889	0.0402	24.8889	0.0277	36.1291
35	1.5446	0.6474	0.0355	28.2079	0.0230	43.5709
40	1.6436	0.6084	0.0319	31.3269	0.0194	51.4896
45	1.7489	0.5718	0.0292	34.2582	0.0167	59.9157
50	1.8610	0.5373	0.0270	37.0129	0.0145	68.8818
55	1.9803	0.5050	0.0253	39.6017	0.0128	78.4225
60	2.1072	0.4746	0.0238	42.0346	0.0113	88.5745
65	2.2422	0.4460	0.0226	44.3210	0.0101	99.3771
70	2.3859	0.4191	0.0215	46.4697	0.0090	110.8720
75	2.5388	0.3939	0.0206	48.4890	0.0081	123.1035
80	2.7015	0.3702	0.0198	50.3867	0.0073	136.1188
85	2.8746	0.3479	0.0192	52.1701	0.0067	149.9682
90	3.0588	0.3269	0.0186	53.8461	0.0061	164.7050
95	3.2548	0.3072	0.0180	55.4211	0.0055	180.3862
100	3.4634	0.2887	0.0176	56.9013	0.0051	197.0723

Table E-6. Discrete compounding factors: i = 1.50%

Year, *n*	To Find *F* Given *P* *F/P*	To Find *P* Given *F* *P/F*	To Find A Given *P* *A/P*	To Find *P* Given A *P/A*	To Find A Given *F* *A/F*	To Find *F* Given A *F/A*
1	1.0150	0.9852	1.0150	0.9852	1.0000	1.0000
2	1.0302	0.9707	0.5113	1.9559	0.4963	2.0150
3	1.0457	0.9563	0.3434	2.9122	0.3284	3.0452
4	1.0614	0.9422	0.2594	3.8544	0.2444	4.0909
5	1.0773	0.9283	0.2091	4.7826	0.1941	5.1523
6	1.0934	0.9145	0.1755	5.6972	0.1605	6.2296
7	1.1098	0.9010	0.1516	6.5982	0.1366	7.3230
8	1.1265	0.8877	0.1336	7.4859	0.1186	8.4328
9	1.1434	0.8746	0.1196	8.3605	0.1046	9.5593
10	1.1605	0.8617	0.1084	9.2222	0.0934	10.7027
11	1.1779	0.8489	0.0993	10.0711	0.0843	11.8633
12	1.1956	0.8364	0.0917	10.9075	0.0767	13.0412
13	1.2136	0.8240	0.0852	11.7315	0.0702	14.2368
14	1.2318	0.8118	0.0797	12.5434	0.0647	15.4504
15	1.2502	0.7999	0.0749	13.3432	0.0599	16.6821
16	1.2690	0.7880	0.0708	14.1313	0.0558	17.9324
17	1.2880	0.7764	0.0671	14.9076	0.0521	19.2014
18	1.3073	0.7649	0.0638	15.6726	0.0488	20.4894
19	1.3270	0.7536	0.0609	16.4262	0.0459	21.7967
20	1.3469	0.7425	0.0582	17.1686	0.0432	23.1237
21	1.3671	0.7315	0.0559	17.9001	0.0409	24.4705
22	1.3876	0.7207	0.0537	18.6208	0.0387	25.8376
23	1.4084	0.7100	0.0517	19.3309	0.0367	27.2251
24	1.4295	0.6995	0.0499	20.0304	0.0349	28.6335
25	1.4509	0.6892	0.0483	20.7196	0.0333	30.0630
26	1.4727	0.6790	0.0467	21.3986	0.0317	31.5140
27	1.4948	0.6690	0.0453	22.0676	0.0303	32.9867
28	1.5172	0.6591	0.0440	22.7267	0.0290	34.4815
29	1.5400	0.6494	0.0428	23.3761	0.0278	35.9987
30	1.5631	0.6398	0.0416	24.0158	0.0266	37.5387
35	1.6839	0.5939	0.0369	27.0756	0.0219	45.5921
40	1.8140	0.5513	0.0334	29.9158	0.0184	54.2679
45	1.9542	0.5117	0.0307	32.5523	0.0157	63.6142
50	2.1052	0.4750	0.0286	34.9997	0.0136	73.6828
55	2.2679	0.4409	0.0268	37.2715	0.0118	84.5296
60	2.4432	0.4093	0.0254	39.3803	0.0104	96.2147
65	2.6320	0.3799	0.0242	41.3378	0.0092	108.8028
70	2.8355	0.3527	0.0232	43.1549	0.0082	122.3638
75	3.0546	0.3274	0.0223	44.8416	0.0073	136.9728
80	3.2907	0.3039	0.0215	46.4073	0.0065	152.7109
85	3.5450	0.2821	0.0209	47.8607	0.0059	169.6652
90	3.8189	0.2619	0.0203	49.2099	0.0053	187.9299
95	4.1141	0.2431	0.0198	50.4622	0.0048	207.6061
100	4.4320	0.2256	0.0194	51.6247	0.0044	228.8030

Table E-7. Discrete compounding factors: i = 1.75%

Year, n	To Find F Given P F/P	To Find P Given F P/F	To Find A Given P A/P	To Find P Given A P/A	To Find A Given F A/F	To Find F Given A F/A
1	1.0175	0.9828	1.0175	0.9828	1.0000	1.0000
2	1.0353	0.9659	0.5132	1.9487	0.4957	2.0175
3	1.0534	0.9493	0.3451	2.8980	0.3276	3.0528
4	1.0719	0.9330	0.2610	3.8309	0.2435	4.1062
5	1.0906	0.9169	0.2106	4.7479	0.1931	5.1781
6	1.1097	0.9011	0.1770	5.6490	0.1595	6.2687
7	1.1291	0.8856	0.1530	6.5346	0.1355	7.3784
8	1.1489	0.8704	0.1350	7.4051	0.1175	8.5075
9	1.1690	0.8554	0.1211	8.2605	0.1036	9.6564
10	1.1894	0.8407	0.1099	9.1012	0.0924	10.8254
11	1.2103	0.8263	0.1007	9.9275	0.0832	12.0148
12	1.2314	0.8121	0.0931	10.7395	0.0756	13.2251
13	1.2530	0.7981	0.0867	11.5376	0.0692	14.4565
14	1.2749	0.7844	0.0812	12.3220	0.0637	15.7095
15	1.2972	0.7709	0.0764	13.0929	0.0589	16.9844
16	1.3199	0.7576	0.0722	13.8505	0.0547	18.2817
17	1.3430	0.7446	0.0685	14.5951	0.0510	19.6016
18	1.3665	0.7318	0.0652	15.3269	0.0477	20.9446
19	1.3904	0.7192	0.0623	16.0461	0.0448	22.3112
20	1.4148	0.7068	0.0597	16.7529	0.0422	23.7016
21	1.4395	0.6947	0.0573	17.4475	0.0398	25.1164
22	1.4647	0.6827	0.0552	18.1303	0.0377	26.5559
23	1.4904	0.6710	0.0532	18.8012	0.0357	28.0207
24	1.5164	0.6594	0.0514	19.4607	0.0339	29.5110
25	1.5430	0.6481	0.0497	20.1088	0.0322	31.0275
26	1.5700	0.6369	0.0482	20.7457	0.0307	32.5704
27	1.5975	0.6260	0.0468	21.3717	0.0293	34.1404
28	1.6254	0.6152	0.0455	21.9870	0.0280	35.7379
29	1.6539	0.6046	0.0443	22.5916	0.0268	37.3633
30	1.6828	0.5942	0.0431	23.1858	0.0256	39.0172
35	1.8353	0.5449	0.0385	26.0073	0.0210	47.7308
40	2.0016	0.4996	0.0350	28.5942	0.0175	57.2341
45	2.1830	0.4581	0.0323	30.9663	0.0148	67.5986
50	2.3808	0.4200	0.0302	33.1412	0.0127	78.9022
55	2.5965	0.3851	0.0285	35.1354	0.0110	91.2302
60	2.8318	0.3531	0.0271	36.9640	0.0096	104.6752
65	3.0884	0.3238	0.0259	38.6406	0.0084	119.3386
70	3.3683	0.2969	0.0249	40.1779	0.0074	135.3308
75	3.6735	0.2722	0.0240	41.5875	0.0065	152.7721
80	4.0064	0.2496	0.0233	42.8799	0.0058	171.7938
85	4.3694	0.2289	0.0227	44.0650	0.0052	192.5393
90	4.7654	0.2098	0.0221	45.1516	0.0046	215.1646
95	5.1972	0.1924	0.0217	46.1479	0.0042	239.8402
100	5.6682	0.1764	0.0212	47.0615	0.0037	266.7518

Table E-10. Discrete compounding factors: i = 4.00%

Year, *n*	To Find *F* Given *P* *F/P*	To Find *P* Given *F* *P/F*	To Find *A* Given *P* *A/P*	To Find *P* Given *A* *P/A*	To Find *A* Given *F* *A/F*	To Find *F* Given *A* *F/A*
1	1.0400	0.9615	1.0400	0.9615	1.0000	1.0000
2	1.0816	0.9246	0.5302	1.8861	0.4902	2.0400
3	1.1249	0.8890	0.3603	2.7751	0.3203	3.1216
4	1.1699	0.8548	0.2755	3.6299	0.2355	4.2465
5	1.2167	0.8219	0.2246	4.4518	0.1846	5.4163
6	1.2653	0.7903	0.1908	5.2421	0.1508	6.6330
7	1.3159	0.7599	0.1666	6.0021	0.1266	7.8983
8	1.3686	0.7307	0.1485	6.7327	0.1085	9.2142
9	1.4233	0.7026	0.1345	7.4353	0.0945	10.5828
10	1.4802	0.6756	0.1233	8.1109	0.0833	12.0061
11	1.5395	0.6496	0.1141	8.7605	0.0741	13.4864
12	1.6010	0.6246	0.1066	9.3851	0.0666	15.0258
13	1.6651	0.6006	0.1001	9.9856	0.0601	16.6268
14	1.7317	0.5775	0.0947	10.5631	0.0547	18.2919
15	1.8009	0.5553	0.0899	11.1184	0.0499	20.0236
16	1.8730	0.5339	0.0858	11.6523	0.0458	21.8245
17	1.9479	0.5134	0.0822	12.1657	0.0422	23.6975
18	2.0258	0.4936	0.0790	12.6593	0.0390	25.6454
19	2.1068	0.4746	0.0761	13.1339	0.0361	27.6712
20	2.1911	0.4564	0.0736	13.5903	0.0336	29.7781
21	2.2788	0.4388	0.0713	14.0292	0.0313	31.9692
22	2.3699	0.4220	0.0692	14.4511	0.0292	34.2480
23	2.4647	0.4057	0.0673	14.8568	0.0273	36.6179
24	2.5633	0.3901	0.0656	15.2470	0.0256	39.0826
25	2.6658	0.3751	0.0640	15.6221	0.0240	41.6459
26	2.7725	0.3607	0.0626	15.9828	0.0226	44.3117
27	2.8834	0.3468	0.0612	16.3296	0.0212	47.0842
28	2.9987	0.3335	0.0600	16.6631	0.0200	49.9676
29	3.1187	0.3207	0.0589	16.9837	0.0189	52.9663
30	3.2434	0.3083	0.0578	17.2920	0.0178	56.0849
35	3.9461	0.2534	0.0536	18.6646	0.0136	73.6522
40	4.8010	0.2083	0.0505	19.7928	0.0105	95.0255
45	5.8412	0.1712	0.0483	20.7200	0.0083	121.0294
50	7.1067	0.1407	0.0466	21.4822	0.0066	152.6671
55	8.6464	0.1157	0.0452	22.1086	0.0052	191.1592
60	10.5196	0.0951	0.0442	22.6235	0.0042	237.9907
65	12.7987	0.0781	0.0434	23.0467	0.0034	294.9684
70	15.5716	0.0642	0.0427	23.3945	0.0027	364.2905
75	18.9453	0.0528	0.0422	23.6804	0.0022	448.6314
80	23.0498	0.0434	0.0418	23.9154	0.0018	551.2450
85	28.0436	0.0357	0.0415	24.1085	0.0015	676.0901
90	34.1193	0.0293	0.0412	24.2673	0.0012	827.9833
95	41.5114	0.0241	0.0410	24.3978	0.0010	1012.7846
100	50.5049	0.0198	0.0408	24.5050	0.0008	1237.6237

Table E-11. Discrete compounding factors: i = 5.00%

Year, n	To Find F Given P F/P	To Find P Given F P/F	To Find A Given P A/P	To Find P Given A P/A	To Find A Given F A/F	To Find F Given A F/A
1	1.0500	0.9524	1.0500	0.9524	1.0000	1.0000
2	1.1025	0.9070	0.5378	1.8594	0.4878	2.0500
3	1.1576	0.8638	0.3672	2.7232	0.3172	3.1525
4	1.2155	0.8227	0.2820	3.5460	0.2320	4.3101
5	1.2763	0.7835	0.2310	4.3295	0.1810	5.5256
6	1.3401	0.7462	0.1970	5.0757	0.1470	6.8019
7	1.4071	0.7107	0.1728	5.7864	0.1228	8.1420
8	1.4775	0.6768	0.1547	6.4632	0.1047	9.5491
9	1.5513	0.6446	0.1407	7.1078	0.0907	11.0266
10	1.6289	0.6139	0.1295	7.7217	0.0795	12.5779
11	1.7103	0.5847	0.1204	8.3064	0.0704	14.2068
12	1.7959	0.5568	0.1128	8.8633	0.0628	15.9171
13	1.8856	0.5303	0.1065	9.3936	0.0565	17.7130
14	1.9799	0.5051	0.1010	9.8986	0.0510	19.5986
15	2.0789	0.4810	0.0963	10.3797	0.0463	21.5786
16	2.1829	0.4581	0.0923	10.8378	0.0423	23.6575
17	2.2920	0.4363	0.0887	11.2741	0.0387	25.8404
18	2.4066	0.4155	0.0855	11.6896	0.0355	28.1324
19	2.5270	0.3957	0.0827	12.0853	0.0327	30.5390
20	2.6533	0.3769	0.0802	12.4622	0.0302	33.0660
21	2.7860	0.3589	0.0780	12.8212	0.0280	35.7193
22	2.9253	0.3418	0.0760	13.1630	0.0260	38.5052
23	3.0715	0.3256	0.0741	13.4886	0.0241	41.4305
24	3.2251	0.3101	0.0725	13.7986	0.0225	44.5020
25	3.3864	0.2953	0.0710	14.0939	0.0210	47.7271
26	3.5557	0.2812	0.0696	14.3752	0.0196	51.1135
27	3.7335	0.2678	0.0683	14.6430	0.0183	54.6691
28	3.9201	0.2551	0.0671	14.8981	0.0171	58.4026
29	4.1161	0.2429	0.0660	15.1411	0.0160	62.3227
30	4.3219	0.2314	0.0651	15.3725	0.0151	66.4388
35	5.5160	0.1813	0.0611	16.3742	0.0111	90.3203
40	7.0400	0.1420	0.0583	17.1591	0.0083	120.7998
45	8.9850	0.1113	0.0563	17.7741	0.0063	159.7002
50	11.4674	0.0872	0.0548	18.2559	0.0048	209.3480
55	14.6356	0.0683	0.0537	18.6335	0.0037	272.7126
60	18.6792	0.0535	0.0528	18.9293	0.0028	353.5837
65	23.8399	0.0419	0.0522	19.1611	0.0022	456.7980
70	30.4264	0.0329	0.0517	19.3427	0.0017	588.5285
75	38.8327	0.0258	0.0513	19.4850	0.0013	756.6537
80	49.5614	0.0202	0.0510	19.5965	0.0010	971.2288
85	63.2544	0.0158	0.0508	19.6838	0.0008	1245.0871
90	80.7304	0.0124	0.0506	19.7523	0.0006	1594.6073
95	103.0347	0.0097	0.0505	19.8059	0.0005	2040.6935
100	131.5013	0.0076	0.0504	19.8479	0.0004	2610.0252

Table E-12. Discrete compounding factors: $i = 6.00\%$

Year, n	To Find F Given P F/P	To Find P Given F P/F	To Find A Given P A/P	To Find P Given A P/A	To Find A Given F A/F	To Find F Given A F/A
1	1.0600	0.9434	1.0600	0.9434	1.0000	1.0000
2	1.1236	0.8900	0.5454	1.8334	0.4854	2.0600
3	1.1910	0.8396	0.3741	2.6730	0.3141	3.1836
4	1.2625	0.7921	0.2886	3.4651	0.2286	4.3746
5	1.3382	0.7473	0.2374	4.2124	0.1774	5.6371
6	1.4185	0.7050	0.2034	4.9173	0.1434	6.9753
7	1.5036	0.6651	0.1791	5.5824	0.1191	8.3938
8	1.5938	0.6274	0.1610	6.2098	0.1010	9.8975
9	1.6895	0.5919	0.1470	6.8017	0.0870	11.4913
10	1.7908	0.5584	0.1359	7.3601	0.0759	13.1808
11	1.8983	0.5268	0.1268	7.8869	0.0668	14.9716
12	2.0122	0.4970	0.1193	8.3838	0.0593	16.8699
13	2.1329	0.4688	0.1130	8.8527	0.0530	18.8821
14	2.2609	0.4423	0.1076	9.2950	0.0476	21.0151
15	2.3966	0.4173	0.1030	9.7122	0.0430	23.2760
16	2.5404	0.3936	0.0990	10.1059	0.0390	25.6725
17	2.6928	0.3714	0.0954	10.4773	0.0354	28.2129
18	2.8543	0.3503	0.0924	10.8276	0.0324	30.9057
19	3.0256	0.3305	0.0896	11.1581	0.0296	33.7600
20	3.2071	0.3118	0.0872	11.4699	0.0272	36.7856
21	3.3996	0.2942	0.0850	11.7641	0.0250	39.9927
22	3.6035	0.2775	0.0830	12.0416	0.0230	43.3923
23	3.8197	0.2618	0.0813	12.3034	0.0213	46.9958
24	4.0489	0.2470	0.0797	12.5504	0.0197	50.8156
25	4.2919	0.2330	0.0782	12.7834	0.0182	54.8645
26	4.5494	0.2198	0.0769	13.0032	0.0169	59.1564
27	4.8223	0.2074	0.0757	13.2105	0.0157	63.7058
28	5.1117	0.1956	0.0746	13.4062	0.0146	68.5281
29	5.4184	0.1846	0.0736	13.5907	0.0136	73.6398
30	5.7435	0.1741	0.0726	13.7648	0.0126	79.0582
35	7.6861	0.1301	0.0690	14.4982	0.0090	111.4348
40	10.2857	0.0972	0.0665	15.0463	0.0065	154.7620
45	13.7646	0.0727	0.0647	15.4558	0.0047	212.7435
50	18.4202	0.0543	0.0634	15.7619	0.0034	290.3359
55	24.6503	0.0406	0.0625	15.9905	0.0025	394.1720
60	32.9877	0.0303	0.0619	16.1614	0.0019	533.1282
65	44.1450	0.0227	0.0614	16.2891	0.0014	719.0829
70	59.0759	0.0169	0.0610	16.3845	0.0010	967.9322
75	79.0569	0.0126	0.0608	16.4558	0.0008	1300.9487
80	105.7960	0.0095	0.0606	16.5091	0.0006	1746.5999
85	141.5789	0.0071	0.0604	16.5489	0.0004	2342.9817
90	189.4645	0.0053	0.0603	16.5787	0.0003	3141.0752
95	253.5463	0.0039	0.0602	16.6009	0.0002	4209.1042
100	339.3021	0.0029	0.0602	16.6175	0.0002	5638.3681

Table E-13. Discrete compounding factors: i = 7.00%

Year, n	To Find F Given P F/P	To Find P Given F P/F	To Find A Given P A/P	To Find P Given A P/A	To Find A Given F A/F	To Find F Given A F/A
1	1.0700	0.9346	1.0700	0.9346	1.0000	1.0000
2	1.1449	0.8734	0.5531	1.8080	0.4831	2.0700
3	1.2250	0.8163	0.3811	2.6243	0.3111	3.2149
4	1.3108	0.7629	0.2952	3.3872	0.2252	4.4399
5	1.4026	0.7130	0.2439	4.1002	0.1739	5.7507
6	1.5007	0.6663	0.2098	4.7665	0.1398	7.1533
7	1.6058	0.6227	0.1856	5.3893	0.1156	8.6540
8	1.7182	0.5820	0.1675	5.9713	0.0975	10.2598
9	1.8385	0.5439	0.1535	6.5152	0.0835	11.9780
10	1.9672	0.5083	0.1424	7.0236	0.0724	13.8164
11	2.1049	0.4751	0.1334	7.4987	0.0634	15.7836
12	2.2522	0.4440	0.1259	7.9427	0.0559	17.8885
13	2.4098	0.4150	0.1197	8.3577	0.0497	20.1406
14	2.5785	0.3878	0.1143	8.7455	0.0443	22.5505
15	2.7590	0.3624	0.1098	9.1079	0.0398	25.1290
16	2.9522	0.3387	0.1059	9.4466	0.0359	27.8881
17	3.1588	0.3166	0.1024	9.7632	0.0324	30.8402
18	3.3799	0.2959	0.0994	10.0591	0.0294	33.9990
19	3.6165	0.2765	0.0968	10.3356	0.0268	37.3790
20	3.8697	0.2584	0.0944	10.5940	0.0244	40.9955
21	4.1406	0.2415	0.0923	10.8355	0.0223	44.8652
22	4.4304	0.2257	0.0904	11.0612	0.0204	49.0057
23	4.7405	0.2109	0.0887	11.2722	0.0187	53.4361
24	5.0724	0.1971	0.0872	11.4693	0.0172	58.1767
25	5.4274	0.1842	0.0858	11.6536	0.0158	63.2490
26	5.8074	0.1722	0.0846	11.8258	0.0146	68.6765
27	6.2139	0.1609	0.0834	11.9867	0.0134	74.4838
28	6.6488	0.1504	0.0824	12.1371	0.0124	80.6977
29	7.1143	0.1406	0.0814	12.2777	0.0114	87.3465
30	7.6123	0.1314	0.0806	12.4090	0.0106	94.4608
35	10.6766	0.0937	0.0772	12.9477	0.0072	138.2369
40	14.9745	0.0668	0.0750	13.3317	0.0050	199.6351
45	21.0025	0.0476	0.0735	13.6055	0.0035	285.7493
50	29.4570	0.0339	0.0725	13.8007	0.0025	406.5289
55	41.3150	0.0242	0.0717	13.9399	0.0017	575.9286
60	57.9464	0.0173	0.0712	14.0392	0.0012	813.5204
65	81.2729	0.0123	0.0709	14.1099	0.0009	1146.7552
70	113.9894	0.0088	0.0706	14.1604	0.0006	1614.1342
75	159.8760	0.0063	0.0704	14.1964	0.0004	2269.6574
80	224.2344	0.0045	0.0703	14.2220	0.0003	3189.0627
85	314.5003	0.0032	0.0702	14.2403	0.0002	4478.5761
90	441.1030	0.0023	0.0702	14.2533	0.0002	6287.1854
95	618.6697	0.0016	0.0701	14.2626	0.0001	8823.8535
100	867.7163	0.0012	0.0701	14.2693	0.0001	12381.6618

Table E-14. Discrete compounding factors: i = 8.00%

Year, n	To Find F Given P F/P	To Find P Given F P/F	To Find A Given P A/P	To Find P Given A P/A	To Find A Given F A/F	To Find F Given A F/A
1	1.0800	0.9259	1.0800	0.9259	1.0000	1.0000
2	1.1664	0.8573	0.5608	1.7833	0.4808	2.0800
3	1.2597	0.7938	0.3880	2.5771	0.3080	3.2464
4	1.3605	0.7350	0.3019	3.3121	0.2219	4.5061
5	1.4693	0.6806	0.2505	3.9927	0.1705	5.8666
6	1.5869	0.6302	0.2163	4.6229	0.1363	7.3359
7	1.7138	0.5835	0.1921	5.2064	0.1121	8.9228
8	1.8509	0.5403	0.1740	5.7466	0.0940	10.6366
9	1.9990	0.5002	0.1601	6.2469	0.0801	12.4876
10	2.1589	0.4632	0.1490	6.7101	0.0690	14.4866
11	2.3316	0.4289	0.1401	7.1390	0.0601	16.6455
12	2.5182	0.3971	0.1327	7.5361	0.0527	18.9771
13	2.7196	0.3677	0.1265	7.9038	0.0465	21.4953
14	2.9372	0.3405	0.1213	8.2442	0.0413	24.2149
15	3.1722	0.3152	0.1168	8.5595	0.0368	27.1521
16	3.4259	0.2919	0.1130	8.8514	0.0330	30.3243
17	3.7000	0.2703	0.1096	9.1216	0.0296	33.7502
18	3.9960	0.2502	0.1067	9.3719	0.0267	37.4502
19	4.3157	0.2317	0.1041	9.6036	0.0241	41.4463
20	4.6610	0.2145	0.1019	9.8181	0.0219	45.7620
21	5.0338	0.1987	0.0998	10.0168	0.0198	50.4229
22	5.4365	0.1839	0.0980	10.2007	0.0180	55.4568
23	5.8715	0.1703	0.0964	10.3711	0.0164	60.8933
24	6.3412	0.1577	0.0950	10.5288	0.0150	66.7648
25	6.8485	0.1460	0.0937	10.6748	0.0137	73.1059
26	7.3964	0.1352	0.0925	10.8100	0.0125	79.9544
27	7.9881	0.1252	0.0914	10.9352	0.0114	87.3508
28	8.6271	0.1159	0.0905	11.0511	0.0105	95.3388
29	9.3173	0.1073	0.0896	11.1584	0.0096	103.9659
30	10.0627	0.0994	0.0888	11.2578	0.0088	113.2832
35	14.7853	0.0676	0.0858	11.6546	0.0058	172.3168
40	21.7245	0.0460	0.0839	11.9246	0.0039	259.0565
45	31.9204	0.0313	0.0826	12.1084	0.0026	386.5056
50	46.9016	0.0213	0.0817	12.2335	0.0017	573.7702
55	68.9139	0.0145	0.0812	12.3186	0.0012	848.9232
60	101.2571	0.0099	0.0808	12.3766	0.0008	1253.2133
65	148.7798	0.0067	0.0805	12.4160	0.0005	1847.2481
70	218.6064	0.0046	0.0804	12.4428	0.0004	2720.0801
75	321.2045	0.0031	0.0802	12.4611	0.0002	4002.5566
80	471.9548	0.0021	0.0802	12.4735	0.0002	5886.9354
85	693.4565	0.0014	0.0801	12.4820	0.0001	8655.7061
90	1018.9151	0.0010	0.0801	12.4877	0.0001	12723.9386
95	1497.1205	0.0007	0.0801	12.4917	0.0001	18701.5069
100	2199.7613	0.0005	0.0800	12.4943	0.0000	27484.5157

Table E-15. Discrete compounding factors: i = 9.00%

Year, n	To Find F Given P F/P	To Find P Given F P/F	To Find A Given P A/P	To Find P Given A P/A	To Find A Given F A/F	To Find F Given A F/A
1	1.0900	0.9174	1.0900	0.9174	1.0000	1.0000
2	1.1881	0.8417	0.5685	1.7591	0.4785	2.0900
3	1.2950	0.7722	0.3951	2.5313	0.3051	3.2781
4	1.4116	0.7084	0.3087	3.2397	0.2187	4.5731
5	1.5386	0.6499	0.2571	3.8897	0.1671	5.9847
6	1.6771	0.5963	0.2229	4.4859	0.1329	7.5233
7	1.8280	0.5470	0.1987	5.0330	0.1087	9.2004
8	1.9926	0.5019	0.1807	5.5348	0.0907	11.0285
9	2.1719	0.4604	0.1668	5.9952	0.0768	13.0210
10	2.3674	0.4224	0.1558	6.4177	0.0658	15.1929
11	2.5804	0.3875	0.1469	6.8052	0.0569	17.5603
12	2.8127	0.3555	0.1397	7.1607	0.0497	20.1407
13	3.0658	0.3262	0.1336	7.4869	0.0436	22.9534
14	3.3417	0.2992	0.1284	7.7862	0.0384	26.0192
15	3.6425	0.2745	0.1241	8.0607	0.0341	29.3609
16	3.9703	0.2519	0.1203	8.3126	0.0303	33.0034
17	4.3276	0.2311	0.1170	8.5436	0.0270	36.9737
18	4.7171	0.2120	0.1142	8.7556	0.0242	41.3013
19	5.1417	0.1945	0.1117	8.9501	0.0217	46.0185
20	5.6044	0.1784	0.1095	9.1285	0.0195	51.1601
21	6.1088	0.1637	0.1076	9.2922	0.0176	56.7645
22	6.6586	0.1502	0.1059	9.4424	0.0159	62.8733
23	7.2579	0.1378	0.1044	9.5802	0.0144	69.5319
24	7.9111	0.1264	0.1030	9.7066	0.0130	76.7898
25	8.6231	0.1160	0.1018	9.8226	0.0118	84.7009
26	9.3992	0.1064	0.1007	9.9290	0.0107	93.3240
27	10.2451	0.0976	0.0997	10.0266	0.0097	102.7231
28	11.1671	0.0895	0.0989	10.1161	0.0089	112.9682
29	12.1722	0.0822	0.0981	10.1983	0.0081	124.1354
30	13.2677	0.0754	0.0973	10.2737	0.0073	136.3075
35	20.4140	0.0490	0.0946	10.5668	0.0046	215.7108
40	31.4094	0.0318	0.0930	10.7574	0.0030	337.8824
45	48.3273	0.0207	0.0919	10.8812	0.0019	525.8587
50	74.3575	0.0134	0.0912	10.9617	0.0012	815.0836
55	114.4083	0.0087	0.0908	11.0140	0.0008	1260.0918
60	176.0313	0.0057	0.0905	11.0480	0.0005	1944.7921
65	270.8460	0.0037	0.0903	11.0701	0.0003	2998.2885
70	416.7301	0.0024	0.0902	11.0844	0.0002	4619.2232
75	641.1909	0.0016	0.0901	11.0938	0.0001	7113.2321
80	986.5517	0.0010	0.0901	11.0998	0.0001	10950.5741
85	1517.9320	0.0007	0.0901	11.1038	0.0001	16854.8003
90	2335.5266	0.0004	0.0900	11.1064	0.0000	25939.1842
95	3593.4971	0.0003	0.0900	11.1080	0.0000	39916.6350
100	5529.0408	0.0002	0.0900	11.1091	0.0000	61422.6755

Table E-16. Discrete compounding factors: i = 10.00%

Year, n	To Find F Given P F/P	To Find P Given F P/F	To Find A Given P A/P	To Find P Given A P/A	To Find A Given F A/F	To Find F Given A F/A
1	1.1000	0.9091	1.1000	0.9091	1.0000	1.0000
2	1.2100	0.8264	0.5762	1.7355	0.4762	2.1000
3	1.3310	0.7513	0.4021	2.4869	0.3021	3.3100
4	1.4641	0.6830	0.3155	3.1699	0.2155	4.6410
5	1.6105	0.6209	0.2638	3.7908	0.1638	6.1051
6	1.7716	0.5645	0.2296	4.3553	0.1296	7.7156
7	1.9487	0.5132	0.2054	4.8684	0.1054	9.4872
8	2.1436	0.4665	0.1874	5.3349	0.0874	11.4359
9	2.3579	0.4241	0.1736	5.7590	0.0736	13.5795
10	2.5937	0.3855	0.1627	6.1446	0.0627	15.9374
11	2.8531	0.3505	0.1540	6.4951	0.0540	18.5312
12	3.1384	0.3186	0.1468	6.8137	0.0468	21.3843
13	3.4523	0.2897	0.1408	7.1034	0.0408	24.5227
14	3.7975	0.2633	0.1357	7.3667	0.0357	27.9750
15	4.1772	0.2394	0.1315	7.6061	0.0315	31.7725
16	4.5950	0.2176	0.1278	7.8237	0.0278	35.9497
17	5.0545	0.1978	0.1247	8.0216	0.0247	40.5447
18	5.5599	0.1799	0.1219	8.2014	0.0219	45.5992
19	6.1159	0.1635	0.1195	8.3649	0.0195	51.1591
20	6.7275	0.1486	0.1175	8.5136	0.0175	57.2750
21	7.4002	0.1351	0.1156	8.6487	0.0156	64.0025
22	8.1403	0.1228	0.1140	8.7715	0.0140	71.4027
23	8.9543	0.1117	0.1126	8.8832	0.0126	79.5430
24	9.8497	0.1015	0.1113	8.9847	0.0113	88.4973
25	10.8347	0.0923	0.1102	9.0770	0.0102	98.3471
26	11.9182	0.0839	0.1092	9.1609	0.0092	109.1818
27	13.1100	0.0763	0.1083	9.2372	0.0083	121.0999
28	14.4210	0.0693	0.1075	9.3066	0.0075	134.2099
29	15.8631	0.0630	0.1067	9.3696	0.0067	148.6309
30	17.4494	0.0573	0.1061	9.4269	0.0061	164.4940
35	28.1024	0.0356	0.1037	9.6442	0.0037	271.0244
40	45.2593	0.0221	0.1023	9.7791	0.0023	442.5926
45	72.8905	0.0137	0.1014	9.8628	0.0014	718.9048
50	117.3909	0.0085	0.1009	9.9148	0.0009	1163.9085
55	189.0591	0.0053	0.1005	9.9471	0.0005	1880.5914
60	304.4816	0.0033	0.1003	9.9672	0.0003	3034.8164
65	490.3707	0.0020	0.1002	9.9796	0.0002	4893.7073
70	789.7470	0.0013	0.1001	9.9873	0.0001	7887.4696
75	1271.8954	0.0008	0.1001	9.9921	0.0001	12708.9537
80	2048.4002	0.0005	0.1000	9.9951	0.0000	20474.0021
85	3298.9690	0.0003	0.1000	9.9970	0.0000	32979.6903
90	5313.0226	0.0002	0.1000	9.9981	0.0000	53120.2261
95	8556.6760	0.0001	0.1000	9.9988	0.0000	85556.7605
100	13780.6123	0.0001	0.1000	9.9993	0.0000	137796.1230

Table E-17. Discrete compounding factors: i = 11.00%

Year, n	To Find F Given P F/P	To Find P Given F P/F	To Find A Given P A/P	To Find P Given A P/A	To Find A Given F A/F	To Find F Given A F/A
1	1.1100	0.9009	1.1100	0.9009	1.0000	1.0000
2	1.2321	0.8116	0.5839	1.7125	0.4739	2.1100
3	1.3676	0.7312	0.4092	2.4437	0.2992	3.3421
4	1.5181	0.6587	0.3223	3.1024	0.2123	4.7097
5	1.6851	0.5935	0.2706	3.6959	0.1606	6.2278
6	1.8704	0.5346	0.2364	4.2305	0.1264	7.9129
7	2.0762	0.4817	0.2122	4.7122	0.1022	9.7833
8	2.3045	0.4339	0.1943	5.1461	0.0843	11.8594
9	2.5580	0.3909	0.1806	5.5370	0.0706	14.1640
10	2.8394	0.3522	0.1698	5.8892	0.0598	16.7220
11	3.1518	0.3173	0.1611	6.2065	0.0511	19.5614
12	3.4985	0.2858	0.1540	6.4924	0.0440	22.7132
13	3.8833	0.2575	0.1482	6.7499	0.0382	26.2116
14	4.3104	0.2320	0.1432	6.9819	0.0332	30.0949
15	4.7846	0.2090	0.1391	7.1909	0.0291	34.4054
16	5.3109	0.1883	0.1355	7.3792	0.0255	39.1899
17	5.8951	0.1696	0.1325	7.5488	0.0225	44.5008
18	6.5436	0.1528	0.1298	7.7016	0.0198	50.3959
19	7.2633	0.1377	0.1276	7.8393	0.0176	56.9395
20	8.0623	0.1240	0.1256	7.9633	0.0156	64.2028
21	8.9492	0.1117	0.1238	8.0751	0.0138	72.2651
22	9.9336	0.1007	0.1223	8.1757	0.0123	81.2143
23	11.0263	0.0907	0.1210	8.2664	0.0110	91.1479
24	12.2392	0.0817	0.1198	8.3481	0.0098	102.1742
25	13.5855	0.0736	0.1187	8.4217	0.0087	114.4133
26	15.0799	0.0663	0.1178	8.4881	0.0078	127.9988
27	16.7386	0.0597	0.1170	8.5478	0.0070	143.0786
28	18.5799	0.0538	0.1163	8.6016	0.0063	159.8173
29	20.6237	0.0485	0.1156	8.6501	0.0056	178.3972
30	22.8923	0.0437	0.1150	8.6938	0.0050	199.0209
35	38.5749	0.0259	0.1129	8.8552	0.0029	341.5896
40	65.0009	0.0154	0.1117	8.9511	0.0017	581.8261
45	109.5302	0.0091	0.1110	9.0079	0.0010	986.6386
50	184.5648	0.0054	0.1106	9.0417	0.0006	1668.7712
55	311.0025	0.0032	0.1104	9.0617	0.0004	2818.2042
60	524.0572	0.0019	0.1102	9.0736	0.0002	4755.0658
65	883.0669	0.0011	0.1101	9.0806	0.0001	8018.7903
70	1488.0191	0.0007	0.1101	9.0848	0.0001	13518.3557
75	2507.3988	0.0004	0.1100	9.0873	0.0000	22785.4434
80	4225.1128	0.0002	0.1100	9.0888	0.0000	38401.0250

Table E-18. Discrete compounding factors: i = 12.00%

Year, n	To Find F Given P F/P	To Find P Given F P/F	To Find A Given P A/P	To Find P Given A P/A	To Find A Given F A/F	To Find F Given A F/A
1	1.1200	0.8929	1.1200	0.8929	1.0000	1.0000
2	1.2544	0.7972	0.5917	1.6901	0.4717	2.1200
3	1.4049	0.7118	0.4163	2.4018	0.2963	3.3744
4	1.5735	0.6355	0.3292	3.0373	0.2092	4.7793
5	1.7623	0.5674	0.2774	3.6048	0.1574	6.3528
6	1.9738	0.5066	0.2432	4.1114	0.1232	8.1152
7	2.2107	0.4523	0.2191	4.5638	0.0991	10.0890
8	2.4760	0.4039	0.2013	4.9676	0.0813	12.2997
9	2.7731	0.3606	0.1877	5.3282	0.0677	14.7757
10	3.1058	0.3220	0.1770	5.6502	0.0570	17.5487
11	3.4785	0.2875	0.1684	5.9377	0.0484	20.6546
12	3.8960	0.2567	0.1614	6.1944	0.0414	24.1331
13	4.3635	0.2292	0.1557	6.4235	0.0357	28.0291
14	4.8871	0.2046	0.1509	6.6282	0.0309	32.3926
15	5.4736	0.1827	0.1468	6.8109	0.0268	37.2797
16	6.1304	0.1631	0.1434	6.9740	0.0234	42.7533
17	6.8660	0.1456	0.1405	7.1196	0.0205	48.8837
18	7.6900	0.1300	0.1379	7.2497	0.0179	55.7497
19	8.6128	0.1161	0.1358	7.3658	0.0158	63.4397
20	9.6463	0.1037	0.1339	7.4694	0.0139	72.0524
21	10.8038	0.0926	0.1322	7.5620	0.0122	81.6987
22	12.1003	0.0826	0.1308	7.6446	0.0108	92.5026
23	13.5523	0.0738	0.1296	7.7184	0.0096	104.6029
24	15.1786	0.0659	0.1285	7.7843	0.0085	118.1552
25	17.0001	0.0588	0.1275	7.8431	0.0075	133.3339
26	19.0401	0.0525	0.1267	7.8957	0.0067	150.3339
27	21.3249	0.0469	0.1259	7.9426	0.0059	169.3740
28	23.8839	0.0419	0.1252	7.9844	0.0052	190.6989
29	26.7499	0.0374	0.1247	8.0218	0.0047	214.5828
30	29.9599	0.0334	0.1241	8.0552	0.0041	241.3327
35	52.7996	0.0189	0.1223	8.1755	0.0023	431.6635
40	93.0510	0.0107	0.1213	8.2438	0.0013	767.0914
45	163.9876	0.0061	0.1207	8.2825	0.0007	1358.2300
50	289.0022	0.0035	0.1204	8.3045	0.0004	2400.0182
55	509.3206	0.0020	0.1202	8.3170	0.0002	4236.0050
60	897.5969	0.0011	0.1201	8.3240	0.0001	7471.6411
65	1581.8725	0.0006	0.1201	8.3281	0.0001	13173.9374
70	2787.7998	0.0004	0.1200	8.3303	0.0000	23223.3319
75	4913.0558	0.0002	0.1200	8.3316	0.0000	40933.7987
80	8658.4831	0.0001	0.1200	8.3324	0.0000	72145.6925

Table E-19. Discrete compounding factors: i = 13.00%

Year, n	To Find F Given P F/P	To Find P Given F P/F	To Find A Given P A/P	To Find P Given A P/A	To Find A Given F A/F	To Find F Given A F/A
1	1.1300	0.8850	1.1300	0.8850	1.0000	1.0000
2	1.2769	0.7831	0.5995	1.6681	0.4695	2.1300
3	1.4429	0.6931	0.4235	2.3612	0.2935	3.4069
4	1.6305	0.6133	0.3362	2.9745	0.2062	4.8498
5	1.8424	0.5428	0.2843	3.5172	0.1543	6.4803
6	2.0820	0.4803	0.2502	3.9975	0.1202	8.3227
7	2.3526	0.4251	0.2261	4.4226	0.0961	10.4047
8	2.6584	0.3762	0.2084	4.7988	0.0784	12.7573
9	3.0040	0.3329	0.1949	5.1317	0.0649	15.4157
10	3.3946	0.2946	0.1843	5.4262	0.0543	18.4197
11	3.8359	0.2607	0.1758	5.6869	0.0458	21.8143
12	4.3345	0.2307	0.1690	5.9176	0.0390	25.6502
13	4.8980	0.2042	0.1634	6.1218	0.0334	29.9847
14	5.5348	0.1807	0.1587	6.3025	0.0287	34.8827
15	6.2543	0.1599	0.1547	6.4624	0.0247	40.4175
16	7.0673	0.1415	0.1514	6.6039	0.0214	46.6717
17	7.9861	0.1252	0.1486	6.7291	0.0186	53.7391
18	9.0243	0.1108	0.1462	6.8399	0.0162	61.7251
19	10.1974	0.0981	0.1441	6.9380	0.0141	70.7494
20	11.5231	0.0868	0.1424	7.0248	0.0124	80.9468
21	13.0211	0.0768	0.1408	7.1016	0.0108	92.4699
22	14.7138	0.0680	0.1395	7.1695	0.0095	105.4910
23	16.6266	0.0601	0.1383	7.2297	0.0083	120.2048
24	18.7881	0.0532	0.1373	7.2829	0.0073	136.8315
25	21.2305	0.0471	0.1364	7.3300	0.0064	155.6196
26	23.9905	0.0417	0.1357	7.3717	0.0057	176.8501
27	27.1093	0.0369	0.1350	7.4086	0.0050	200.8406
28	30.6335	0.0326	0.1344	7.4412	0.0044	227.9499
29	34.6158	0.0289	0.1339	7.4701	0.0039	258.5834
30	39.1159	0.0256	0.1334	7.4957	0.0034	293.1992
35	72.0685	0.0139	0.1318	7.5856	0.0018	546.6808
40	132.7816	0.0075	0.1310	7.6344	0.0010	1013.7042
45	244.6414	0.0041	0.1305	7.6609	0.0005	1874.1646
50	450.7359	0.0022	0.1303	7.6752	0.0003	3459.5071
55	830.4517	0.0012	0.1302	7.6830	0.0002	6380.3979
60	1530.0535	0.0007	0.1301	7.6873	0.0001	11761.9498
65	2819.0243	0.0004	0.1300	7.6896	0.0000	21677.1103
70	5193.8696	0.0002	0.1300	7.6908	0.0000	39945.1510

Table E-20. Discrete compounding factors: i = 14.00%

Year, *n*	To Find *F* Given *P* *F/P*	To Find *P* Given *F* *P/F*	To Find *A* Given *P* *A/P*	To Find *P* Given *A* *P/A*	To Find *A* Given *F* *A/F*	To Find *F* Given *A* *F/A*
1	1.1400	0.8772	1.1400	0.8772	1.0000	1.0000
2	1.2996	0.7695	0.6073	1.6467	0.4673	2.1400
3	1.4815	0.6750	0.4307	2.3216	0.2907	3.4396
4	1.6890	0.5921	0.3432	2.9137	0.2032	4.9211
5	1.9254	0.5194	0.2913	3.4331	0.1513	6.6101
6	2.1950	0.4556	0.2572	3.8887	0.1172	8.5355
7	2.5023	0.3996	0.2332	4.2883	0.0932	10.7305
8	2.8526	0.3506	0.2156	4.6389	0.0756	13.2328
9	3.2519	0.3075	0.2022	4.9464	0.0622	16.0853
10	3.7072	0.2697	0.1917	5.2161	0.0517	19.3373
11	4.2262	0.2366	0.1834	5.4527	0.0434	23.0445
12	4.8179	0.2076	0.1767	5.6603	0.0367	27.2707
13	5.4924	0.1821	0.1712	5.8424	0.0312	32.0887
14	6.2613	0.1597	0.1666	6.0021	0.0266	37.5811
15	7.1379	0.1401	0.1628	6.1422	0.0228	43.8424
16	8.1372	0.1229	0.1596	6.2651	0.0196	50.9804
17	9.2765	0.1078	0.1569	6.3729	0.0169	59.1176
18	10.5752	0.0946	0.1546	6.4674	0.0146	68.3941
19	12.0557	0.0829	0.1527	6.5504	0.0127	78.9692
20	13.7435	0.0728	0.1510	6.6231	0.0110	91.0249
21	15.6676	0.0638	0.1495	6.6870	0.0095	104.7684
22	17.8610	0.0560	0.1483	6.7429	0.0083	120.4360
23	20.3616	0.0491	0.1472	6.7921	0.0072	138.2970
24	23.2122	0.0431	0.1463	6.8351	0.0063	158.6586
25	26.4619	0.0378	0.1455	6.8729	0.0055	181.8708
26	30.1666	0.0331	0.1448	6.9061	0.0048	208.3327
27	34.3899	0.0291	0.1442	6.9352	0.0042	238.4993
28	39.2045	0.0255	0.1437	6.9607	0.0037	272.8892
29	44.6931	0.0224	0.1432	6.9830	0.0032	312.0937
30	50.9502	0.0196	0.1428	7.0027	0.0028	356.7868
35	98.1002	0.0102	0.1414	7.0700	0.0014	693.5727
40	188.8835	0.0053	0.1407	7.1050	0.0007	1342.0251
45	363.6791	0.0027	0.1404	7.1232	0.0004	2590.5648
50	700.2330	0.0014	0.1402	7.1327	0.0002	4994.5213
55	1348.2388	0.0007	0.1401	7.1376	0.0001	9623.1343
60	2595.9187	0.0004	0.1401	7.1401	0.0001	18535.1333
65	4998.2196	0.0002	0.1400	7.1414	0.0000	35694.4260
70	9623.6450	0.0001	0.1400	7.1421	0.0000	68733.1785

Table E-21. Discrete compounding factors: i = 15.00%

Year, n	To Find F Given P F/P	To Find P Given F P/F	To Find A Given P A/P	To Find P Given A P/A	To Find A Given F A/F	To Find F Given A F/A
1	1.1500	0.8696	1.1500	0.8696	1.0000	1.0000
2	1.3225	0.7561	0.6151	1.6257	0.4651	2.1500
3	1.5209	0.6575	0.4380	2.2832	0.2880	3.4725
4	1.7490	0.5718	0.3503	2.8550	0.2003	4.9934
5	2.0114	0.4972	0.2983	3.3522	0.1483	6.7424
6	2.3131	0.4323	0.2642	3.7845	0.1142	8.7537
7	2.6600	0.3759	0.2404	4.1604	0.0904	11.0668
8	3.0590	0.3269	0.2229	4.4873	0.0729	13.7268
9	3.5179	0.2843	0.2096	4.7716	0.0596	16.7858
10	4.0456	0.2472	0.1993	5.0188	0.0493	20.3037
11	4.6524	0.2149	0.1911	5.2337	0.0411	24.3493
12	5.3503	0.1869	0.1845	5.4206	0.0345	29.0017
13	6.1528	0.1625	0.1791	5.5831	0.0291	34.3519
14	7.0757	0.1413	0.1747	5.7245	0.0247	40.5047
15	8.1371	0.1229	0.1710	5.8474	0.0210	47.5804
16	9.3576	0.1069	0.1679	5.9542	0.0179	55.7175
17	10.7613	0.0929	0.1654	6.0472	0.0154	65.0751
18	12.3755	0.0808	0.1632	6.1280	0.0132	75.8364
19	14.2318	0.0703	0.1613	6.1982	0.0113	88.2118
20	16.3665	0.0611	0.1598	6.2593	0.0098	102.4436
21	18.8215	0.0531	0.1584	6.3125	0.0084	118.8101
22	21.6447	0.0462	0.1573	6.3587	0.0073	137.6316
23	24.8915	0.0402	0.1563	6.3988	0.0063	159.2764
24	28.6252	0.0349	0.1554	6.4338	0.0054	184.1678
25	32.9190	0.0304	0.1547	6.4641	0.0047	212.7930
26	37.8568	0.0264	0.1541	6.4906	0.0041	245.7120
27	43.5353	0.0230	0.1535	6.5135	0.0035	283.5688
28	50.0656	0.0200	0.1531	6.5335	0.0031	327.1041
29	57.5755	0.0174	0.1527	6.5509	0.0027	377.1697
30	66.2118	0.0151	0.1523	6.5660	0.0023	434.7451
35	133.1755	0.0075	0.1511	6.6166	0.0011	881.1702
40	267.8635	0.0037	0.1506	6.6418	0.0006	1779.0903
45	538.7693	0.0019	0.1503	6.6543	0.0003	3585.1285
50	1083.6574	0.0009	0.1501	6.6605	0.0001	7217.7163
55	2179.6222	0.0005	0.1501	6.6636	0.0001	14524.1479
60	4383.9987	0.0002	0.1500	6.6651	0.0000	29219.9916
65	8817.7874	0.0001	0.1500	6.6659	0.0000	58778.5826
70	17735.7200	0.0001	0.1500	6.6663	0.0000	118231.4670

Table E-22. Discrete compounding factors: i = 16.00%

Year, *n*	To Find F Given P F/P	To Find P Given F P/F	To Find A Given P A/P	To Find P Given A P/A	To Find A Given F A/F	To Find F Given A F/A
1	1.1600	0.8621	1.1600	0.8621	1.0000	1.0000
2	1.3456	0.7432	0.6230	1.6052	0.4630	2.1600
3	1.5609	0.6407	0.4453	2.2459	0.2853	3.5056
4	1.8106	0.5523	0.3574	2.7982	0.1974	5.0665
5	2.1003	0.4761	0.3054	3.2743	0.1454	6.8771
6	2.4364	0.4104	0.2714	3.6847	0.1114	8.9775
7	2.8262	0.3538	0.2476	4.0386	0.0876	11.4139
8	3.2784	0.3050	0.2302	4.3436	0.0702	14.2401
9	3.8030	0.2630	0.2171	4.6065	0.0571	17.5185
10	4.4114	0.2267	0.2069	4.8332	0.0469	21.3215
11	5.1173	0.1954	0.1989	5.0286	0.0389	25.7329
12	5.9360	0.1685	0.1924	5.1971	0.0324	30.8502
13	6.8858	0.1452	0.1872	5.3423	0.0272	36.7862
14	7.9875	0.1252	0.1829	5.4675	0.0229	43.6720
15	9.2655	0.1079	0.1794	5.5755	0.0194	51.6595
16	10.7480	0.0930	0.1764	5.6685	0.0164	60.9250
17	12.4677	0.0802	0.1740	5.7487	0.0140	71.6730
18	14.4625	0.0691	0.1719	5.8178	0.0119	84.1407
19	16.7765	0.0596	0.1701	5.8775	0.0101	98.6032
20	19.4608	0.0514	0.1687	5.9288	0.0087	115.3797
21	22.5745	0.0443	0.1674	5.9731	0.0074	134.8405
22	26.1864	0.0382	0.1664	6.0113	0.0064	157.4150
23	30.3762	0.0329	0.1654	6.0442	0.0054	183.6014
24	35.2364	0.0284	0.1647	6.0726	0.0047	213.9776
25	40.8742	0.0245	0.1640	6.0971	0.0040	249.2140
26	47.4141	0.0211	0.1634	6.1182	0.0034	290.0883
27	55.0004	0.0182	0.1630	6.1364	0.0030	337.5024
28	63.8004	0.0157	0.1625	6.1520	0.0025	392.5028
29	74.0085	0.0135	0.1622	6.1656	0.0022	456.3032
30	85.8499	0.0116	0.1619	6.1772	0.0019	530.3117
35	180.3141	0.0055	0.1609	6.2153	0.0009	1120.7130
40	378.7212	0.0026	0.1604	6.2335	0.0004	2360.7572
45	795.4438	0.0013	0.1602	6.2421	0.0002	4965.2739
50	1670.7038	0.0006	0.1601	6.2463	0.0001	10435.6488

Table E-23. Discrete compounding factors: i = 18.00%

Year, n	To Find F Given P F/P	To Find P Given F P/F	To Find A Given P A/P	To Find P Given A P/A	To Find A Given F A/F	To Find F Given A F/A
1	1.1800	0.8475	1.1800	0.8475	1.0000	1.0000
2	1.3924	0.7182	0.6387	1.5656	0.4587	2.1800
3	1.6430	0.6086	0.4599	2.1743	0.2799	3.5724
4	1.9388	0.5158	0.3717	2.6901	0.1917	5.2154
5	2.2878	0.4371	0.3198	3.1272	0.1398	7.1542
6	2.6996	0.3704	0.2859	3.4976	0.1059	9.4420
7	3.1855	0.3139	0.2624	3.8115	0.0824	12.1415
8	3.7589	0.2660	0.2452	4.0776	0.0652	15.3270
9	4.4355	0.2255	0.2324	4.3030	0.0524	19.0859
10	5.2338	0.1911	0.2225	4.4941	0.0425	23.5213
11	6.1759	0.1619	0.2148	4.6560	0.0348	28.7551
12	7.2876	0.1372	0.2086	4.7932	0.0286	34.9311
13	8.5994	0.1163	0.2037	4.9095	0.0237	42.2187
14	10.1472	0.0985	0.1997	5.0081	0.0197	50.8180
15	11.9737	0.0835	0.1964	5.0916	0.0164	60.9653
16	14.1290	0.0708	0.1937	5.1624	0.0137	72.9390
17	16.6722	0.0600	0.1915	5.2223	0.0115	87.0680
18	19.6733	0.0508	0.1896	5.2732	0.0096	103.7403
19	23.2144	0.0431	0.1881	5.3162	0.0081	123.4135
20	27.3930	0.0365	0.1868	5.3527	0.0068	146.6280
21	32.3238	0.0309	0.1857	5.3837	0.0057	174.0210
22	38.1421	0.0262	0.1848	5.4099	0.0048	206.3448
23	45.0076	0.0222	0.1841	5.4321	0.0041	244.4868
24	53.1090	0.0188	0.1835	5.4509	0.0035	289.4945
25	62.6686	0.0160	0.1829	5.4669	0.0029	342.6035
26	73.9490	0.0135	0.1825	5.4804	0.0025	405.2721
27	87.2598	0.0115	0.1821	5.4919	0.0021	479.2211
28	102.9666	0.0097	0.1818	5.5016	0.0018	566.4809
29	121.5005	0.0082	0.1815	5.5098	0.0015	669.4475
30	143.3706	0.0070	0.1813	5.5168	0.0013	790.9480
35	327.9973	0.0030	0.1806	5.5386	0.0006	1816.6516
40	750.3783	0.0013	0.1802	5.5482	0.0002	4163.2130
45	1716.6839	0.0006	0.1801	5.5523	0.0001	9531.5771
50	3927.3569	0.0003	0.1800	5.5541	0.0000	21813.0937

Table E-24. Discrete compounding factors: i = 20.00%

Year, *n*	To Find F Given P F/P	To Find P Given F P/F	To Find A Given P A/P	To Find P Given A P/A	To Find A Given F A/F	To Find F Given A F/A
1	1.2000	0.8333	1.2000	0.8333	1.0000	1.0000
2	1.4400	0.6944	0.6545	1.5278	0.4545	2.2000
3	1.7280	0.5787	0.4747	2.1065	0.2747	3.6400
4	2.0736	0.4823	0.3863	2.5887	0.1863	5.3680
5	2.4883	0.4019	0.3344	2.9906	0.1344	7.4416
6	2.9860	0.3349	0.3007	3.3255	0.1007	9.9299
7	3.5832	0.2791	0.2774	3.6046	0.0774	12.9159
8	4.2998	0.2326	0.2606	3.8372	0.0606	16.4991
9	5.1598	0.1938	0.2481	4.0310	0.0481	20.7989
10	6.1917	0.1615	0.2385	4.1925	0.0385	25.9587
11	7.4301	0.1346	0.2311	4.3271	0.0311	32.1504
12	8.9161	0.1122	0.2253	4.4392	0.0253	39.5805
13	10.6993	0.0935	0.2206	4.5327	0.0206	48.4966
14	12.8392	0.0779	0.2169	4.6106	0.0169	59.1959
15	15.4070	0.0649	0.2139	4.6755	0.0139	72.0351
16	18.4884	0.0541	0.2114	4.7296	0.0114	87.4421
17	22.1861	0.0451	0.2094	4.7746	0.0094	105.9306
18	26.6233	0.0376	0.2078	4.8122	0.0078	128.1167
19	31.9480	0.0313	0.2065	4.8435	0.0065	154.7400
20	38.3376	0.0261	0.2054	4.8696	0.0054	186.6880
21	46.0051	0.0217	0.2044	4.8913	0.0044	225.0256
22	55.2061	0.0181	0.2037	4.9094	0.0037	271.0307
23	66.2474	0.0151	0.2031	4.9245	0.0031	326.2369
24	79.4968	0.0126	0.2025	4.9371	0.0025	392.4842
25	95.3962	0.0105	0.2021	4.9476	0.0021	471.9811
26	114.4755	0.0087	0.2018	4.9563	0.0018	567.3773
27	137.3706	0.0073	0.2015	4.9636	0.0015	681.8528
28	164.8447	0.0061	0.2012	4.9697	0.0012	819.2233
29	197.8136	0.0051	0.2010	4.9747	0.0010	984.0680
30	237.3763	0.0042	0.2008	4.9789	0.0008	1181.8816
35	590.6682	0.0017	0.2003	4.9915	0.0003	2948.3411
40	1469.7716	0.0007	0.2001	4.9966	0.0001	7343.8578
45	3657.2620	0.0003	0.2001	4.9986	0.0001	18281.3099
50	9100.4382	0.0001	0.2000	4.9995	0.0000	45497.1908

Table E-25. Discrete compounding factors: i = 25.00%

Year, *n*	To Find *F* Given *P* *F/P*	To Find *P* Given *F* *P/F*	To Find *A* Given *P* *A/P*	To Find *P* Given *A* *P/A*	To Find *A* Given *F* *A/F*	To Find *F* Given *A* *F/A*
1	1.2500	0.8000	1.2500	0.8000	1.0000	1.0000
2	1.5625	0.6400	0.6944	1.4400	0.4444	2.2500
3	1.9531	0.5120	0.5123	1.9520	0.2623	3.8125
4	2.4414	0.4096	0.4234	2.3616	0.1734	5.7656
5	3.0518	0.3277	0.3718	2.6893	0.1218	8.2070
6	3.8147	0.2621	0.3388	2.9514	0.0888	11.2588
7	4.7684	0.2097	0.3163	3.1611	0.0663	15.0735
8	5.9605	0.1678	0.3004	3.3289	0.0504	19.8419
9	7.4506	0.1342	0.2888	3.4631	0.0388	25.8023
10	9.3132	0.1074	0.2801	3.5705	0.0301	33.2529
11	11.6415	0.0859	0.2735	3.6564	0.0235	42.5661
12	14.5519	0.0687	0.2684	3.7251	0.0184	54.2077
13	18.1899	0.0550	0.2645	3.7801	0.0145	68.7596
14	22.7374	0.0440	0.2615	3.8241	0.0115	86.9495
15	28.4217	0.0352	0.2591	3.8593	0.0091	109.6868
16	35.5271	0.0281	0.2572	3.8874	0.0072	138.1085
17	44.4089	0.0225	0.2558	3.9099	0.0058	173.6357
18	55.5112	0.0180	0.2546	3.9279	0.0046	218.0446
19	69.3889	0.0144	0.2537	3.9424	0.0037	273.5558
20	86.7362	0.0115	0.2529	3.9539	0.0029	342.9447
21	108.4202	0.0092	0.2523	3.9631	0.0023	429.6809
22	135.5253	0.0074	0.2519	3.9705	0.0019	538.1011
23	169.4066	0.0059	0.2515	3.9764	0.0015	673.6264
24	211.7582	0.0047	0.2512	3.9811	0.0012	843.0329
25	264.6978	0.0038	0.2509	3.9849	0.0009	1054.7912
26	330.8722	0.0030	0.2508	3.9879	0.0008	1319.4890
27	413.5903	0.0024	0.2506	3.9903	0.0006	1650.3612
28	516.9879	0.0019	0.2505	3.9923	0.0005	2063.9515
29	646.2349	0.0015	0.2504	3.9938	0.0004	2580.9394
30	807.7936	0.0012	0.2503	3.9950	0.0003	3227.1743
35	2465.1903	0.0004	0.2501	3.9984	0.0001	9856.7613
40	7523.1638	0.0001	0.2500	3.9995	0.0000	30088.6554
45	22958.8740	0.0000	0.2500	3.9998	0.0000	91831.4962
50	70064.9232	0.0000	0.2500	3.9999	0.0000	280255.693

Table E-26. Discrete compounding factors: i = 30.00%

Year, n	To Find F Given P F/P	To Find P Given F P/F	To Find A Given P A/P	To Find P Given A P/A	To Find A Given F A/F	To Find F Given A F/A
1	1.3000	0.7692	1.3000	0.7692	1.0000	1.0000
2	1.6900	0.5917	0.7348	1.3609	0.4348	2.3000
3	2.1970	0.4552	0.5506	1.8161	0.2506	3.9900
4	2.8561	0.3501	0.4616	2.1662	0.1616	6.1870
5	3.7129	0.2693	0.4106	2.4356	0.1106	9.0431
6	4.8268	0.2072	0.3784	2.6427	0.0784	12.7560
7	6.2749	0.1594	0.3569	2.8021	0.0569	17.5828
8	8.1573	0.1226	0.3419	2.9247	0.0419	23.8577
9	10.6045	0.0943	0.3312	3.0190	0.0312	32.0150
10	13.7858	0.0725	0.3235	3.0915	0.0235	42.6195
11	17.9216	0.0558	0.3177	3.1473	0.0177	56.4053
12	23.2981	0.0429	0.3135	3.1903	0.0135	74.3270
13	30.2875	0.0330	0.3102	3.2233	0.0102	97.6250
14	39.3738	0.0254	0.3078	3.2487	0.0078	127.9125
15	51.1859	0.0195	0.3060	3.2682	0.0060	167.2863
16	66.5417	0.0150	0.3046	3.2832	0.0046	218.4722
17	86.5042	0.0116	0.3035	3.2948	0.0035	285.0139
18	112.4554	0.0089	0.3027	3.3037	0.0027	371.5180
19	146.1920	0.0068	0.3021	3.3105	0.0021	483.9734
20	190.0496	0.0053	0.3016	3.3158	0.0016	630.1655
21	247.0645	0.0040	0.3012	3.3198	0.0012	820.2151
22	321.1839	0.0031	0.3009	3.3230	0.0009	1067.2796
23	417.5391	0.0024	0.3007	3.3254	0.0007	1388.4635
24	542.8008	0.0018	0.3006	3.3272	0.0006	1806.0026
25	705.6410	0.0014	0.3004	3.3286	0.0004	2348.8033
26	917.3333	0.0011	0.3003	3.3297	0.0003	3054.4443
27	1192.5333	0.0008	0.3003	3.3305	0.0003	3971.7776
28	1550.2933	0.0006	0.3002	3.3312	0.0002	5164.3109
29	2015.3813	0.0005	0.3001	3.3317	0.0001	6714.6042
30	2619.9956	0.0004	0.3001	3.3321	0.0001	8729.9855
35	9727.8604	0.0001	0.3000	3.3330	0.0000	32422.8681

Table E-27. Discrete compounding factors: i = 35.00%

Year, n	To Find F Given P F/P	To Find P Given F P/F	To Find A Given P A/P	To Find P Given A P/A	To Find A Given F A/F	To Find F Given A F/A
1	1.3500	0.7407	1.3500	0.7407	1.0000	1.0000
2	1.8225	0.5487	0.7755	1.2894	0.4255	2.3500
3	2.4604	0.4064	0.5897	1.6959	0.2397	4.1725
4	3.3215	0.3011	0.5008	1.9969	0.1508	6.6329
5	4.4840	0.2230	0.4505	2.2200	0.1005	9.9544
6	6.0534	0.1652	0.4193	2.3852	0.0693	14.4384
7	8.1722	0.1224	0.3988	2.5075	0.0488	20.4919
8	11.0324	0.0906	0.3849	2.5982	0.0349	28.6640
9	14.8937	0.0671	0.3752	2.6653	0.0252	39.6964
10	20.1066	0.0497	0.3683	2.7150	0.0183	54.5902
11	27.1439	0.0368	0.3634	2.7519	0.0134	74.6967
12	36.6442	0.0273	0.3598	2.7792	0.0098	101.8406
13	49.4697	0.0202	0.3572	2.7994	0.0072	138.4848
14	66.7841	0.0150	0.3553	2.8144	0.0053	187.9544
15	90.1585	0.0111	0.3539	2.8255	0.0039	254.7385
16	121.7139	0.0082	0.3529	2.8337	0.0029	344.8970
17	164.3138	0.0061	0.3521	2.8398	0.0021	466.6109
18	221.8236	0.0045	0.3516	2.8443	0.0016	630.9247
19	299.4619	0.0033	0.3512	2.8476	0.0012	852.7483
20	404.2736	0.0025	0.3509	2.8501	0.0009	1152.2103
21	545.7693	0.0018	0.3506	2.8519	0.0006	1556.4838
22	736.7886	0.0014	0.3505	2.8533	0.0005	2102.2532
23	994.6646	0.0010	0.3504	2.8543	0.0004	2839.0418
24	1342.7973	0.0007	0.3503	2.8550	0.0003	3833.7064
25	1812.7763	0.0006	0.3502	2.8556	0.0002	5176.5037
26	2447.2480	0.0004	0.3501	2.8560	0.0001	6989.2800
27	3303.7848	0.0003	0.3501	2.8563	0.0001	9436.5280
28	4460.1095	0.0002	0.3501	2.8565	0.0001	12740.3128
29	6021.1478	0.0002	0.3501	2.8567	0.0001	17200.4222
30	8128.5495	0.0001	0.3500	2.8568	0.0000	23221.5700

Table E-28. Discrete compounding factors: i = 40.00%

Year, n	To Find F Given P F/P	To Find P Given F P/F	To Find A Given P A/P	To Find P Given A P/A	To Find A Given F A/F	To Find F Given A F/A
1	1.4000	0.7143	1.4000	0.7143	1.0000	1.0000
2	1.9600	0.5102	0.8167	1.2245	0.4167	2.4000
3	2.7440	0.3644	0.6294	1.5889	0.2294	4.3600
4	3.8416	0.2603	0.5408	1.8492	0.1408	7.1040
5	5.3782	0.1859	0.4914	2.0352	0.0914	10.9456
6	7.5295	0.1328	0.4613	2.1680	0.0613	16.3238
7	10.5414	0.0949	0.4419	2.2628	0.0419	23.8534
8	14.7579	0.0678	0.4291	2.3306	0.0291	34.3947
9	20.6610	0.0484	0.4203	2.3790	0.0203	49.1526
10	28.9255	0.0346	0.4143	2.4136	0.0143	69.8137
11	40.4957	0.0247	0.4101	2.4383	0.0101	98.7391
12	56.6939	0.0176	0.4072	2.4559	0.0072	139.2348
13	79.3715	0.0126	0.4051	2.4685	0.0051	195.9287
14	111.1201	0.0090	0.4036	2.4775	0.0036	275.3002
15	155.5681	0.0064	0.4026	2.4839	0.0026	386.4202
16	217.7953	0.0046	0.4018	2.4885	0.0018	541.9883
17	304.9135	0.0033	0.4013	2.4918	0.0013	759.7837
18	426.8789	0.0023	0.4009	2.4941	0.0009	1064.6971
19	597.6304	0.0017	0.4007	2.4958	0.0007	1491.5760
20	836.6826	0.0012	0.4005	2.4970	0.0005	2089.2064
21	1171.3556	0.0009	0.4003	2.4979	0.0003	2925.8889
22	1639.8978	0.0006	0.4002	2.4985	0.0002	4097.2445
23	2295.8569	0.0004	0.4002	2.4989	0.0002	5737.1423
24	3214.1997	0.0003	0.4001	2.4992	0.0001	8032.9993
25	4499.8796	0.0002	0.4001	2.4994	0.0001	11247.1990
26	6299.8314	0.0002	0.4001	2.4996	0.0001	15747.0785
27	8819.7640	0.0001	0.4000	2.4997	0.0000	22046.9099
28	12347.6696	0.0001	0.4000	2.4998	0.0000	30866.6739
29	17286.7374	0.0001	0.4000	2.4999	0.0000	43214.3435
30	24201.4324	0.0000	0.4000	2.4999	0.0000	60501.0809

Table E-29. Discrete compounding factors: i = 50.00%

Year, n	To Find F Given P F/P	To Find P Given F P/F	To Find A Given P A/P	To Find P Given A P/A	To Find A Given F A/F	To Find F Given A F/A
1	1.5000	0.6667	1.5000	0.6667	1.0000	1.0000
2	2.2500	0.4444	0.9000	1.1111	0.4000	2.5000
3	3.3750	0.2963	0.7105	1.4074	0.2105	4.7500
4	5.0625	0.1975	0.6231	1.6049	0.1231	8.1250
5	7.5938	0.1317	0.5758	1.7366	0.0758	13.1875
6	11.3906	0.0878	0.5481	1.8244	0.0481	20.7813
7	17.0859	0.0585	0.5311	1.8829	0.0311	32.1719
8	25.6289	0.0390	0.5203	1.9220	0.0203	49.2578
9	38.4434	0.0260	0.5134	1.9480	0.0134	74.8867
10	57.6650	0.0173	0.5088	1.9653	0.0088	113.3301
11	86.4976	0.0116	0.5058	1.9769	0.0058	170.9951
12	129.7463	0.0077	0.5039	1.9846	0.0039	257.4927
13	194.6195	0.0051	0.5026	1.9897	0.0026	387.2390
14	291.9293	0.0034	0.5017	1.9931	0.0017	581.8585
15	437.8939	0.0023	0.5011	1.9954	0.0011	873.7878
16	656.8408	0.0015	0.5008	1.9970	0.0008	1311.6817
17	985.2613	0.0010	0.5005	1.9980	0.0005	1968.5225
18	1477.8919	0.0007	0.5003	1.9986	0.0003	2953.7838
19	2216.8378	0.0005	0.5002	1.9991	0.0002	4431.6756
20	3325.2567	0.0003	0.5002	1.9994	0.0002	6648.5135
21	4987.8851	0.0002	0.5001	1.9996	0.0001	9973.7702
22	7481.8276	0.0001	0.5001	1.9997	0.0001	14961.6553
23	11222.7415	0.0001	0.5000	1.9998	0.0000	22443.4829
24	16834.1122	0.0001	0.5000	1.9999	0.0000	33666.2244
25	25251.1683	0.0000	0.5000	1.9999	0.0000	50500.3366

Appendix F

Review Question Answers

CHAPTER 48: PERSONAL EFFECTIVENESS

48.1) Content listening requires a person to focus on the content as opposed to the delivery style.
48.2) Proposals are generally persuasive in nature as opposed to objectively communicating factual information.
48.3) Line and bar charts
48.4) Best alternative to a negotiated agreement
48.5) Confrontation implies addressing or bringing issues to another person's attention where conflict implies disagreement, arguing, and hurt feelings.
48.6) Withdrawal

CHAPTER 49: MACHINING PROCESSES ANALYSIS

49.1) Carbide and ceramic
49.2) Inscribed circle (IC)
49.3) 10,000 rpm
49.4) No
49.5) Rubber
49.6) Grinding ratio
49.7) Harder materials require a lower thread percentage

CHAPTER 50: FORMING PROCESSES ANALYSIS

50.1) The ratio of the billet's cross-sectional area to the cross-sectional area of the final extrusion
50.2) Below
50.3) Location inside the material where there is neither tension nor compression
50.4) Overbending or bottoming (setting)
50.5) Circle grid analysis (CGA)

CHAPTER 51: JOINING AND FASTENING ANALYSIS

51.1) The flux alloys the weld pool, forms a protective slag, and burns off, forming a protective gas.
51.2) Wire speed adjusts the amperage
51.3) To initiate the arc
51.4) By using low-hydrogen stick electrodes and/or preheating and postheating
51.5) The distance parallel to the axis from any point on a screw thread to a corresponding point on the next thread
51.6) Blind rivets (pop rivets)

CHAPTER 52: DEBURRING AND FINISHING ANALYSIS

52.1) Abrasive flow machining
52.2) Honing
52.3) Hard anodizing builds two layers on the workpiece where regular anodizing builds only one layer
52.4) Polymer binder, pigments, solvent, and additives
52.5) Volatile organic compound
52.6) Electrostatic spraying, fluidized bed, and flame spraying

CHAPTER 53: FIXTURE AND JIG DESIGN

53.1) To hold, grip, or chuck a workpiece while it is being machined or processed by other manufacturing operations
53.2) Assembled locators are replaceable
53.3) Locators and supports
53.4) Short lead times, infrequent production runs, and small production quantities
53.5) Jigs guide the cutting tool into the workpiece throughout the cutting cycle in addition to holding the workpiece in place while it is being processed.

CHAPTER 54: ADVANCED QUALITY ANALYSIS

54.1) $UCL_p = 0.08$
$LCL_p = 0$
54.2) 0.9099 = 90.99%
54.3) Reproducibility
54.4) 16
54.5) Variability
54.6) ISO 9000 is a generic quality management standard and QS-9000 is a set of quality management requirements for DaimlerChrysler, General Motors, and Ford Motor Company suppliers.
54.7) Green belt
54.8) Ultrasonic testing

CHAPTER 55: ENGINEERING ECONOMICS ANALYSIS

55.1) a) $76,000
b) $60,727
c) $51,200
d) $42,000
55.2) 1,400 units per month = 70% utilization

CHAPTER 56: MANAGEMENT THEORY AND PRACTICE

56.1) Matrix structure
56.2) Normative
56.3) No
56.4) Benchmarking
56.5 Focus group
56.6) Not novel
56.7) Warranty of merchantability

CHAPTER 57: INDUSTRIAL SAFETY, HEALTH, AND ENVIRONMENTAL MANAGEMENT

57.1) One or more
57.2) Letters 6 in. (152 mm) tall and illuminated by a continuous power source that includes battery backup
57.3) Combustible metals
57.4) 85 dB
57.5) They do not prevent the worker from accessing the hazard. They only warn of the situation.
57.6) Where they can be accessed at all times by persons exposed to the chemicals
57.7) Behavior of humans and the environment where the accident takes place
57.8) >2,200 lb (>1,000 kg)
57.9) It permits the tracking of waste from the point of generation to the point of ultimate treatment, storage, or disposal.

Bibliography

Bedeian, A. 1989. *Management*, Second Edition. New York: Dryden Press.

Bhateja, C. and Lindsay, R. 1982. "Principles of Grinding." *Grinding: Theory, Techniques, and Troubleshooting*. Dearborn MI: Society of Manufacturing Engineers.

Bhote, Keki R. 2002. *The Ultimate Six Sigma: Beyond Quality Excellence to Total Business Excellence*. New York: AMACOM/ American Management Association.

Breyfogle III, Forrest W. 1999. *Implementing Six Sigma: Smarter Solutions Using Statistical Methods*. New York: John Wiley & Sons, Inc.

Diamond, W. J. 2001. *Practical Experiment Designs*, Third Edition. New York: John Wiley & Sons.

George, Michael L. 2002. *Lean Six Sigma: Combining Six Sigma Quality with Lean Speed*. New York: McGraw-Hill.

Harry, Mikel. 2000. "Framework for Business Leadership." *Quality Progress* 33, 4: 80-83.

Howes, N. 2001. *Modern Project Management*. New York: American Management Association.

Hunger, J. and Wheelen, T. 2001. *Essentials of Strategic Management*, Second Edition. Upper Saddle River, NJ: Prentice-Hall.

Hunter, W.G. and Hunter J.S. 1978. *Statistics for Experimenters*. New York: John Wiley & Sons.

International Forum for Management Systems. December 27, 2002. "Why is ISO so Important?" http://www.informintl.com/iso9000. Durham, NC: INFORM.

Meredith, J. and Mantel, S. 1989. *Project Management: A Managerial Approach*, Second Edition. New York: John Wiley & Sons.

Montgomery, D.C. 2001. *Design and Analysis of Experiments*, Fifth Edition. New York: John Wiley & Sons.

Occupational Safety and Health Administration. 1994. *OSHA Field Inspection Reference Manual* (FIRM). No. CPL2.103. Washington, DC: OSHA.

Occupational Safety and Health Administration. 2001. *OSHA Personal Protective Equipment Guide*. Washington, DC: OSHA.

Occupational Safety and Health Administration. 2000. *Process Safety Management*. OSHA publication No. 3132. Washington, DC: OSHA.

Occupational Safety and Health Administration. 1994. *Process Safety Management Guidelines for Compliance*. OSHA publication No. 3133. Washington, DC: OSHA.

Porter, M. 1998. *Competitive Advantage: Creating and Sustaining Superior Performance*. New York: The Free Press.

Ross, P. J. 1995. *Taguchi Techniques for Quality Engineering*, Second Edition. New York: McGraw-Hill.

Veilleux, R. and Petro, L., eds. 1988. *Tool and Manufacturing Engineers Handbook*, Fourth Edition. Volume 5: *Manufacturing Management*. Dearborn, MI: Society of Manufacturing Engineers.

Index

D

E

N

O

P

Q

R

S

T

U

V

W